VOLUME EIGHTY SEVEN

ADVANCES IN VIRUS RESEARCH

VOLUME EIGHTY SEVEN

ADVANCES IN VIRUS RESEARCH

Edited by

KARL MARAMOROSCH
Rutgers University, New Brunswick, New Jersey, USA

FREDERICK A. MURPHY
University of Texas Medical Branch, Galveston, Texas, USA

AMSTERDAM • BOSTON • HEIDELBERG • LONDON
NEW YORK • OXFORD • PARIS • SAN DIEGO
SAN FRANCISCO • SINGAPORE • SYDNEY • TOKYO
Academic Press is an imprint of Elsevier

Academic Press is an imprint of Elsevier
525 B Street, Suite 1800, San Diego, CA 92101-4495, USA
225 Wyman Street, Waltham, MA 02451, USA
The Boulevard, Langford Lane, Kidlington, Oxford, OX5 1GB, UK
32 Jamestown Road, London NW1 7BY, UK
Radarweg 29, PO Box 211, 1000 AE Amsterdam, The Netherlands

First edition 2013

Library of Congress Cataloging-in-Publication Data
A catalog record for this book is available from the Library of Congress

British Library Cataloguing-in-Publication Data
A catalogue record for this book is available from the British Library

ISBN: 978-0-12-407698-3
ISSN: 0065-3527

Printed and bound in United States of America
13 14 15 16 11 10 9 8 7 6 5 4 3 2 1

CONTENTS

CONTRIBUTORS

Frederic Aparicio
Instituto de Biología Molecular y Celular de Plantas (IBMCP), Universidad Politécnica de Valencia-Consejo Superior de Investigaciones Científicas, Valencia, Spain

Karl W. Boehme
Department of Pediatrics, and Elizabeth B. Lamb Center for Pediatric Research, Vanderbilt University School of Medicine, Nashville, Tennessee, USA

Satinder Dahiya
Department of Microbiology and Immunology, Center for Molecular Virology and Translational Neuroscience, Institute for Molecular Medicine and Infectious Disease, Drexel University College of Medicine, Philadelphia, Pennsylvania, USA

Terence S. Dermody
Department of Pediatrics; Elizabeth B. Lamb Center for Pediatric Research, and Department of Pathology, Microbiology, and Immunology, Vanderbilt University School of Medicine, Nashville, Tennessee, USA

Mari C. Herranz
Instituto de Biología Molecular y Celular de Plantas (IBMCP), Universidad Politécnica de Valencia-Consejo Superior de Investigaciones Científicas, Valencia, Spain

Chuji Hiruki
Department of Agricultural, Food, and Nutritional Science, University of Alberta, Edmonton, Alberta, Canada

Bryan P. Irish
Department of Microbiology and Immunology, Center for Molecular Virology and Translational Neuroscience, Institute for Molecular Medicine and Infectious Disease, Drexel University College of Medicine, Philadelphia, Pennsylvania, USA

Kook-Hyung Kim
Department of Agricultural Biotechnology, Plant Genomics and Breeding Institute, and Research Institute for Agriculture and Life Sciences, Seoul National University, Seoul, Republic of Korea

Caroline M. Lai
Elizabeth B. Lamb Center for Pediatric Research, and Department of Pathology, Microbiology, and Immunology, Vanderbilt University School of Medicine, Nashville, Tennessee, USA

Jaime Martín-Benito
Centro Nacional de Biotecnología (CSIC), Madrid, Spain

Michael R. Nonnemacher
Department of Microbiology and Immunology, Center for Molecular Virology and Translational Neuroscience, Institute for Molecular Medicine and Infectious Disease, Drexel University College of Medicine, Philadelphia, Pennsylvania, USA

Tetsuro Okuno
Laboratory of Plant Pathology, Graduate School of Agriculture, Kyoto University, Sakyo-ku, Kyoto, Japan

Juan Ortín
Centro Nacional de Biotecnología (CSIC), and CIBER de Enfermedades Respiratorias (ISCIII), Madrid, Spain

Vicente Pallas
Instituto de Biología Molecular y Celular de Plantas (IBMCP), Universidad Politécnica de Valencia-Consejo Superior de Investigaciones Científicas, Valencia, Spain

Mi-Ri Park
Department of Agricultural Biotechnology, Plant Genomics and Breeding Institute, and Research Institute for Agriculture and Life Sciences, Seoul National University, Seoul, Republic of Korea

Jesus A. Sanchez-Navarro
Instituto de Biología Molecular y Celular de Plantas (IBMCP), Universidad Politécnica de Valencia-Consejo Superior de Investigaciones Científicas, Valencia, Spain

Simon W. Scott
Department of Biological Sciences, Clemson University, Clemson, South Carolina, USA

Jang-Kyun Seo
Department of Agricultural Biotechnology, Plant Genomics and Breeding Institute; Research Institute for Agriculture and Life Sciences, Seoul National University, Seoul, and Crop Protection Division, National Academy of Agricultural Science, Rural Development Administration, Suwon, Republic of Korea

Brian Wigdahl
Department of Microbiology and Immunology, Center for Molecular Virology and Translational Neuroscience, Institute for Molecular Medicine and Infectious Disease, Drexel University College of Medicine, Philadelphia, Pennsylvania, USA

CHAPTER ONE

Mechanisms of Reovirus Bloodstream Dissemination

Karl W. Boehme[*,†,1], **Caroline M. Lai**[†,‡], **Terence S. Dermody**[*,†,‡,2]
[*]Department of Pediatrics, Vanderbilt University School of Medicine, Nashville, Tennessee, USA
[†]Elizabeth B. Lamb Center for Pediatric Research, Vanderbilt University School of Medicine, Nashville, Tennessee, USA
[‡]Department of Pathology, Microbiology, and Immunology, Vanderbilt University School of Medicine, Nashville, Tennessee, USA
[1]Current address: Department of Microbiology and Immunology, University of Arkansas for Medical Sciences, Little Rock, Arkansas, USA.
[2]Corresponding author: e-mail address: terry.dermody@vanderbilt.edu

Contents

Abstract

Many viruses cause disease within an infected host after spread from an initial portal of entry to sites of secondary replication. Viruses can disseminate via the bloodstream or through nerves. Mammalian orthoreoviruses (reoviruses) are neurotropic viruses that use both bloodborne and neural pathways to spread systemically within their hosts

Advances in Virus Research, Volume 87
ISSN 0065-3527
http://dx.doi.org/10.1016/B978-0-12-407698-3.00001-6

to cause disease. Using a robust mouse model and a dynamic reverse genetics system, we have identified a viral receptor and a viral nonstructural protein that are essential for hematogenous reovirus dissemination. Junctional adhesion molecule-A (JAM-A) is a member of the immunoglobulin superfamily expressed in tight junctions and on hematopoietic cells that serves as a receptor for all reovirus serotypes. Expression of JAM-A is required for infection of endothelial cells and development of viremia in mice, suggesting that release of virus into the bloodstream from infected endothelial cells requires JAM-A. Nonstructural protein σ1s is implicated in cell cycle arrest and apoptosis in reovirus-infected cells but is completely dispensable for reovirus replication in cultured cells. Surprisingly, a recombinant σ1s-null reovirus strain fails to spread hematogenously in infected mice, suggesting that σ1s facilitates apoptosis of reovirus-infected intestinal epithelial cells. It is possible that apoptotic bodies formed as a consequence of σ1s expression lead to reovirus uptake by dendritic cells for subsequent delivery to the mesenteric lymph node and the blood. Thus, both host and viral factors are required for efficient hematogenous dissemination of reovirus. Understanding mechanisms of reovirus bloodborne spread may shed light on how microbial pathogens invade the bloodstream to disseminate and cause disease in infected hosts.

1. INTRODUCTION

Many pathogenic human and animal viruses disseminate from mucosal sites to peripheral tissues where they cause organ-specific disease (Nathanson & Tyler, 1997). The capacity of a virus to spread systemically can correlate with increased virulence (de Jong et al., 2006; Gu et al., 2007; Kuiken et al., 2003; Pallansch & Roos, 2001). Systemic dissemination requires that the virus effectively navigate diverse intracellular and extracellular environments to infect, replicate, and evade immune detection in multiple cell types and tissues (Adair et al., 2012; Antar et al., 2009; Boehme, Frierson, Konopka, Kobayashi, & Dermody, 2011; Boehme, Guglielmi, & Dermody, 2009). Although some general principles of virus dissemination are understood, little is known about the precise viral and cellular determinants that govern virus spread. Defining mechanisms by which viruses disseminate within their hosts is of fundamental importance to an understanding of viral pathogenesis.

Mammalian orthoreoviruses (reoviruses) are highly tractable models for studies of viral pathogenesis. Studies of reovirus neural spread have provided important information about mechanisms by which neurotropic viruses cause disease in the central nervous system (CNS). The recent identification of new viral and host determinants that govern reovirus spread by the blood provides new insights into how hematogenous dissemination contributes to viral disease.

1.1. Reoviruses

Viruses of the *Reoviridae* family infect a wide range of host organisms, including mammals, birds, insects, and plants (Dermody, Parker, & Sherry, 2013). The *Reoviridae* includes rotaviruses, the most common diarrheal pathogen among children (Parashar, Bresee, Gentsch, & Glass, 1998), orbiviruses, which are economically important pathogens of sheep, cattle, and horses (Coetzee et al., 2012), and reoviruses. Three reovirus serotypes (T1, T2, and T3) currently circulate in humans and other mammals. The serotypes are distinguished on the basis of antibody-mediated neutralization of infectivity and inhibition of hemagglutination. Each serotype is represented by a prototype strain isolated from a human host: type 1 Lang (T1L), type 2 Jones (T2J), and type 3 Dearing (T3D). These strains differ dramatically in host cell tropism, mechanisms of cell killing, modes of dissemination, and CNS disease. In particular, studies of T1 and T3 reoviruses have generated foundational knowledge about strategies used by viruses to replicate and cause neural injury. Development of a plasmid-based reverse genetics system allows introduction of mutations into the viral genome to test specific hypotheses about the structure and function of viral proteins and RNAs (Kobayashi et al., 2007; Kobayashi, Ooms, Ikizler, Chappell, & Dermody, 2010). In concert with an experimentally facile mouse model of infection (Fields, 1992; Parashar, Tarlow, & McCrae, 1992), reovirus is an ideal experimental platform for studies of virus–host interactions.

Reoviruses are nonenveloped, icosahedral viruses that contain a genome consisting of 10 segments of double-stranded (ds) RNA (Fig. 1.1; Dermody et al., 2013). There are three large (L1, L2, L3), three medium (M1, M2, M3), and four small (S1, S2, S3, S4) dsRNA segments that are packaged in an equimolar stoichiometric relationship with one copy of each per virion. With the exception of the M3 and S1 gene segments, each of the reovirus gene segments is monocistronic. Reovirus virions are composed of two concentric protein shells, the outer capsid and core (Fig. 1.1; Dryden et al., 1993). The outer capsid consists of heterohexameric complexes of the μ1 (encoded by M2) and σ3 (encoded by S4) proteins. At each of the icosahedral fivefold symmetry axes, the attachment protein σ1 (encoded by S1) extends from turret-like structures formed by pentamers of λ2 (encoded by L2) protein. The inner core shell is formed by parallel asymmetric dimers of λ1 (encoded by L3) protein that are stabilized by σ2 (encoded by S2) protein. The λ3 (encoded by L1) and μ2 (encoded by M1) proteins are anchored to the inner surface of the core via interactions

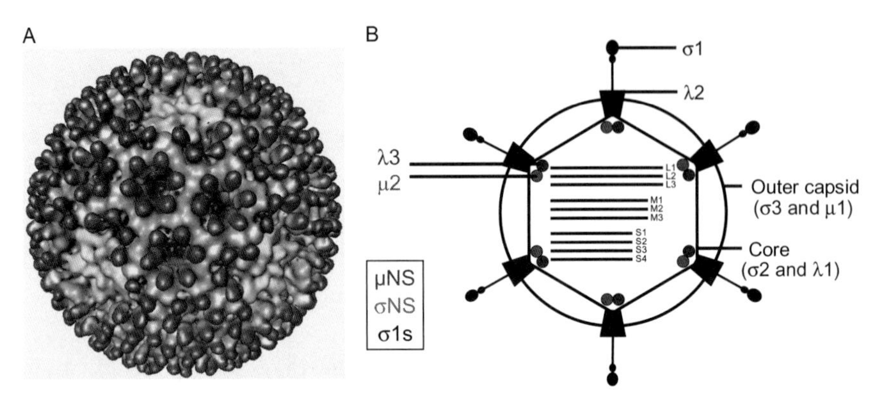

Figure 1.1 The reovirus virion. (A) Cryoelectron micrograph image reconstruction of a reovirus virion. Outer-capsid protein σ3 (blue) is the initial target for virion disassembly in infected cells. Pentameric λ2 protein (yellow) forms an insertion pedestal for σ1, which is the viral attachment protein. (B) Schematic of a reovirus virion. Reovirus particles are formed from two concentric protein shells, the outer capsid and core. The core contains the viral genome, which consists of 10 dsRNA segments. Reovirus also encodes nonstructural proteins, σNS, μNS, and σ1s. *Copyright © American Society for Microbiology, Nason et al. (2001).* (For interpretation of the references to color in this figure legend, the reader is referred to the online version of this chapter.)

with λ1. Lastly, the M3 gene segment encodes nonstructural protein μNS, the S3 gene segment encodes nonstructural protein σNS, and the S1 gene segment encodes nonstructural protein σ1s.

Viral attachment protein σ1 is a long filamentous molecule with head-and-tail morphology (Fig. 1.2A; Chappell, Prota, Dermody, & Stehle, 2002; Fraser et al., 1990; Mercier et al., 2004; Reiter et al., 2011). The σ1 protein is comprised of three distinct structural domains: an N-terminal α-helical coiled-coil tail, a central β-spiral body, and a C-terminal globular head (Chappell et al., 2002; Reiter et al., 2011). Short regions of undefined structure partition each domain and are hypothesized to permit molecular flexibility required to engage cellular receptors during viral entry (Bokiej et al., 2012; Chappell et al., 2002; Fraser et al., 1990; Reiter et al., 2011). Attachment of the σ1 protein to cell-surface receptors initiates reovirus infection of susceptible host cells (Lee, Hayes, & Joklik, 1981; Weiner, Ault, & Fields, 1980; Weiner, Powers, & Fields, 1980). The σ1 protein of T3 reovirus targets two different receptors, α-linked sialic acid (SA) (Armstrong, Paul, & Lee, 1984; Dermody, Nibert, Bassel-Duby, & Fields, 1990a; Pacitti & Gentsch, 1987; Paul, Choi, & Lee, 1989; Paul & Lee, 1987) and junctional adhesion molecule-A (JAM-A) (Barton, Forrest, et al., 2001; Campbell et al., 2005; Prota et al., 2003). Residues in the T3 σ1 β-spiral body domain bind SA

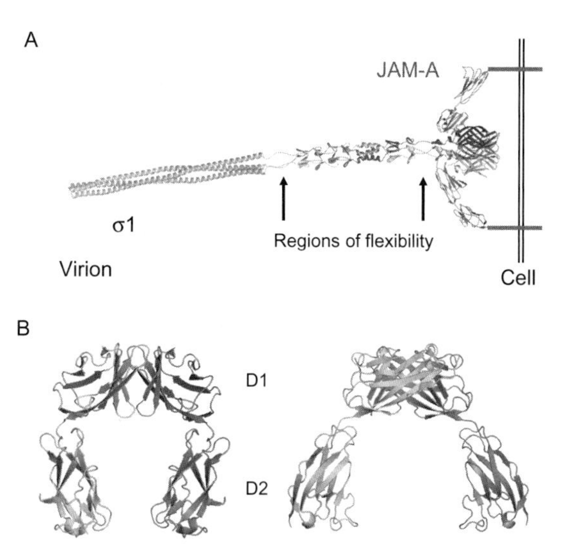

Figure 1.2 Structure of σ1 and JAM-A. (A) Full-length model of attachment protein σ1 bound to JAM-A. A model of full-length σ1 extending from the virion is shown as a ribbon drawing, with the known structure of the C-terminus (Reiter et al., 2011) in tricolor and the predicted structure of the N-terminus in gray. Arrows indicate predicted regions of flexibility. A model of full-length JAM-A is shown in green as a ribbon drawing of the known structure of the extracellular domain (Prota et al., 2003) and a schematic representation of the transmembrane and intracellular domains. For clarity, only two JAM-A monomers are shown bound to σ1. (B) Structure of human JAM-A D1 and D2 domains. Ribbon drawings of a JAM-A homodimer, with one monomer shown in yellow and the other in green. Two orthogonal views are displayed. *(A) Adapted from Kirchner, Guglielmi, Strauss, Dermody, and Stehle (2008, Fig. 1). (B) Adapted from Prota et al. (2003). Copyright (2003) National Academy of Sciences, USA.* (For interpretation of the references to color in this figure legend, the reader is referred to the online version of this chapter.)

(Chappell, Duong, Wright, & Dermody, 2000; Reiter et al., 2011), whereas sequences in the σ1 globular head domain engage JAM-A (Barton, Forrest, et al., 2001; Kirchner et al., 2008).

After receptor binding, virions are internalized into endosomes via a process dependent on β1 integrin (Maginnis et al., 2006) and distributed to organelles marked by Rab7 and Rab9 where viral disassembly takes place (Mainou & Dermody, 2012). During viral disassembly, outer-capsid protein σ3 is degraded by cathepsin proteases, attachment protein σ1 undergoes a conformational change, and outer-capsid protein μ1 is cleaved to form infectious subvirion particles (ISVPs) (Danthi et al., 2010). The μ1 cleavage fragments undergo conformational rearrangement to facilitate endosome penetration and delivery of transcriptionally active core particles into the cytoplasm (Nibert, Odegard, Agosto, Chandran, & Schiff, 2005; Odegard et al., 2004). Primary transcription occurs within the viral core, and nascent RNAs are translated or encapsidated into new viral cores, where they serve as templates for negative-strand synthesis. Within new viral cores, secondary rounds of transcription occur. Outer-capsid proteins are added to nascent cores, which silences viral transcription and yields progeny viral particles. Reovirus release from host cells is hypothesized to occur via a lytic mechanism, but the egress pathway is not understood (Dermody et al., 2013).

1.2. Junctional adhesion molecule-A

JAM-A is the only known proteinaceous receptor for reovirus. It mediates entry of prototype and field-isolated strains of all three reovirus serotypes (Barton, Forrest, et al., 2001; Campbell et al., 2005). JAM-A is a member of the immunoglobulin (Ig) superfamily of proteins that functions in cell–cell adhesion (Bazzoni, 2003). It is expressed on the surface of endothelial and epithelial cells as a component of tight junctions that maintain the integrity of barriers formed between polarized cells (Martin-Padura et al., 1998; Woodfin et al., 2007). JAM-A also is expressed on hematopoietic cells, where it mediates leukocyte extravasation (Corada et al., 2005; Ghislin et al., 2011), and on platelets, where it functions in platelet activation during blood clot formation (Bazzoni, 2003; Sobocka et al., 2004). JAM-A contains three distinct structural domains: an N-terminal ectodomain, a single-span transmembrane anchor, and a C-terminal cytoplasmic tail (Fig. 1.2B; Prota et al., 2003). The ectodomain consists of two Ig-like domains, a membrane-distal D1 domain and a membrane-proximal D2 domain. The cytoplasmic tail terminates in a PDZ-binding domain that interacts with intracellular tight junction components (Bazzoni et al., 2005; Nomme et al., 2011). JAM-A participates in homotypic interactions between D1 domains on opposing monomers (Prota et al., 2003). An interaction between two JAM-A monomers on adjacent cells promotes cell adhesion (Iden et al., 2012; Mandell, Babbin, Nusrat, & Parkos, 2005; Ostermann et al., 2005).

The σ1 protein interacts with the JAM-A D1 domain to adhere reovirus virions to the surface of target cells (Kirchner et al., 2008). Interestingly, the σ1–JAM-A interaction is substantially stronger (approximately 1000-fold) than the interaction between JAM-A monomers (Kirchner et al., 2008). Consequently, σ1 binding to JAM-A likely disrupts JAM-A homodimers. Studies using JAM-A-deficient mice indicate that JAM-A is required for the establishment of viremia, which is essential for dissemination and disease in newborn mice following peroral inoculation of reovirus (Antar et al., 2009). JAM-A is not required for reovirus replication in the murine CNS or development of encephalitis (Antar et al., 2009). These findings suggest that reovirus utilizes other cell-surface receptors to mediate entry into specific cell types.

1.3. Reovirus pathogenesis

Reoviruses are highly virulent in newborn mice and cause injury to a variety of host organs, including the CNS, heart, and liver (Dermody et al., 2013). T1 and T3 reovirus strains invade the CNS but use different routes and produce distinct pathologic consequences following peroral or intramuscular inoculation. T1 reoviruses spread by hematogenous routes and infect ependymal cells, causing nonlethal hydrocephalus (Tyler, McPhee, & Fields, 1986; Weiner, Drayna, Averill, & Fields, 1977; Weiner, Powers, et al., 1980). T3 reoviruses spread to the CNS by both hematogenous and neural routes, and infect neurons (Antar et al., 2009; Boehme et al., 2011; Tyler et al., 1986). In the brain, T3 reoviruses induce neuronal apoptosis, which results in fatal encephalitis (Morrison, Sidman, & Fields, 1991; Tyler et al., 1986; Weiner et al., 1977; Weiner, Powers, et al., 1980). Studies using T1L × T3D reassortant viruses mapped the major determinant of CNS pathology to the viral S1 gene (Dichter & Weiner, 1984; Tardieu & Weiner, 1982), which encodes attachment protein σ1 and nonstructural protein σ1s (Sarkar et al., 1985; Weiner, Ault, et al., 1980). Because of its role in viral attachment and entry, these serotype-specific differences in dissemination and disease have largely been ascribed to the σ1 protein. However, σ1s plays a critical role in promoting reovirus spread by the bloodstream (Boehme et al., 2011, 2009).

1.4. Nonstructural protein σ1s

Protein σ1s is a 14 kDa nonstructural protein encoded by the viral S1 gene segment (Cashdollar, Chmelo, Wiener, & Joklik, 1985; Ernst & Shatkin, 1985; Sarkar et al., 1985). The σ1s open-reading frame (ORF) completely overlaps the σ1 coding sequence; however, σ1s lies in a different reading frame (Cashdollar et al., 1985; Cenatiempo et al., 1984; Dermody, Nibert, Bassel-Duby, & Fields, 1990b; Ernst & Shatkin, 1985; Sarkar et al., 1985). Although every reovirus strain sequenced to date contains a σ1s ORF, little amino acid sequence identity exists between the σ1s proteins from the different reovirus serotypes (Cashdollar et al., 1985;

Dermody et al., 1990b). The only conserved sequence among σ1s proteins is a cluster of basic amino acids near the amino terminus (Cashdollar et al., 1985; Dermody et al., 1990b). The basic cluster from T3 σ1s functions as an autonomous nuclear localization signal that can redirect an appended heterologous protein to the nucleus (Hoyt, Bouchard, & Tyler, 2004). While the majority of native σ1s localizes to the nucleus (Rodgers, Connolly, Chappell, & Dermody, 1998), it is not known whether the basic cluster mediates nuclear translocation in the context of reovirus infection. Functionally, the σ1s protein is implicated in reovirus-induced cell cycle arrest at the G_2/M boundary (Poggioli, Dermody, & Tyler, 2001; Poggioli, Keefer, Connolly, Dermody, & Tyler, 2000) and may influence reovirus neurovirulence by promoting reovirus-induced apoptosis in the murine CNS (Hoyt et al., 2005). Initial studies to define the function of σ1s in reovirus pathogenesis were complicated by the use of nonisogenic σ1s-null mutant and parental virus strains (Rodgers et al., 1998).

Development of a plasmid-based reverse genetics system for mammalian reovirus (Kobayashi et al., 2007, 2010) made it possible to elucidate the function of σ1s in reovirus replication and pathogenesis. Recombinant reoviruses deficient in σ1s expression were engineered by incorporating a single nucleotide change (AUG to ACG) to disrupt the σ1s translational start site into the plasmid containing the cDNA encoding the S1 gene segment. Importantly, the mutation does not affect the coding sequence of the overlapping σ1 ORF. Thus, except for σ1s expression, the resultant viruses are isogenic with the parental strain. Viruses deficient in σ1s expression have been generated in the T1 and T3 S1 gene backgrounds. In both cases, the σ1s-null viruses are viable and replicate with equivalent kinetics and produce yields of progeny virus comparable to the corresponding wild-type viruses, indicating that the σ1s protein is dispensable for reovirus replication in cultured cells (Boehme et al., 2011, 2009). These viruses were used to uncover a role for σ1s in promoting hematogenous reovirus dissemination.

2. DYNAMICS OF REOVIRUS INFECTION IN THE INTESTINE AND LUNG

Reoviruses infect their hosts by the fecal-oral and respiratory routes. Virus enters the host by ingestion of contaminated food or inhalation of virus-containing aerosols. At both portals of entry, reoviruses infect epithelial cells and disseminate to peripheral sites where they cause disease.

2.1. Infection via the gastrointestinal tract

Reoviruses have been isolated from the stools of healthy (Ramos-Alvarez & Sabin, 1954, 1956) and ill (Ramos-Alvarez & Sabin, 1958) children as well as a variety of animals (Ramos-Alvarez & Sabin, 1958). These findings suggest that reovirus is ingested into and shed from the gastrointestinal tract. The dynamics of reovirus infection *in vivo* have largely been elucidated using experimental mouse and rat model systems. Following entry into the gastrointestinal tract of rodents, intestinal proteases rapidly convert reovirus virions to ISVPs, suggesting that the form of the reovirus particle that initiates infection in the intestine is the ISVP (Bass et al., 1990; Bodkin, Nibert, & Fields, 1989; Chappell et al., 1998). In newborn mice, cells at the tips of microvilli are readily infected, whereas cells in the intestinal crypts are spared (Antar et al., 2009; Boehme et al., 2009). In contrast, intestinal crypt cells are infected in adult mice, and cells at the villus tips are uninfected (Rubin, Kornstein, & Anderson, 1985). Infectious reovirus can be recovered following peroral inoculation from the duodenum, jejunum, ileum, and colon (Rubin, Eaton, & Anderson, 1986; Rubin et al., 1985). However, the vast majority of virus is produced in the ileum. This differential production of virus may be due to the capacity of reovirus to infect Peyer patches. Reoviruses are thought to penetrate the intestinal barrier via transport across microfold (M) cells, which are specialized cells of the follicle-associated epithelium (FAE) that overlay the Peyer patches (Amerongen, Wilson, Fields, & Neutra, 1994; Wolf, Dambrauskas, Sharpe, & Trier, 1987; Wolf et al., 1983, 1981). M cells transfer antigens from the intestinal lumen to lymphoid cells of the gut-associated lymphoid tissue (van de Pavert & Mebius, 2010) and serve to monitor luminal contents by exposing Peyer patch lymphoid cells to food antigens, the intestinal microbiota, and invading pathogens. This process is essential for induction of oral tolerance and activation of immune responses to pathogenic microorganisms (van de Pavert & Mebius, 2010). The preferential targeting of crypt cells observed in adult mice is hypothesized to result from transcytosis of virus across M cells and subsequent infection of crypt cells via the basolateral surface (Rubin, 1987). However, M cells also take up reovirus in neonatal mice (Antar et al., 2009; Boehme et al., 2009; Wolf et al., 1981), suggesting that viral transcytosis across M cells is unlikely to explain the difference in intestinal cell tropism observed in adult and newborn mice. It is possible that the proliferative status of stem cells in the crypts of adult mice may recapitulate the cellular environment of neonatal intestinal cells, thereby facilitating reovirus infection of intestinal crypt cells.

2.2. Infection via the respiratory tract

Reovirus also infects the respiratory tract (Sabin, 1959). In rats, both T1 and T3 reovirus strains cause a pneumonia that is characterized by destruction of type 1 alveolar epithelial cells and infiltration of leukocytes into the alveolar spaces (Morin, Warner, & Fields, 1996). The pathology associated with reovirus infection closely mimics disease progression in bronchiolitis obliterans organizing pneumonia, which is notable for fibrous extensions into alveolar spaces in the context of an organizing pneumonia (Bellum et al., 1997). Following inoculation into the respiratory tract, lung proteases convert reovirus virions to ISVPs (Golden & Schiff, 2005; Nygaard, Golden, & Schiff, 2012). Similar to infection in the intestine, reovirus infects the lung by transcytosis through M cells that overlie the bronchus-associated lymphoid tissue (Morin, Warner, & Fields, 1994; Morin et al., 1996).

3. HEMATOGENOUS SPREAD OF REOVIRUS

3.1. Transport of reovirus from the intestine to the bloodstream

Systemic reovirus infection is thought to originate from infected lymphoid cells in the Peyer patch. From the Peyer patch, reovirus transits intestinal lymphatics to the mesenteric lymph node (MLN) and ultimately enters the bloodstream via the thoracic duct (Antar et al., 2009; Boehme et al., 2009; Wolf et al., 1981). Many pathogens that cause systemic disease, including poliovirus (Bodian, 1955; Sabin, 1956) and *Salmonella* (Carter & Collins, 1974; Galan & Curtiss, 1989; Jones, Ghori, & Falkow, 1994), initiate extraintestinal infection and access the bloodstream via this route.

Reovirus reaches the Peyer patches early after infection; viral antigen is detected in Peyer patches within 24 h after peroral inoculation (Antar et al., 2009; Bass, Trier, Dambrauskas, & Wolf, 1988; Boehme et al., 2009; Wolf et al., 1987, 1983, 1981). However, the mechanism by which reovirus infects Peyer patch cells is not known. It is possible that dendritic cells in the Peyer patches take up reovirus virions immediately following viral transcytosis across M cells. This is the most direct route from the intestinal lumen to the Peyer patch and the primary pathway used for processing of intestinal antigens for immune surveillance. A second possibility is that progeny virions released from the basolateral surface of infected FAE cells are taken up by lymphoid cells in Peyer patches. Both viral structural and nonstructural proteins are detected in FAE cells (Fleeton et al., 2004), indicating that active viral replication occurs in these cells. However, it is not known whether FAE cells produce virus. A third possibility is that dendritic cells in Peyer patches take up apoptotic fragments from infected FAE cells, which undergo apoptosis following reovirus infection (Fleeton et al., 2004). Dendritic cells in the underlying Peyer patches immediately adjacent to apoptotic FAE cells contain both active caspase-3 and reovirus structural proteins (Fleeton et al., 2004). These observations suggest that Peyer patch dendritic cells take up apoptotic bodies from infected FAE cells. Additionally, apoptosis induction in the FAE may signal Peyer patch cells to phagocytose the apoptotic remnants, along with reovirus particles.

Regardless of the mechanism by which reovirus accesses Peyer patches, reovirus antigen is detected in the MLN 24 h after peroral inoculation. Little is known about the cell types that support reovirus replication within the intestine and dissemination to the MLN. In adult mice, $CD11c^+$ dendritic cells harbor reovirus antigen, but these cells are not thought to be actively infected (Fleeton et al., 2004). Viral nonstructural proteins are not present in these cells (Fleeton et al., 2004), suggesting that active replication does not occur. $CD11c^+$ dendritic cells are present in neonatal animals (Muthukkumar, Goldstein, & Stein, 2000), but it is not known whether these cells internalize reovirus following peroral inoculation of newborn mice. Reovirus productively infects bulk splenocytes isolated from newborn mice (Tardieu, Powers, & Weiner, 1983), suggesting that reovirus can replicate in primary lymphoid cells. Reovirus cannot productively infect splenocytes explanted after the mouse reaches 7 days of age (Tardieu et al., 1983). Thus, the lack of viral replication in Peyer patch cells in older animals may contribute to the age restriction to reovirus infection.

From Peyer patches, reovirus is hypothesized to traffic via afferent lymphatics to the MLN, then through efferent lymphatics to the blood. It is possible that infected lymphoid cells or lymphoid cells harboring virus mediate transport from the Peyer patches to the bloodstream. However, migrating dendritic cells rarely exit lymph nodes once they enter and present antigen to B and T cells (Iwasaki, 2007). Thus, the cells responsible for transport of reovirus from the Peyer patch are likely retained in the MLN. Reovirus titers in the MLN increase rapidly after peroral inoculation (Antar et al., 2009; Boehme et al., 2009), suggesting that active viral replication occurs in the MLN. However, it is also possible that the increase in viral load in the MLN represents migration of infected lymphoid cells from the Peyer patches. Dissemination from the MLN to the bloodstream may occur as free virus or within another lymphoid cell subset.

An alternative mechanism for accessing the blood is direct uptake of viral particles from the gut. $CD18^+$ phagocytes extend cellular processes between enterocytes to directly sample luminal contents. Dendritic cells also extend processes through the epithelial monolayer, while maintaining barrier integrity to sample gut pathogens (Rescigno et al., 2001). A number of pathogens, including *Salmonella* (Vazquez-Torres et al., 1999) and *Yersinia* (Isberg & Barnes, 2001), use macrophages or dendritic cells to invade the bloodstream and cause extraintestinal infection. Following uptake of luminal pathogens, $CD18^+$ phagocytes traffic across the lamina propria and directly into the blood allowing for rapid entry of the pathogen into the bloodstream.

3.2. Reovirus viremia

Although virus is detected in the blood of infected animals, it is not known whether reovirus virions within the blood are free in the plasma or associated with hematopoietic cells. Other *Reoviridae* family members, including bluetongue virus (BTV) and Colorado tick fever virus, produce cell-associated viremia during infection. BTV infects and replicates in mononuclear cells, lymphocytes, and endothelial cells (Barratt-Boyes & MacLachlan, 1994; Ellis et al., 1993; MacLachlan, Jagels, Rossitto, Moore, & Heidner, 1990; Mahrt & Osburn, 1986; Veronesi et al., 2009). Colorado tick fever virus is detected in mature erythrocytes (Oshiro, Dondero, Emmons, & Lennette, 1978). However, arthropod vectors transmit BTV and Colorado tick fever virus, making viremia a necessary part of the viral infectious cycle in nature. Mammalian reoviruses are not transmitted by arthropod vectors and may produce a distinctly different type of viremia. Studies in which oncolytic reovirus was delivered intravenously to persons with cancer revealed that virus is largely found in hematopoietic cells, specifically mononuclear cells, granulocytes, and platelets (Adair et al., 2012). Each of these cell types express JAM-A (Martin-Padura et al., 1998; Naik, Naik, Eckfeld, Martin-DeLeon, & Spychala, 2001; Sobocka et al., 2000), suggesting that reovirus associates with or infects blood cells to disseminate through the blood to target organs. However, in these studies, virus was delivered directly into the bloodstream by intravenous inoculation. It is not known how reovirus spreads systemically following infection from a natural portal, such as the intestine or lung.

3.3. Role of receptors in reovirus dissemination

Interactions between viral attachment proteins and host cell receptors play a pivotal role in viral pathogenesis. Receptor engagement is a primary mechanism to define cells targeted by viruses. Therefore, patterns of receptor expression are a key determinant of viral disease. Reoviruses engage two types of cellular receptors: cell-surface carbohydrate (Paul et al., 1989) and JAM-A (Barton, Forrest, et al., 2001). Both T1 and T3 (Dermody et al., 1990a; Pacitti & Gentsch, 1987; Paul et al., 1989) reoviruses bind cell-surface SA (Armstrong et al., 1984; Dermody et al., 1990a; Pacitti & Gentsch, 1987; Paul et al., 1989; Paul & Lee, 1987). However, the domains of σ1 that engage glycans differ between the serotypes (Dermody et al., 1990a; Chappell et al., 2000), as do the specific glycans bound (Reiss et al., 2012).

SA engagement enhances reovirus infection through an adhesion-strengthening mechanism in which viral particles are tethered to the cell surface via a low-affinity interaction with the carbohydrate (Barton, Connolly, Forrest, Chappell, & Dermody, 2001). This interaction maintains the virus on the cell surface and increases the opportunity to engage JAM-A. SA-binding reovirus strains have an increased capacity to infect cells compared with non-SA-binding viruses; pretreatment of cells with neuraminidase to remove cell-surface SA eliminates this advantage (Barton, Connolly, et al., 2001). SA engagement also enhances reovirus tropism for bile duct epithelial cells in mice following peroral inoculation (Barton et al., 2003). The resulting disease closely mimics biliary atresia in human infants (Barton et al., 2003), an illness epidemiologically associated with reovirus (Richardson, Bishop, & Smith, 1994; Tyler et al., 1998).

Reovirus strains circulating in nature vary in the capacity to bind SA (Dermody et al., 1990a, 1990b). This finding suggests that SA binding comes with a fitness cost. Accordingly, SA binding appears to inhibit the capacity of reovirus to establish infection at mucosal portals of entry. Non-SA-binding viruses infect primary human airway epithelial cells substantially more efficiently than SA-binding strains (Excoffon et al., 2008). Moreover, infection of primary human airway epithelial cells by SA-binding viruses is enhanced by removal of cell-surface SA with neuraminidase. Mucosal surfaces are covered with a glycocalyx consisting of polysaccharides and glycoproteins that are rich in SA (Excoffon et al., 2008). SA-binding viruses may be trapped by SA within the glycocalyx and incapable of reaching the underlying epithelium (Excoffon et al., 2008). However, once infection is established, SA binding may enhance the capacity of reovirus to cause disease. In addition to the capacity to target bile duct epithelium, SA-binding strains are more neurovirulent than non-SA-binding viruses following intracranial inoculation (Barton et al., 2003; Frierson et al., 2012). This increase in virulence is likely due to more efficient infection of neurons, which results in neuronal apoptosis and encephalitis. The function of SA binding in reovirus hematogenous spread remains to be determined.

Although all reoviruses bind JAM-A, T1, and T3 reoviruses infect distinct cells and cause serotype-specific patterns of pathologic injury within the CNS. These observations suggest that JAM-A binding does not influence serotype-specific differences in reovirus neural tropism and CNS disease. Following peroral inoculation, reovirus produces similar viral titers in the intestine of wild-type and JAM-A-deficient mice, suggesting that JAM-A is not required for reovirus replication in the mouse gastrointestinal tract (Antar et al., 2009). In sharp contrast, viral titers at all sites of secondary replication are significantly lower in JAM-A-deficient animals compared with wild-type controls (Fig. 1.3). Viral loads are comparable within the brains of wild-type and JAM-A-deficient animals after intracranial inoculation, suggesting that JAM-A is not required for viral replication at this site of secondary replication (Antar et al., 2009). These results suggest that JAM-A is required for dissemination of the virus from the intestine to replication sites in target organs.

How might JAM-A promote hematogenous dissemination? Substantially lower reovirus titers are detected in the blood of JAM-A-deficient mice compared with wild-type mice (Fig. 1.4), suggesting that JAM-A is involved in the establishment of viremia (Antar et al., 2009). Diminished viremia is detected in mice inoculated with either T1 or T3 reovirus, indicating that JAM-A functions in promoting bloodborne spread of T1 viruses that disseminate by strictly hematogenous mechanisms as well as neurotropic T3 reoviruses. Primary pulmonary endothelial cells isolated from JAM-A-deficient mice are refractory to reovirus infection compared with those harvested from wild-type mice (Fig. 1.5). These data suggest that reovirus engages JAM-A to infect endothelial cells, likely in the lymphatics or vasculature of the gastrointestinal tract. It is possible that virus released from endothelial cells invades the bloodstream to disseminate to peripheral target organs either free in the plasma or associated with hematopoietic cells.

3.4. Function of nonstructural protein σ1s in reovirus dissemination

Studies using T1 σ1s-null virus uncovered a role for σ1s in promoting bloodborne reovirus spread (Boehme et al., 2009). The σ1s protein is not required for the initial establishment of reovirus infection in the gut.

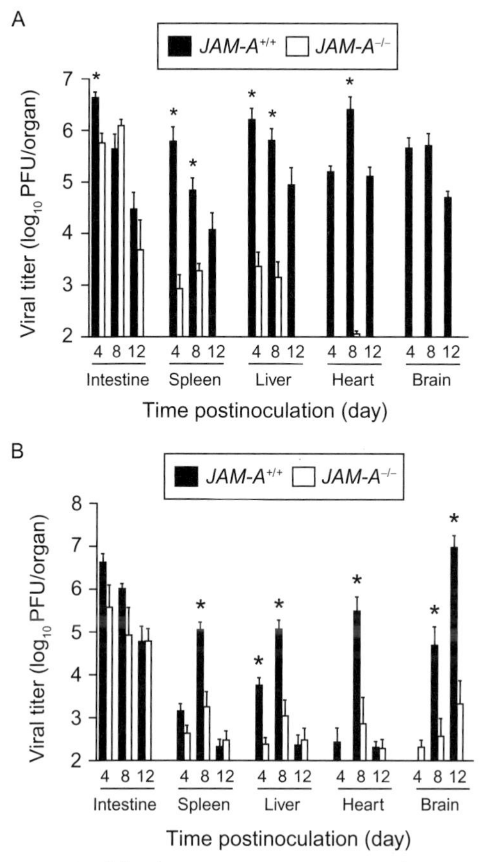

Figure 1.3 JAM-A is required for hematogenous reovirus dissemination. (A) Newborn C57/BL6 *JAM-A*$^{+/+}$ and *JAM-A*$^{-/-}$ mice were inoculated perorally with 10^6 PFU of strain T1L. At days 4, 8, and 12 after inoculation, mice were euthanized, organs were resected, and viral titers were determined by plaque assay. Results are expressed as mean viral titers for six animals for each time point. Error bars indicate SD. $^*P<0.005$ by Student's *t*-test. When all values are less than the limit of detection (spleen, liver, heart, and brain in *JAM-A*$^{-/-}$ mice), a Student's *t*-test *P* value cannot be calculated. (B) Newborn *JAM-A*$^{+/+}$ and *JAM-A*$^{-/-}$ mice were inoculated perorally with 10^4 PFU of strain T3SA−. At days 4, 8, and 12 after inoculation, mice were euthanized, organs were resected, and viral titers were determined by plaque assay. Results are expressed as mean viral titers for 6–13 animals for each time point. Error bars indicate SD. $^*P<0.05$ by Student's *t*-test in comparison to *JAM-A*$^{-/-}$ mice. *Reprinted from Antar et al. (2009), Copyright (2009), with permission from Elsevier.*

Wild-type and σ1s-null viruses replicate to comparable levels in the gastrointestinal tract following peroral inoculation (Fig. 1.6). Reovirus antigen is evident in the intestinal epithelium and Peyer patches of mice inoculated with wild-type or σ1s-null virus, indicating that σ1s does not influence

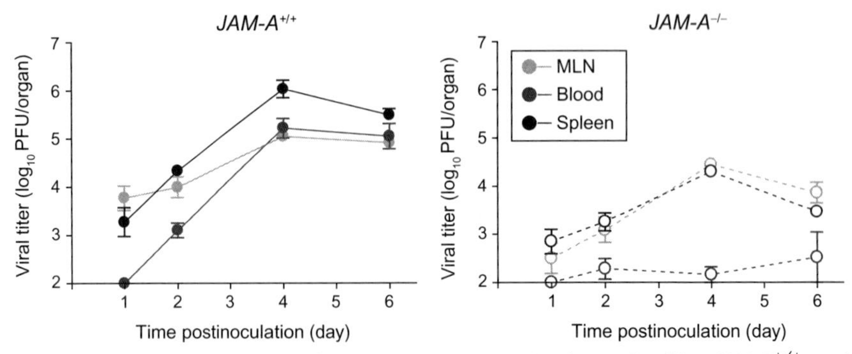

Figure 1.4 JAM-A is required for reovirus viremia. Newborn C57/BL6 *JAM-A*$^{+/+}$ and *JAM-A*$^{-/-}$ mice were inoculated perorally with 10^8 PFU of T1L. At days 1, 2, 4, and 6 after inoculation, mice were euthanized, mesenteric lymph node (MLN), blood, and spleen were collected, and viral titers were determined by plaque assay. Results are expressed as mean viral titers for three to eight animals for each time point. Error bars indicate SD. *Reprinted from Antar et al. (2009), Copyright (2009), with permission from Elsevier.* (For color version of this figure, the reader is referred to the online version of this chapter.)

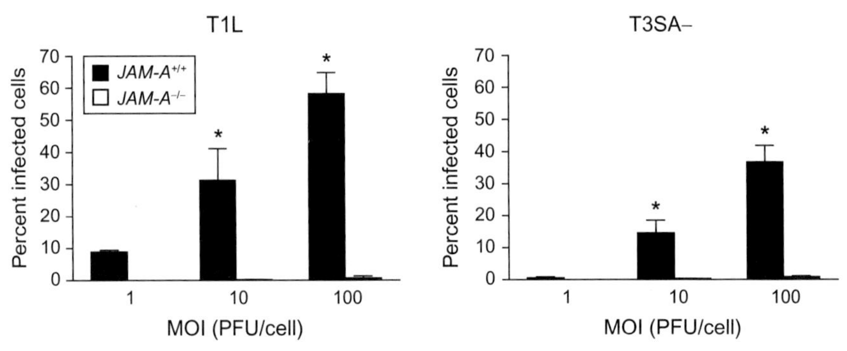

Figure 1.5 JAM-A is required for efficient reovirus infection of endothelial cells. *JAM-A*$^{+/+}$ and *JAM-A*$^{-/-}$ primary endothelial cells were adsorbed with T1L or T3SA– at MOIs of 1, 10, or 100 PFU/cell and incubated for 20 h. The percentage of infected cells was quantified by dividing the number of cells exhibiting reovirus staining by the total number of cell nuclei exhibiting DAPI staining in whole 96 wells for triplicate experiments. Wells contained between 200 and 1600 nuclei. Error bars indicate SD. $^*P < 0.05$ as determined by Student's *t*-test in comparison to *JAM-A*$^{-/-}$ endothelial cells inoculated at the same MOI. *Reprinted from Antar et al. (2009), with permission from Elsevier.*

reovirus tropism in the intestine. In contrast to wild-type virus, the σ1s-null mutant fails to produce substantial titers in the MLN. The σ1s-null virus is detected at low titer in the MLN, but viral titers do not increase over the course of infection. These findings indicate that σ1s either is essential for transit through lymphatic channels to the MLN or serves to promote

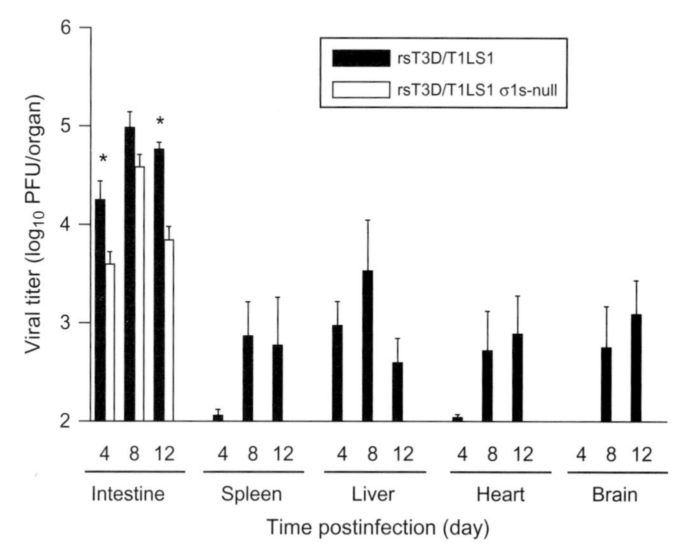

Figure 1.6 The σ1s protein is required for systemic reovirus dissemination following peroral inoculation. Newborn C57/BL6 mice were inoculated perorally with 10^4 PFU of wild-type or σ1s-null reovirus. At days 4, 8, and 12 postinoculation, viral titers in the organs shown were determined by plaque assay. Error bars indicate SEM. $^*P < 0.05$ as determined by Mann–Whitney test in comparison to wild-type virus. When all values are less than the limit of detection, a Mann–Whitney test P value cannot be calculated. *Adapted from Boehme et al. (2009). Copyright (2009) National Academy of Sciences, USA.*

replication in MLN cells. Wild-type virus is detected in the blood and sites of secondary viral replication, including the brain, heart, liver, and spleen (Fig. 1.6). The σ1s-null virus is not detected in the blood or any of the target organs examined. This difference is probably not due to a requirement for σ1s in mediating replication at these sites, as wild-type and σ1s-null viruses produce comparable titers in the brain following intracranial inoculation. Together, these findings suggest that the σ1s protein performs a function that is essential for reovirus to spread from the gut through intestinal lymphatics and ultimately to the blood where it gains access peripheral organs.

In contrast to T1 reoviruses that spread by strictly hematogenous mechanisms, T3 reoviruses disseminate by both hematogenous and neural pathways. The amino acid sequences of the σ1s proteins from the different reovirus serotypes differ markedly (Dermody et al., 1990b). Therefore, it is possible that the σ1s proteins perform serotype-specific functions. In nature, reovirus infects by the peroral route and spreads to the CNS in infant animals resulting in neuropathology. However, infectivity of T3 prototype strain T3D is diminished significantly within the gastrointestinal tract (Bodkin & Fields, 1989; Bodkin et al., 1989) due to cleavage of its σ1 protein by intestinal proteases (Bodkin & Fields, 1989; Bodkin et al., 1989; Nibert, Chappell, & Dermody, 1995). Consequently, intramuscular inoculation into the hindlimb is used to assess mechanisms of T3D dissemination. Inoculation of Type 3 reoviruses intramuscularly leads to invasion of the brain by neural routes (Tyler et al., 1986). Following intramuscular inoculation, wild-type T3D produces substantially higher titers than the σ1s-null virus in peripheral organs including the heart, liver, intestine, and spleen, similar to results obtained using wild-type and σ1s-null T1 viruses (Boehme et al., 2011). Moreover, wild-type T3D but not the T3D σ1s-null mutant virus is detected in the blood. Together, these data suggest that σ1s functions to promote the establishment of reovirus viremia in a serotype-independent manner, which ultimately leads to infection of peripheral target tissues.

In contrast to its function in hematogenous spread, the σ1s protein is dispensable for reovirus spread to the CNS by neural routes. Both wild-type and σ1s-null viruses produce comparable titers in the spinal cord following inoculation into the hindlimb muscle (Boehme et al., 2011). Both viruses also produce comparable titers in the brain following direct intracranial inoculation and in cultured primary neurons. Together, these findings indicate that σ1s is not required for reovirus neural spread or replication in the murine CNS. Thus, although T3 viruses spread via neural and hematogenous mechanisms, the T3 σ1s protein only influences the efficiency of hematogenous dissemination.

4. NEURAL DISSEMINATION OF REOVIRUS

In addition to bloodborne spread, T3 reoviruses use neural circuits to disseminate to the CNS (Boehme et al., 2011; Tyler et al., 1986). Spread via neural routes is a fundamental mechanism of reovirus pathogenesis that is essential for development of reovirus-induced encephalitis (Boehme et al., 2011; Tyler et al., 1986). Direct infection of neurons at peripheral sites provides the virus with access to the CNS and serves as a conduit to the brain. Although the importance of neural spread in reovirus pathogenesis is well appreciated, the cellular and molecular mechanisms that underlie neuronal reovirus trafficking are not well understood.

In contrast to hematogenous spread, JAM-A is dispensable for neural dissemination. Although JAM-A is expressed in the brain, the cell types on which it is present have not been defined. JAM-A is found on NG2-glia cells, which are a subset of stem cells that give rise to oligodendrocytes (Nomme et al., 2011). It is unclear whether JAM-A is expressed on peripheral or CNS neurons. Viral titers in the brains of wild-type and JAM-A-deficient mice are comparable after intracranial inoculation (Antar et al., 2009). Viral tropism in the brain for hippocampal, thalamic, and cortical regions also does not differ between wild-type and JAM-A-deficient mice. Concordantly, primary cortical neurons isolated from wild-type and JAM-A-deficient mice are equally susceptible to reovirus infection and produce equivalent yields of viral progeny (Antar et al., 2009). Together, these data indicate JAM-A is not required for reovirus infection of neural tissue and suggest that JAM-A is dispensable for reovirus spread by neural routes. These findings further suggest that a cellular receptor distinct from JAM-A mediates reovirus infection of neurons.

Some evidence exists about the means by which reovirus traverses neural pathways. Treatment of animals with colchicine to inhibit fast axonal transport impairs reovirus spread to the spinal cord following hindlimb inoculation (Sjostrand & Karlsson, 1969; Tyler et al., 1986). However, treatment with β-β′-iminodipropionitrile to inhibit slow axonal transport does not affect reovirus dissemination to the spinal cord (Hansson, Kristensson, Olsson, & Sjostrand, 1971; Mahrt & Osburn, 1986). These findings suggest that reovirus traffics in neurons by fast axonal transport. However, these inhibitors may act nonspecifically to impair other aspects of viral replication. It also is not known whether reovirus uses afferent or efferent neurons to traffic to the CNS or whether virions can travel using both retrograde and anterograde pathways within neurons. Finally, it is not known where or how progeny virions exit neurons. Much work is required to fully elucidate how reoviruses replicate and traffic in neurons.

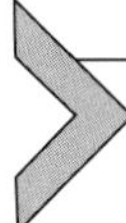

5. FUNCTION OF HEMATOGENOUS AND NEURAL SPREAD IN REOVIRUS PATHOGENESIS

T1 reoviruses disseminate to sites of secondary viral replication solely by hematogenous pathways. Following peroral inoculation of JAM-A-deficient mice, T1 reovirus does not reach the blood or peripheral organs (Antar et al., 2009). Similarly, T1 σ1s-null virus fails to disseminate from the intestine to sites of secondary replication (Boehme et al., 2009). Because T1 reoviruses utilize a single mechanism to spread within its host, inhibiting that mode of dissemination prevents virus-induced systemic disease.

T3 reoviruses, in contrast, disseminate to peripheral organs using a combination of hematogenous and neural mechanisms. Neural spread is essential for maximal neural injury induced by T3 reovirus (Tyler et al., 1986). Following peroral inoculation or inoculation into the hindlimb muscle, T3 reovirus infects peripheral neurons and travels along nerve fibers to infect the CNS and cause disease (Morrison et al., 1991; Tyler et al., 1986). Inhibiting neural spread by sectioning the sciatic nerve prior to hindlimb inoculation prevents virus spread to the spinal cord (Tyler et al., 1986). This finding indicates that T3 reovirus spreads along neural routes to the CNS and suggests that neural dissemination is essential for reovirus neuropathogenesis.

The importance of hematogenous spread in reovirus neuropathogenesis is evident from studies that identified host and viral factors that mediate reovirus transport through the blood. JAM-A-deficient mice are completely resistant to reovirus-induced disease following peroral inoculation with T3 reovirus, whereas wild-type mice succumb to infection (Antar et al., 2009). Viral titers in the brains of JAM-A-deficient mice are substantially reduced in comparison to those in wild-type controls. Concordantly, viral loads in the blood of JAM-A-deficient mice are lower than those detected in wild-type mice (Antar et al., 2009). However, following intracranial inoculation, wild-type and JAM-A-deficient mice are equally susceptible to reovirus disease, and equivalent viral yields are produced in the brains of wild-type and JAM-A-deficient mice (Antar et al., 2009). These results indicate that reduced reovirus virulence in JAM-A-deficient mice following peroral inoculation is not the result of differences in reovirus replication in the brain.

Studies of T3 σ1s-null viruses also highlight the requirement of hematogenous dissemination for reovirus neuropathogenesis. Wild-type T3 reovirus is substantially more virulent than the T3 σ1s-null virus following hindlimb inoculation (Fig. 1.7). Approximately 75% of animals inoculated with wild-type virus succumb to infection compared with 25% of mice inoculated with the σ1s-null virus (Boehme et al., 2011). Wild-type and σ1s-null T3 reoviruses induced 100% mortality following intracranial inoculation, although animals inoculated with wild-type virus succumbed to CNS disease with slightly faster kinetics than those inoculated with the σ1s-null virus. Wild-type and σ1s-null viruses also produce equivalent titers in the brain following intracranial inoculation, indicating that σ1s is dispensable for viral replication in the murine CNS. Thus, the disparity in virulence between wild-type and σ1s-null viruses following intramuscular inoculation does not result from differences in replication in the CNS between the two viral strains.

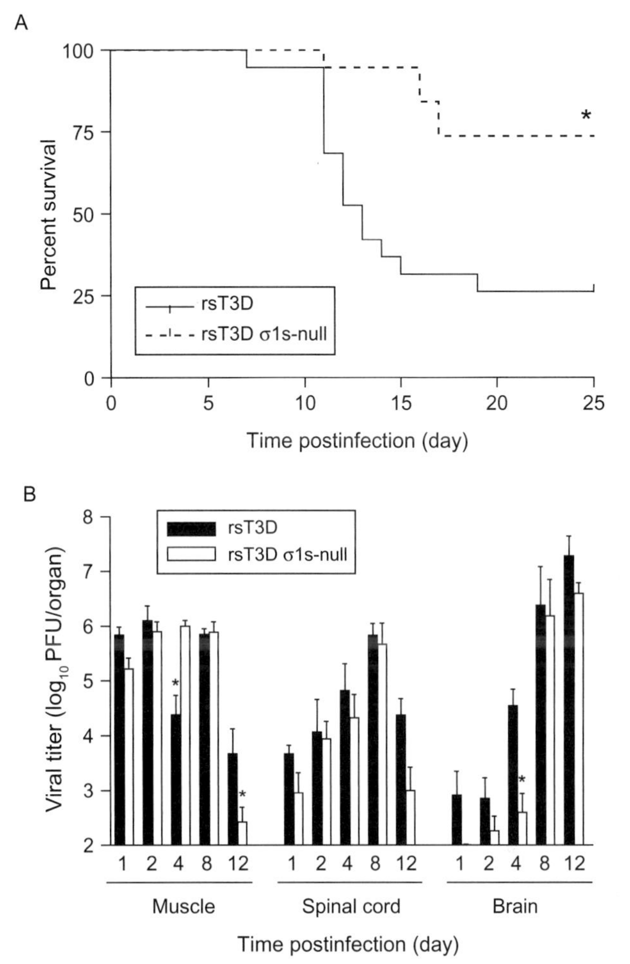

Figure 1.7 The σ1s protein enhances reovirus virulence following intramuscular inoculation. (A) Newborn C57/BL6 mice were inoculated in the left hindlimb with 10^6 PFU of wild-type or σ1s-null T3 reovirus. Mice ($n = 19$ for each virus strain) were monitored for survival for 25 days. $^*P < 0.001$ as determined by log-rank test in comparison to wild-type T3 reovirus. (B) The σ1s protein is not required for reovirus spread by neural routes. Newborn C57/BL6 mice were inoculated in the left hindlimb with 10^6 PFU of wild-type or σ1s-null T3 reovirus. At days 1, 2, 4, 8, and 12 postinoculation, mice were euthanized, hindlimb muscle, spinal cord, and brain were resected, and viral titers were determined by plaque assay. Results are expressed as mean viral titers for six to nine animals for each time point. Error bars indicate SEM. $^*P < 0.05$ as determined by Mann–Whitney test in comparison to wild-type T3 reovirus. *Copyright © American Society for Microbiology, Boehme et al. (2011).*

Following hindlimb inoculation, wild-type virus is detected in the brain 1 day after infection (Fig. 1.7). In contrast, the σ1s-null virus is not found in the brain until 2 days after inoculation. At days 2 and 4 postinoculation, viral titers in the brains of animals inoculated with wild-type virus are markedly higher than those observed in mice inoculated with the σ1s-null mutant. This finding correlates with significantly higher loads of wild-type virus in the blood of infected animals at early times postinoculation compared with the σ1s-null virus (Boehme et al., 2011). Comparable titers of wild-type and σ1s-null viruses are found in the spinal cord at days 1, 2, and 4 post-inoculation. This observation suggests that transport of the σ1s-null virus to the CNS by neural pathways is not impaired. At day 8 postinoculation, titers of wild-type and σ1s-null viruses in the brain are equivalent, possibly reflecting delivery of virus to the brain via neural routes. Collectively, these findings suggest that hematogenous spread is required for reovirus transport to the brain at early times after infection. These results also suggest that the timing of reovirus delivery to the brain is critical for neuropathogenesis. Viral transport by neural routes does not differ between wild-type and σ1s-null viruses, and both virus strains produce equivalent peak titers in the brain. However, peak titers of the σ1s-null virus appear to be achieved after the mice reach the age-imposed limit to reovirus infection, and these animals are no longer susceptible to reovirus-induced CNS disease (Mann, Knipe, Fischbach, & Fields, 2002). Thus, reovirus transport to the brain by the blood at early times after infection is critical for neuropathogenesis.

Reovirus spreads to the spinal cord via the sciatic nerve following intramuscular inoculation into the hindlimb (Tyler et al., 1986). Transection of the sciatic nerve prior to inoculation inhibits neural transmission of the virus to the spinal cord; however, viral dissemination by the blood is unaffected (Tyler et al., 1986). T3 reovirus retains the capacity to spread to the brain after sciatic nerve section (Boehme et al., 2011), suggesting that reovirus can access the brain even in the absence of neural spread, likely via the bloodstream (Fig. 1.8). In addition, almost no virus is detected in the brain following hindlimb inoculation with the σ1s-null virus when the sciatic nerve is sectioned. Thus, virus cannot access the brain when both hematogenous and neural pathways of spread are inhibited.

Together, these findings suggest that (i) spread by neural routes alone is not sufficient to cause reovirus CNS disease, (ii) bloodborne spread is required for delivery of reovirus to the brain at early times postinfection, (iii) hematogenous viral dissemination to the brain is an essential mechanism of reovirus neuropathogenesis, and (iv) virus must be delivered to the brain by the blood early after inoculation for full reovirus neurovirulence.

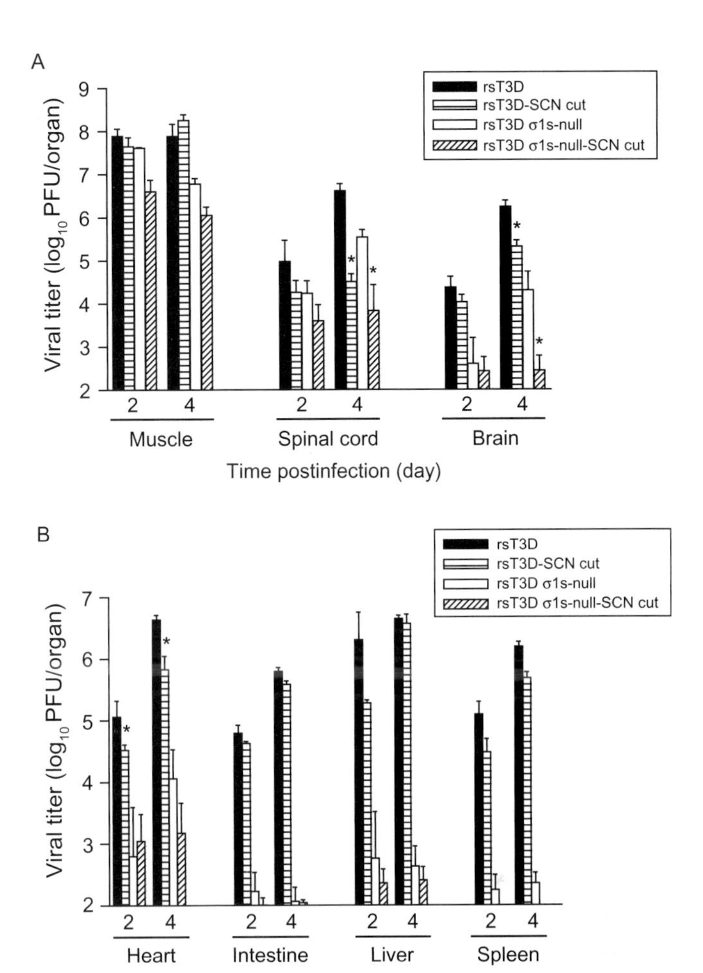

Figure 1.8 Reovirus disseminates to the CNS by hematogenous and neural routes. The left sciatic nerve of newborn C57/BL6 mice was sectioned prior to inoculation in the left hindlimb with 10^6 PFU of wild-type or σ1s-null T3 reovirus. In parallel, mice in which the left sciatic nerve was not sectioned were inoculated in the left hindlimb with 10^6 PFU of wild-type or σ1s-null T3 reovirus. At days 2 and 4 postinoculation, mice were euthanized: (A) hindlimb muscle, spinal cord, and brain and (B) heart, intestine, liver, and spleen were resected, and viral titers were determined by plaque assay. Results are expressed as mean viral titers for six animals for each time point. Error bars indicate SEM. $^*P < 0.05$ as determined by Mann–Whitney test in comparison to animals in which the sciatic nerve was not sectioned. *Copyright © American Society for Microbiology, Boehme et al. (2011).*

6. UNANSWERED QUESTIONS AND FUTURE DIRECTIONS

We have identified host and viral factors essential for the hematogenous dissemination of reovirus. However, many unanswered questions remain. Because viruses capable of bloodstream spread may share similar mechanisms of dissemination, understanding how reovirus spreads in the infected host may aid in the development of therapeutics that target this critical step in viral pathogenesis.

6.1. How does reovirus enter and exit the bloodstream?

To spread to peripheral organs by hematogenous pathways, reovirus must first enter the bloodstream. Studies of reovirus pathogenesis suggest that following peroral inoculation, reovirus infects Peyer patch lymphoid cells that transport virus to the bloodstream (Fig. 1.9). However, reovirus also

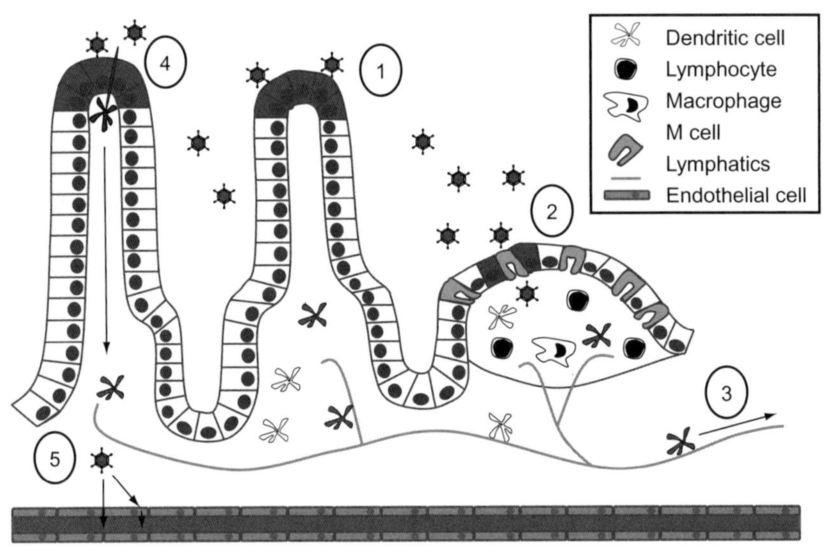

Figure 1.9 Model of reovirus hematogenous spread from the intestine. (1) Following peroral inoculation, reovirus infects intestinal epithelial cells (2) and is taken up by lymphoid cells in the Peyer patch. (3) Infected dendritic cells or lymphocytes carry reovirus from the Peyer patch through the lymphatics and finally to the blood. (4) Phagocytic cells that extend processes into the lumen of the intestine also might be infected for subsequent transport of virus through the lymphatics. (5) Reovirus may enter directly into the blood by passing between endothelial cells or via release into the bloodstream from infected cells. (For color version of this figure, the reader is referred to the online version of this chapter.)

disseminates hematogenously following intracranial inoculation (Boehme et al., 2011, 2009). This observation suggests that reovirus has the capacity to cross endothelial barriers to enter the blood. Little is known about how reovirus infects polarized cells, such as those that constitute the endothelium. JAM-A localizes to tight junctions linking endothelial cells and functions in maintaining the barrier between the tissue and blood compartments. JAM-A is required for hematogenous spread of reovirus and infection of primary cultures of pulmonary vascular endothelial cells (Antar et al., 2009). We envision several possible mechanisms to explain how reovirus uses JAM-A to facilitate entry into the blood. First, JAM-A may function as a gatekeeper for reovirus entry into the bloodstream (Fig. 1.9). Although reovirus infection does not change the barrier function of primary human airway epithelial cells (Excoffon et al., 2008), it is unclear whether the same phenomenon occurs during reovirus infection of the endothelium. Free reovirus virions may interact with JAM-A in endothelial cell tight junctions, transiently disrupt these structures, and cause focal breaches of the endothelial barrier to allow viral invasion of the bloodstream. Other viruses are known to disrupt polarized cell barriers during infection. Mouse adenovirus-1 infection of endothelial cells reduces tight junction protein expression and decreases barrier function in polarized endothelial cell monolayers (Gralinski et al., 2009). Coxsackieviruses engage decay-accelerating factor, an apically distributed protein of polarized epithelial cells, to disrupt tight junctions (Coyne, Shen, Turner, & Bergelson, 2007). In doing so, coxsackieviruses gain access to the basolaterally located coxsackievirus and adenovirus receptor (Coyne et al., 2007). HIV-1 gp120 diminishes expression of tight junction proteins and increases vascular permeability (Kanmogne et al., 2005).

Second, it is possible that reovirus infects endothelial cells to allow progeny virus to be released directly into the blood (Fig. 1.9). Endothelial cells function as sites of amplification for many viruses that spread via the bloodstream. Murine cytomegalovirus dissemination occurs after an episode of secondary viremia that requires viral replication in endothelial cells (Sacher et al., 2008). It is possible that reovirus productively infects endothelium from the basolateral surface on the abluminal side of the endothelium and is released from the apical surface into the blood. Many viruses that infect polarized cells egress apically (Roberts, 1995). This mechanism is common for respiratory viruses, in which release from the apical surface of infected respiratory epithelial cells ensures that the virus will be shed into the respiratory tract to facilitate transmission to susceptible hosts

(Brock et al., 2003; Gerl et al., 2012; Rodriguez & Sabatini, 1978). Studies of reovirus infection of polarized endothelial cells will shed light on mechanisms used by the virus to traverse endothelial monolayers.

Third, reovirus spread may involve infection or association with hematopoietic cells (Fig. 1.9). Hematopoietic cells express JAM-A as an adhesin to allow monocyte extravasation across endothelial barriers (Martin-Padura et al., 1998; Williams, Martin-Padura, Dejana, Hogg, & Simmons, 1999). It is not known whether hematopoietic cells are infected or whether infected blood cells transport reovirus systemically following infection by a natural route of inoculation. However, in cancer patients treated with an intravenous infusion of reovirus, virions associate with mononuclear cells, granulocytes, and platelets to allow dissemination to tumors localized in the viscera (Adair et al., 2012). If hematopoietic cells are responsible for hematogenous reovirus dissemination, age-dependent restriction of reovirus replication in these cells may be one mechanism to explain the limitation of reovirus disease to newborn animals (Tardieu et al., 1983).

Reovirus exit from the bloodstream is required for infection and replication in target tissues and development of organ-specific disease. After peroral inoculation of reovirus, high viral titers are found in virtually all organs (Antar et al., 2009; Boehme et al., 2011, 2009). Mechanisms similar to those that facilitate reovirus entry into the vasculature may mediate reovirus escape from the blood. Reovirus interactions with JAM-A may induce localized perturbations of tight junction integrity that permit virus escape into tissues. Reovirus virions in the blood may infect endothelial cells from the apical surface and progeny virions may be released basolaterally. Finally, infected hematopoietic cells may transport virus from the blood into target organs. None of these possibilities is mutually exclusive; reovirus may use multiple strategies to enter and exit the bloodstream. Studies using mice with tissue-specific expression of JAM-A may help to elucidate mechanisms by which JAM-A facilitates reovirus spread through the bloodstream.

6.2. How does σ1s promote hematogenous spread?

Mechanisms by which σ1s promotes dissemination have not been determined. The σ1s protein is required for reovirus-induced cell cycle arrest at the G_2/M boundary (Poggioli et al., 2002, 2001) and has been implicated in apoptosis *in vivo* (Hoyt et al., 2005). It not known whether inhibition of cell cycle progression is related to the induction of apoptosis following reovirus infection. Cells respond to replication stress or DNA damage by activating checkpoints that arrest the cell cycle. For cells in which genomic damage cannot be repaired, apoptosis is induced to ensure that only faithfully replicated DNA is passed to daughter cells. The relationship between cell cycle arrest and apoptosis in the context of reovirus infection has not been examined. It is possible that σ1s-mediated cell cycle arrest contributes to reovirus-induced apoptosis. Interaction of σ1s with components of the host cell cycle machinery that inhibit normal cell cycle progression could cause the cell to undergo apoptosis.

It is not known whether σ1s-dependent cell cycle arrest and apoptosis are responsible for σ1s-mediated reovirus dissemination. It is possible that σ1s-dependent apoptosis in intestinal epithelial cells promotes reovirus uptake by phagocytic cells at the site of inoculation, and these cells in turn traffic virus to the bloodstream where the virus has access to JAM-A. Although σ1s is not required for reovirus replication in cultured cells (Boehme et al., 2011, 2009; Rodgers et al., 1998), it is possible that σ1s is necessary for efficient reovirus replication in specific cell types that are required for viral dissemination. Defining the cell types used by reovirus to spread through the blood may help uncover how σ1s promotes hematogenous spread. Finally, σ1s may mediate evasion of the host immune response, thereby allowing viral spread. Differences in viral dissemination between wild-type and σ1s are evident at early times postinoculation. This suggests that σ1s would dampen host innate immune mechanisms, as opposed to adaptive responses that develop at later times after infection. Determining how σ1s promotes hematogenous reovirus spread is essential to understand how an enteric, neurotropic virus transits from the intestine to the CNS.

6.3. Clinical implications

Defining factors that govern reovirus dissemination in the blood is essential for optimum use of reovirus in clinical applications. Reovirus efficiently replicates in and kills cancer cells (Adair et al., 2012; Karapanagiotou et al., 2012). Phase II and III clinical trials are underway to test the efficacy of reovirus as an adjunct to conventional cancer therapies (Adair et al., 2012; Karapanagiotou et al., 2012; Kottke et al., 2011). Following intravenous administration, reovirus must navigate and exit the bloodstream to infect solid organ tumors. Intratumoral injection of reovirus may allow for enhanced replication in tumor cells and subsequent spread through the blood to target metastatic tumor foci. Thus, defining viral and cellular determinants underlying how reoviruses gain access to the blood compartment, spread within the bloodstream, and exit from the circulation may aid in oncolytic design. Use of the reverse genetics system may allow engineering of reovirus therapeutics with mutations that increase vector potency or safety by manipulating dissemination determinants (Kobayashi et al., 2007).

We have uncovered a central role for hematogenous dissemination in reovirus neuropathogenesis and elucidated molecular mechanisms that govern reovirus spread by the blood. However, we have much more to learn. Understanding mechanisms of reovirus dissemination will provide broader insight into events at the pathogen–host interface that lead to systemic disease and may aid in the development of therapeutics that target this critical step in viral pathogenesis.

ACKNOWLEDGMENTS

We thank members of our laboratories and Dr. J. Craig Forrest for many useful discussions. This research was supported by Public Health Service awards K22 AI094079 (K. W. B.), F31 NS074596 (C. M. L.), R37 AI38296 (T. S. D.), and the Elizabeth B. Lamb Center for Pediatric Research.

REFERENCES

Adair, R. A., Roulstone, V., Scott, K. J., Morgan, R., Nuovo, G. J., Fuller, M., et al. (2012). Cell carriage, delivery, and selective replication of an oncolytic virus in tumor in patients. *Science Translational Medicine*, *4*, 138ra177.

Amerongen, H. M., Wilson, G. A. R., Fields, B. N., & Neutra, M. R. (1994). Proteolytic processing of reovirus is required for adherence to intestinal M cells. *Journal of Virology*, *68*, 8428–8432.

Antar, A. A. R., Konopka, J. L., Campbell, J. A., Henry, R. A., Perdigoto, A. L., Carter, B. D., et al. (2009). Junctional adhesion molecule-A is required for hematogenous dissemination of reovirus. *Cell Host & Microbe*, *5*, 59–71.

Armstrong, G. D., Paul, R. W., & Lee, P. W. (1984). Studies on reovirus receptors of L cells: Virus binding characteristics and comparison with reovirus receptors of erythrocytes. *Virology*, *138*, 37–48.

Barratt-Boyes, S. M., & MacLachlan, N. J. (1994). Dynamics of viral spread in bluetongue virus infected calves. *Veterinary Microbiology*, *40*, 361–371.

Barton, E. S., Connolly, J. L., Forrest, J. C., Chappell, J. D., & Dermody, T. S. (2001). Utilization of sialic acid as a coreceptor enhances reovirus attachment by multistep adhesion strengthening. *The Journal of Biological Chemistry*, *276*, 2200–2211.

Barton, E. S., Forrest, J. C., Connolly, J. L., Chappell, J. D., Liu, Y., Schnell, F., et al. (2001). Junction adhesion molecule is a receptor for reovirus. *Cell*, *104*, 441–451.

Barton, E. S., Youree, B. E., Ebert, D. H., Forrest, J. C., Connolly, J. L., Valyi-Nagy, T., et al. (2003). Utilization of sialic acid as a coreceptor is required for reovirus-induced biliary disease. *The Journal of Clinical Investigation*, *111*, 1823–1833.

Bass, D. M., Bodkin, D., Dambrauskas, R., Trier, J. S., Fields, B. N., & Wolf, J. L. (1990). Intraluminal proteolytic activation plays an important role in replication of type 1 reovirus in the intestines of neonatal mice. *Journal of Virology*, *64*, 1830–1833.

Bass, D. M., Trier, J. S., Dambrauskas, R., & Wolf, J. L. (1988). Reovirus type 1 infection of small intestinal epithelium in suckling mice and its effect on M cells. *Laboratory Investigations, 58*, 226–235.

Bazzoni, G. (2003). The JAM family of junctional adhesion molecules. *Current Opinion in Cell Biology, 15*, 525–530.

Bazzoni, G., Tonetti, P., Manzi, L., Cera, M. R., Balconi, G., & Dejana, E. (2005). Expression of junctional adhesion molecule-A prevents spontaneous and random motility. *Journal of Cell Science, 118*, 623–632.

Bellum, S. C., Dove, D., Harley, R. A., Greene, W. B., Judson, M. A., London, L., et al. (1997). Respiratory reovirus 1/L induction of intraluminal fibrosis. A model for the study of bronchiolitis obliterans organizing pneumonia. *The American Journal of Pathology, 150*, 2243–2254.

Bodian, D. (1955). Emerging concept of poliomyelitis infection. *Science, 122*, 105–108.

Bodkin, D. K., & Fields, B. N. (1989). Growth and survival of reovirus in intestinal tissue: Role of the L2 and S1 genes. *Journal of Virology, 63*, 1188–1193.

Bodkin, D. K., Nibert, M. L., & Fields, B. N. (1989). Proteolytic digestion of reovirus in the intestinal lumens of neonatal mice. *Journal of Virology, 63*, 4676–4681.

Boehme, K. W., Frierson, J. M., Konopka, J. L., Kobayashi, T., & Dermody, T. S. (2011). The reovirus sigma1s protein is a determinant of hematogenous but not neural virus dissemination in mice. *Journal of Virology, 85*, 11781–11790.

Boehme, K. W., Guglielmi, K. M., & Dermody, T. S. (2009). Reovirus nonstructural protein σ1s is required for establishment of viremia and systemic dissemination. *Proceedings of the National Academy of Sciences of the United States of America, 106*, 19986–19991.

Bokiej, M., Ogden, K. M., Ikizler, M., Reiter, D. M., Stehle, T., & Dermody, T. S. (2012). Optimum length and flexibility of reovirus attachment protein sigma1 are required for efficient viral infection. *Journal of Virology, 86*, 10270–10280.

Brock, S. C., Goldenring, J. R., & Crowe, J. E., Jr. (2003). Apical recycling systems regulate directional budding of respiratory syncytial virus from polarized epithelial cells. *Proceedings of the National Academy of Sciences of the United States of America, 100*, 15143–15148.

Campbell, J. A., Shelling, P., Wetzel, J. D., Johnson, E. M., Wilson, G. A. R., Forrest, J. C., et al. (2005). Junctional adhesion molecule-A serves as a receptor for prototype and field-isolate strains of mammalian reovirus. *Journal of Virology, 79*, 7967–7978.

Carter, P. B., & Collins, F. M. (1974). The route of enteric infection in normal mice. *The Journal of Experimental Medicine, 139*, 1189–1203.

Cashdollar, L. W., Chmelo, R. A., Wiener, J. R., & Joklik, W. K. (1985). Sequences of the S1 genes of the three serotypes of reovirus. *Proceedings of the National Academy of Sciences of the United States of America, 82*, 24–28.

Cenatiempo, Y., Twardowski, T., Shoeman, R., Ernst, H., Brot, N., Weissbach, H., et al. (1984). Two initiation sites detected in the small s1 species of reovirus mRNA by dipeptide synthesis in vitro. *Proceedings of the National Academy of Sciences of the United States of America, 81*, 1084–1088.

Chappell, J. D., Barton, E. S., Smith, T. H., Baer, G. S., Duong, D. T., Nibert, M. L., et al. (1998). Cleavage susceptibility of reovirus attachment protein σ1 during proteolytic disassembly of virions is determined by a sequence polymorphism in the σ1 neck. *Journal of Virology, 72*, 8205–8213.

Chappell, J. D., Duong, J. L., Wright, B. W., & Dermody, T. S. (2000). Identification of carbohydrate-binding domains in the attachment proteins of type 1 and type 3 reoviruses. *Journal of Virology, 74*, 8472–8479.

Chappell, J. D., Prota, A., Dermody, T. S., & Stehle, T. (2002). Crystal structure of reovirus attachment protein σ1 reveals evolutionary relationship to adenovirus fiber. *The EMBO Journal, 21*, 1–11.

Coetzee, P., Van Vuuren, M., Stokstad, M., Myrmel, M., & Venter, E. H. (2012). Bluetongue virus genetic and phenotypic diversity: towards identifying the molecular determinants that influence virulence and transmission potential. *Veterinary Microbiology, 161*, 1–12.

Corada, M., Chimenti, S., Cera, M. R., Vinci, M., Salio, M., Fiordaliso, F., et al. (2005). Junctional adhesion molecule-A-deficient polymorphonuclear cells show reduced diapedesis in peritonitis and heart ischemia-reperfusion injury. *Proceedings of the National Academy of Sciences of the United States of America, 102*, 10634–10639.

Coyne, C. B., Shen, L., Turner, J. R., & Bergelson, J. M. (2007). Coxsackievirus entry across epithelial tight junctions requires occludin and the small GTPases Rab34 and Rab5. *Cell Host & Microbe, 2*, 181–192.

Danthi, P., Guglielmi, K. M., Kirchner, E., Mainou, B., Stehle, T., & Dermody, T. S. (2010). From touchdown to transcription: The reovirus cell entry pathway. *Current Topics in Microbiology and Immunology, 343*, 91–119.

de Jong, M. D., Simmons, C. P., Thanh, T. T., Hien, V. M., Smith, G. J., Chau, T. N., et al. (2006). Fatal outcome of human influenza A (H5N1) is associated with high viral load and hypercytokinemia. *Nature Medicine, 12*, 1203–1207.

Dermody, T. S., Nibert, M. L., Bassel-Duby, R., & Fields, B. N. (1990a). A σ1 region important for hemagglutination by serotype 3 reovirus strains. *Journal of Virology, 64*, 5173–5176.

Dermody, T. S., Nibert, M. L., Bassel-Duby, R., & Fields, B. N. (1990b). Sequence diversity in S1 genes and S1 translation products of 11 serotype 3 reovirus strains. *Journal of Virology, 64*, 4842–4850.

Dermody, T. S., Parker, J. S., & Sherry, B. (2013). Orthoreoviruses. In: Knipe, D. M., and P. M. Howley (eds.): Fields Virology. Sixth Edition. Lippincott Williams & Wilkins, Philadelphia, pp. 1304–1346.

Dichter, M. A., & Weiner, H. L. (1984). Infection of neuronal cell cultures with reovirus mimics in vitro patterns of neurotropism. *Annals of Neurology, 16*, 603–610.

Dryden, K. A., Wang, G., Yeager, M., Nibert, M. L., Coombs, K. M., Furlong, D. B., et al. (1993). Early steps in reovirus infection are associated with dramatic changes in supramolecular structure and protein conformation: Analysis of virions and subviral particles by cryoelectron microscopy and image reconstruction. *The Journal of Cell Biology, 122*, 1023–1041.

Ellis, J. A., Coen, M. L., MacLachlan, N. J., Wilson, W. C., Williams, E. S., & Leudke, A. J. (1993). Prevalence of bluetongue virus expression in leukocytes from experimentally infected ruminants. *American Journal of Veterinary Research, 54*, 1452–1456.

Ernst, H., & Shatkin, A. J. (1985). Reovirus hemagglutinin mRNA codes for two polypeptides in overlapping reading frames. *Proceedings of the National Academy of Sciences of the United States of America, 82*, 48–52.

Excoffon, K. J. D. A., Guglielmi, K. M., Wetzel, J. D., Gansemer, N. D., Campbell, J. A., Dermody, T. S., et al. (2008). Reovirus preferentially infects the basolateral surface and is released from the apical surface of polarized human respiratory epithelial cells. *The Journal of Infectious Diseases, 197*, 1189–1197.

Fields, B. N. (1992). Studies of reovirus pathogenesis reveal potential sites for antiviral intervention. *Advances in Experimental Medicine and Biology, 312*, 1–14.

Fleeton, M., Contractor, N., Leon, F., Wetzel, J. D., Dermody, T. S., & Kelsall, B. (2004). Peyer's patch dendritic cells process viral antigen from apoptotic epithelial cells in the intestine of reovirus-infected mice. *The Journal of Experimental Medicine, 200*, 235–245.

Fraser, R. D. B., Furlong, D. B., Trus, B. L., Nibert, M. L., Fields, B. N., & Steven, A. C. (1990). Molecular structure of the cell-attachment protein of reovirus: Correlation of computer-processed electron micrographs with sequence-based predictions. *Journal of Virology, 64*, 2990–3000.

Frierson, J. M., Pruijssers, A. J., Konopka, J. L., Reiter, D. M., Abel, T. W., Stehle, T., et al. (2012). Utilization of sialylated glycans as coreceptors enhances the neurovirulence of serotype 3 reovirus. *Journal of Virology, 86*, 13164–13173.

Galan, J. E., & Curtiss, R., 3rd. (1989). Cloning and molecular characterization of genes whose products allow Salmonella typhimurium to penetrate tissue culture cells. *Proceedings of the National Academy of Sciences of the United States of America, 86*, 6383–6387.

Gerl, M. J., Sampaio, J. L., Urban, S., Kalvodova, L., Verbavatz, J. M., Binnington, B., et al. (2012). Quantitative analysis of the lipidomes of the influenza virus envelope and MDCK cell apical membrane. *Journal of Cell Biology, 196*, 213–221.

Ghislin, S., Obino, D., Middendorp, S., Boggetto, N., Alcaide-Loridan, C., & Deshayes, F. (2011). Junctional adhesion molecules are required for melanoma cell lines transendothelial migration in vitro. *Pigment Cell & Melanoma Research, 24*, 504–511.

Golden, J. W., & Schiff, L. A. (2005). Neutrophil elastase, an acid-independent serine protease, facilitates reovirus uncoating and infection in U937 promonocyte cells. *Virology Journal, 2*, 48.

Gralinski, L. E., Ashley, S. L., Dixon, S. D., & Spindler, K. R. (2009). Mouse adenovirus type 1-induced breakdown of the blood-brain barrier. *Journal of Virology, 83*, 9398–9410.

Gu, J., Xie, Z., Gao, Z., Liu, J., Korteweg, C., Ye, J., et al. (2007). H5N1 infection of the respiratory tract and beyond: A molecular pathology study. *The Lancet, 370*, 1137–1145.

Hansson, G., Kristensson, K., Olsson, Y., & Sjostrand, J. (1971). Embryonal and postnatal development of mast cells in rat peripheral nerve. *Acta Neuropathologica, 17*, 139–149.

Hoyt, C. C., Bouchard, R. J., & Tyler, K. L. (2004). Novel nuclear herniations induced by nuclear localization of a viral protein. *Journal of Virology, 78*, 6360–6369.

Hoyt, C. C., Richardson-Burns, S. M., Goody, R. J., Robinson, B. A., Debiasi, R. L., & Tyler, K. L. (2005). Nonstructural protein σ1s is a determinant of reovirus virulence and influences the kinetics and severity of apoptosis induction in the heart and central nervous system. *Journal of Virology, 79*, 2743–2753.

Iden, S., Misselwitz, S., Peddibhotla, S. S. D., Tuncay, H., Rehder, D., Gerke, V., et al. (2012). aPKC phosphorylates JAM-A at Ser285 to promote cell contact maturation and tight junction formation. *The Journal of Cell Biology, 196*, 623–639.

Isberg, R. R., & Barnes, P. (2001). Subversion of integrins by enteropathogenic Yersinia. *Journal of Cell Science, 114*, 21–28.

Iwasaki, A. (2007). Mucosal dendritic cells. *Annual Review of Immunology, 25*, 381–418.

Jones, B. D., Ghori, N., & Falkow, S. (1994). Salmonella typhimurium initiates murine infection by penetrating and destroying the specialized epithelial M cells of the Peyer's patches. *The Journal of Experimental Medicine, 180*, 15–23.

Kanmogne, G. D., Primeaux, C., & Grammas, P. (2005). HIV-1 gp120 proteins alter tight junction protein expression and brain endothelial cell permeability: implications for the pathogenesis of HIV-associated dementia. *Journal of Neuropathology and Experimental Neurology, 64*, 498–505.

Karapanagiotou, E. M., Roulstone, V., Twigger, K., Ball, M., Tanay, M., Nutting, C., et al. (2012). Phase I/II trial of carboplatin and paclitaxel chemotherapy in combination with intravenous oncolytic reovirus in patients with advanced malignancies. *Clinical Cancer Research, 18*, 2080–2089.

Kirchner, E., Guglielmi, K. M., Strauss, H. M., Dermody, T. S., & Stehle, T. (2008). Structure of reovirus σ1 in complex with its receptor junctional adhesion molecule-A. *PLoS Pathology, 4*, e1000235.

Kobayashi, T., Antar, A. A. R., Boehme, K. W., Danthi, P., Eby, E. A., Guglielmi, K. M., et al. (2007). A plasmid-based reverse genetics system for animal double-stranded RNA viruses. *Cell Host & Microbe, 1*, 147–157.

Kobayashi, T., Ooms, L. S., Ikizler, M., Chappell, J. D., & Dermody, T. S. (2010). An improved reverse genetics system for mammalian orthoreoviruses. *Virology, 2*, 194–200.

Kottke, T., Chester, J., Ilett, E., Thompson, J., Diaz, R., Coffey, M., et al. (2011). Precise scheduling of chemotherapy primes VEGF-producing tumors for successful systemic oncolytic virotherapy. *Molecular Therapy, 19,* 1802–1812.
Kuiken, T., Fouchier, R. A., Schutten, M., Rimmelzwaan, G. F., van Amerongen, G., van Riel, D., et al. (2003). Newly discovered coronavirus as the primary cause of severe acute respiratory syndrome. *The Lancet, 362,* 263–270.
Lee, P. W., Hayes, E. C., & Joklik, W. K. (1981). Protein sigma 1 is the reovirus cell attachment protein. *Virology, 108,* 156–163.
MacLachlan, N. J., Jagels, G., Rossitto, P. V., Moore, P. F., & Heidner, H. W. (1990). The pathogenesis of experimental bluetongue virus infection of calves. *Veterinary Pathology, 27,* 223–229.
Maginnis, M. S., Forrest, J. C., Kopecky-Bromberg, S. A., Dickeson, S. K., Santoro, S. A., Zutter, M. M., et al. (2006). Beta1 integrin mediates internalization of mammalian reovirus. *Journal of Virology, 80,* 2760–2770.
Mahrt, C. R., & Osburn, B. I. (1986). Experimental bluetongue virus infection of sheep; effect of vaccination: Pathologic, immunofluorescent, and ultrastructural studies. *American Journal of Veterinary Research, 47,* 1198–1203.
Mainou, B. A., & Dermody, T. S. (2012). Transport to late endosomes is required for efficient reovirus infection. *Journal of Virology, 86,* 8346–8358.
Mandell, K. J., Babbin, B. A., Nusrat, A., & Parkos, C. A. (2005). Junctional adhesion molecule-1 (JAM1) regulates epithelial cell morphology through effects on β1 integrins and Rap1 activity. *The Journal of Biological Chemistry, 280,* 11665–11674.
Mann, M. A., Knipe, D. M., Fischbach, G. D., & Fields, B. N. (2002). Type 3 reovirus neuroinvasion after intramuscular inoculation: Direct invasion of nerve terminals and age-dependent pathogenesis. *Virology, 303,* 222–231.
Martin-Padura, I., Lostaglio, S., Schneemann, M., Williams, L., Romano, M., Fruscella, P., et al. (1998). Junctional adhesion molecule, a novel member of the immunoglobulin superfamily that distributes at intercellular junctions and modulates monocyte transmigration. *The Journal of Cell Biology, 142,* 117–127.
Mercier, G. T., Campbell, J. A., Chappell, J. D., Stehle, T., Dermody, T. S., & Barry, M. A. (2004). A chimeric adenovirus vector encoding reovirus attachment protein σ1 targets cells expressing junctional adhesion molecule 1. *Proceedings of the National Academy of Sciences of the United States of America, 101,* 6188–6193.
Morin, M. J., Warner, A., & Fields, B. N. (1994). A pathway for entry of reoviruses into the host through M cells of the respiratory tract. *The Journal of Experimental Medicine, 180,* 1523–1527.
Morin, M. J., Warner, A., & Fields, B. N. (1996). Reovirus infection in rat lungs as a model to study the pathogenesis of viral pneumonia. *Journal of Virology, 70,* 541–548.
Morrison, L. A., Sidman, R. L., & Fields, B. N. (1991). Direct spread of reovirus from the intestinal lumen to the central nervous system through vagal autonomic nerve fibers. *Proceedings of the National Academy of Sciences of the United States of America, 88,* 3852–3856.
Muthukkumar, S., Goldstein, J., & Stein, K. E. (2000). The ability of B cells and dendritic cells to present antigen increases during ontogeny. *The Journal of Immunology, 165,* 4803–4813.
Naik, U. P., Naik, M. U., Eckfeld, K., Martin-DeLeon, P., & Spychala, J. (2001). Characterization and chromosomal localization of JAM-1, a platelet receptor for a stimulatory monoclonal antibody. *Journal of Cell Science, 114,* 539–547.
Nason, E. L., Wetzel, J. D., Mukherjee, S. K., Barton, E. S., Prasad, B. V. V., & Dermody, T. S. (2001). A monoclonal antibody specific for reovirus outer-capsid protein σ3 inhibits σ1-mediated hemagglutination by steric hindrance. *Journal of Virology, 75,* 6625–6634.
Nathanson, N., & Tyler, K. L. (1997). Entry, dissemination, shedding, and transmission of viruses. In N. Nathanson (Ed.), *Viral pathogenesis* (pp. 13–33). Philadelphia: Lippincott-Raven.

Nibert, M. L., Chappell, J. D., & Dermody, T. S. (1995). Infectious subvirion particles of reovirus type 3 Dearing exhibit a loss in infectivity and contain a cleaved σ1 protein. *Journal of Virology, 69*, 5057–5067.

Nibert, M. L., Odegard, A. L., Agosto, M. A., Chandran, K., & Schiff, L. A. (2005). Putative autocleavage of reovirus mu1 protein in concert with outer-capsid disassembly and activation for membrane permeabilization. *Journal of Molecular Biology, 345*, 461–474.

Nomme, J., Fanning, A. S., Caffrey, M., Lye, M. F., Anderson, J. M., & Lavie, A. (2011). The Src homology 3 domain is required for junctional adhesion molecule binding to the third PDZ domain of the scaffolding protein ZO-1. *The Journal of Biological Chemistry, 286*, 43352–43360.

Nygaard, R., Golden, J. W., & Schiff, L. A. (2012). Impact of host proteases on reovirus infection in the respiratory tract. *Journal of Virology, 86*, 1238–1243.

Odegard, A. L., Chandran, K., Zhang, X., Parker, J. S., Baker, T. S., & Nibert, M. L. (2004). Putative autocleavage of outer capsid protein micro1, allowing release of myristoylated peptide micro1N during particle uncoating, is critical for cell entry by reovirus. *Journal of Virology, 78*, 8732–8745.

Oshiro, L. S., Dondero, D. V., Emmons, R. W., & Lennette, E. H. (1978). The development of Colorado tick fever virus within cells of the haemopoietic system. *The Journal of General Virology, 39*, 73–79.

Ostermann, G., Fraemohs, L., Baltus, T., Schober, A., Lietz, M., Zernecke, A., et al. (2005). Involvement of JAM-A in mononuclear cell recruitment on inflamed or atherosclerotic endothelium: Inhibition by soluble JAM-A. *Arteriosclerosis, Thrombosis, and Vascular Biology, 25*, 729–735.

Pacitti, A., & Gentsch, J. R. (1987). Inhibition of reovirus type 3 binding to host cells by sialylated glycoproteins is mediated through the viral attachment protein. *Journal of Virology, 61*, 1407–1415.

Pallansch, M. A., & Roos, R. P. (2001). Enteroviruses: Polioviruses, coxsacieviruses, echoviruses. In D. M. Knipe & P. M. Howley (Eds.), *Fields virology* (pp. 723–775). Philadelphia: Lippincott-Raven Press.

Parashar, U. D., Bresee, J. S., Gentsch, J. R., & Glass, R. I. (1998). Rotavirus. *Emerging Infectious Diseases, 4*, 561–570.

Parashar, K., Tarlow, M. J., & McCrae, M. A. (1992). Experimental reovirus type 3-induced murine biliary tract disease. *Journal of Pediatric Surgery, 27*, 843–847.

Paul, R. W., Choi, A. H., & Lee, P. W. K. (1989). The α-anomeric form of sialic acid is the minimal receptor determinant recognized by reovirus. *Virology, 172*, 382–385.

Paul, R. W., & Lee, P. W. K. (1987). Glycophorin is the reovirus receptor on human erythrocytes. *Virology, 159*, 94–101.

Poggioli, G. J., DeBiasi, R. L., Bickel, R., Jotte, R., Spalding, A., Johnson, G. L., et al. (2002). Reovirus-induced alterations in gene expression related to cell cycle regulation. *Journal of Virology, 76*, 2585–2594.

Poggioli, G. J., Dermody, T. S., & Tyler, K. L. (2001). Reovirus-induced σ1s-dependent G2/M cell cycle arrest results from inhibition of p34cdc2. *Journal of Virology, 75*, 7429–7434.

Poggioli, G. J., Keefer, C. J., Connolly, J. L., Dermody, T. S., & Tyler, K. L. (2000). Reovirus-induced G2/M cell cycle arrest requires σ1s and occurs in the absence of apoptosis. *Journal of Virology, 74*, 9562–9570.

Prota, A. E., Campbell, J. A., Schelling, P., Forrest, J. C., Peters, T. R., Watson, M. J., et al. (2003). Crystal structure of human junctional adhesion molecule 1: Implications for reovirus binding. *Proceedings of the National Academy of Sciences of the United States of America, 100*, 5366–5371.

Ramos-Alvarez, M., & Sabin, A. B. (1954). Characteristics of poliomyelitis and other enteric viruses recovered in tissue culture from healthy American children. *Proceedings of the Society for Experimental Biology and Medicine, 87*, 655–661.

Ramos-Alvarez, M., & Sabin, A. B. (1956). Intestinal flora of healthy children demonstrable by monkey kidney tissue culture. *American Journal of Public Health, 46*, 295–299.

Ramos-Alvarez, M., & Sabin, A. B. (1958). Enteropathogenic viruses and bacteria. Role in summer diarrheal diseases of infancy and early childhood. *JAMA: The Journal of the American Medical Association, 167*, 147–158.

Reiss, K., Stencel, J. E., Liu, Y., Blaum, B. S., Reiter, D. M., Feizi, T., et al. (2012). The GM2 glycan serves as a functional coreceptor for serotype 1 reovirus. *PLoS Pathogens, 8*, e1003078.

Reiter, D. M., Frierson, J. M., Halvorson, E. E., Kobayashi, T., Dermody, T. S., & Stehle, T. (2011). Crystal structure of reovirus attachment protein σ1 in complex with sialylated oligosaccharides. *PLoS Pathology*, 7, e1002166.

Rescigno, M., Rotta, G., Valzasina, B., & Ricciardi-Castagnoli, P. (2001). Dendritic cells shuttle microbes across gut epithelial monolayers. *Immunobiology, 204*, 572–581.

Richardson, S. C., Bishop, R. F., & Smith, A. L. (1994). Reovirus serotype 3 infection in infants with extrahepatic biliary atresia or neonatal hepatitis. *Journal of Gastroenterology and Hepatology, 9*, 264–268.

Roberts, S. R., Compans, R. W., & Wertz, G. W. (1995). Respiratory syncytial virus matures at the apical surfaces of polarized epithelial cells. *Journal of Virology, 69*, 2667–2673.

Rodgers, S. E., Connolly, J. L., Chappell, J. D., & Dermody, T. S. (1998). Reovirus growth in cell culture does not require the full complement of viral proteins: Identification of a σ1s-null mutant. *Journal of Virology, 72*, 8597–8604.

Rodriguez Boulan, E., & Sabatini, D. D. (1978). Asymmetric budding of viruses in epithelial monolayers: a model system for study of epithelial polarity. *Proceedings of the National Academy of Sciences of the United States of America, 75*, 5071–5075.

Rubin, D. H. (1987). Reovirus serotype 1 binds to the basolateral membrane of intestinal epithelial cells. *Microbial Pathogenesis, 3*, 215–220.

Rubin, D. H., Eaton, M. A., & Anderson, A. O. (1986). Reovirus infection in adult mice: The virus hemagglutinin determines the site of intestinal disease. *Microbial Pathogenesis, 1*, 79–87.

Rubin, D. H., Kornstein, M. J., & Anderson, A. O. (1985). Reovirus serotype 1 intestinal infection: A novel replicative cycle with ileal disease. *Journal of Virology, 53*, 391–398.

Sabin, A. B. (1956). Pathogenesis of poliomyelitis; reappraisal in the light of new data. *Science, 123*, 1151–1157.

Sabin, A. B. (1959). Reoviruses: A new group of respiratory and enteric viruses formerly classified as ECHO type 10 is described. *Science, 130*, 1387–1389.

Sacher, T., Podlech, J., Mohr, C. A., Jordan, S., Ruzsics, Z., Reddehase, M. J., et al. (2008). The major virus-producing cell type during murine cytomegalovirus infection, the hepatocyte, is not the source of virus dissemination in the host. *Cell Host & Microbe, 3*, 263–272.

Sarkar, G., Pelletier, J., Bassel-Duby, R., Jayasuriya, A., Fields, B. N., & Sonenberg, N. (1985). Identification of a new polypeptide coded by reovirus gene S1. *Journal of Virology, 54*, 720–725.

Sjostrand, J., & Karlsson, J. O. (1969). Axoplasmic transport in the optic nerve and tract of the rabbit: A biochemical and radioautographic study. *Journal of Neurochemistry, 16*, 833–844.

Sobocka, M. B., Sobocki, T., Babinska, A., Hartwig, J., Li, M., Ehrlich, Y. H., et al. (2004). Signaling pathways of the F11 receptor (F11R; a.k.a. JAM-1, JAM-A) in human platelets: F11R dimerization, phosphorylation and complex formation with the integrin GPIIIa. *Journal of Receptors and Signal Transduction Research, 24*, 85–105.

Sobocka, M. B., Sobocki, T., Banerjee, P., Weiss, C., Rushbrook, J. I., Norin, A. J., et al. (2000). Cloning of the human platelet F11 receptor: A cell adhesion molecule member of the immunoglobulin superfamily involved in platelet aggregation. *Blood, 95*, 2600–2609.

Tardieu, M., Powers, M. L., & Weiner, H. L. (1983). Age-dependent susceptibility to reovirus type 3 encephalitis: Role of viral and host factors. *Annals of Neurology*, *13*, 602–607.

Tardieu, M., & Weiner, H. L. (1982). Viral receptors on isolated murine and human ependymal cells. *Science*, *215*, 419–421.

Tyler, K. L., McPhee, D. A., & Fields, B. N. (1986). Distinct pathways of viral spread in the host determined by reovirus S1 gene segment. *Science*, *233*, 770–774.

Tyler, K. L., Sokol, R. J., Oberhaus, S. M., Le, M., Karrer, F. M., Narkewicz, M. R., et al. (1998). Detection of reovirus RNA in hepatobiliary tissues from patients with extrahepatic biliary atresia and choledochal cysts. *Hepatology*, *27*, 1475–1482.

van de Pavert, S. A., & Mebius, R. E. (2010). New insights into the development of lymphoid tissues. *Nature Reviews. Immunology*, *10*, 664–674.

Vazquez-Torres, A., Jones-Carson, J., Baumler, A. J., Falkow, S., Valdivia, R., Brown, W., et al. (1999). Extraintestinal dissemination of Salmonella by CD18-expressing phagocytes. *Nature*, *401*, 804–808.

Veronesi, E., Darpel, K. E., Hamblin, C., Carpenter, S., Takamatsu, H. H., Anthony, S. J., et al. (2009). Viraemia and clinical disease in Dorset Poll sheep following vaccination with live attenuated bluetongue virus vaccines serotypes 16 and 4. *Vaccine*, *28*, 1397–1403.

Weiner, H. L., Ault, K. A., & Fields, B. N. (1980). Interaction of reovirus with cell surface receptors. I. Murine and human lymphocytes have a receptor for the hemagglutinin of reovirus type 3. *The Journal of Immunology*, *124*, 2143–2148.

Weiner, H. L., Drayna, D., Averill, D. R., Jr., & Fields, B. N. (1977). Molecular basis of reovirus virulence: Role of the S1 gene. *Proceedings of the National Academy of Sciences of the United States of America*, *74*, 5744–5748.

Weiner, H. L., Powers, M. L., & Fields, B. N. (1980). Absolute linkage of virulence and central nervous system cell tropism of reoviruses to viral hemagglutinin. *The Journal of Infectious Diseases*, *141*, 609–616.

Williams, L. A., Martin-Padura, I., Dejana, E., Hogg, N., & Simmons, D. L. (1999). Identification and characterisation of human junctional adhesion molecule (JAM). *Molecular Immunology*, *36*, 1175–1188.

Wolf, J. L., Dambrauskas, R., Sharpe, A. H., & Trier, J. S. (1987). Adherence to and penetration of the intestinal epithelium by reovirus type 1 in neonatal mice. *Gastroenterology*, *92*, 82–91.

Wolf, J. L., Kauffman, R. S., Finberg, R., Dambrauskas, R., Fields, B. N., & Trier, J. S. (1983). Determinants of reovirus interaction with the intestinal M cells and absorptive cells of murine intestine. *Gastroenterology*, *85*, 291–300.

Wolf, J. L., Rubin, D. H., Finberg, R., Kaufman, R. S., Sharpe, A. H., Trier, J. S., et al. (1981). Intestinal M cells: A pathway of entry of reovirus into the host. *Science*, *212*, 471–472.

Woodfin, A., Reichel, C. A., Khandoga, A., Corada, M., Voisin, M. B., Scheiermann, C., et al. (2007). JAM-A mediates neutrophil transmigration in a stimulus-specific manner in vivo: Evidence for sequential roles for JAM-A and PECAM-1 in neutrophil transmigration. *Blood*, *110*, 1848–1856.

CHAPTER TWO

Molecular Biology and Epidemiology of Dianthoviruses

Tetsuro Okuno[*,1], **Chuji Hiruki**[†]

[*]Laboratory of Plant Pathology, Graduate School of Agriculture, Kyoto University, Sakyo-ku, Kyoto, Japan
[†]Department of Agricultural, Food, and Nutritional Science, University of Alberta, Edmonton, Alberta, Canada
[1]Corresponding author: e-mail address: okuno@kais.kyoto-u.ac.jp

Contents

Advances in Virus Research, Volume 87
ISSN 0065-3527
http://dx.doi.org/10.1016/B978-0-12-407698-3.00002-8

Abstract

The genus *Dianthovirus* is one of eight genera in the family *Tombusviridae*. All the genera have monopartite positive-stranded RNA genomes, except the dianthoviruses which have bipartite genomes. The dianthoviruses are distributed worldwide. Although they share common structural features with the other *Tombusviridae* viruses in their virions and the terminal structure of the genomic RNAs, the bipartite nature of the dianthovirus genome offers an ideal experimental system with which to study basic issues of virology. The two genomic RNAs seem to use distinct strategies to regulate their translation, transcription, genome replication, genome packaging, and cell-to-cell movement during infection. This review summarizes the current state of our knowledge of the dianthoviruses, with its main emphasis on the molecular biology of the virus, including the viral and host factors required for its infection of host plants. The epidemiology of the virus and the possible viral impacts on agriculture and the environment are also discussed.

1. INTRODUCTION

Since the dianthoviruses were established as a new plant virus group by the International Committee on Taxonomy of Viruses (ICTV) in 1981 (Matthews, 1982), much information has been collected not only on their basic characteristics but also on the effects of their infection of various plant species and the behavior of the viruses in different environments. While members of this group are small in number, they cause characteristic necrosis, both local and systemic, which is often associated with severe dwarfing, a frequent cause of serious reductions in crop yields. The number of *Dianthovirus* species may increase as the molecular genetic characterization of suspected viruses' advances. The viral particles are stable and abundant in reproduction, and are therefore suitable for experimental studies in many aspects.

Our knowledge of the molecular mechanisms underlying the infection processes of the dianthoviruses has increased over the past two decades, after their infectious cDNA clones became available (Ge, Hiruki, & Roy, 1993; Sit, Haikal, Callaway, & Lommel, 2001; Xiong & Lommel, 1991) since the complete nucleotide sequences of their genomes were determined (Ge, Hiruki, & Roy, 1992; Ge et al., 1993; Kendall & Lommel, 1992; Lommel, Weston-Fina, Xiong, & Lomonossoff, 1988; Ryabov, Generozov, Kendall, Lommel, & Zavriev, 1994; Xiong & Lommel, 1989).

This review summarizes the current state of our knowledge of the dianthoviruses, with its main emphasis on recent advances in our understanding of the molecular mechanisms underlying the genome strategies used for translation, replication, and cell-to-cell movement during infection. The epidemiology and possible impacts of the dianthoviruses on agriculture and the environment are also discussed.

2. DIANTHOVIRUSES

The group consists of four viruses: the type member is *Carnation ringspot virus* (CRSV) and the other members are *Red clover necrotic mosaic virus* (RCNMV), *Sweet clover necrotic mosaic virus* (SCNMV) (Hiruki, 1987); *Furcraea necrotic streak virus* (FNSV) is a tentative member of the genus at present (Morales, Castaño, Calvert, & Arroyave, 1992).

2.1. CRSV

Symptoms such as leaf mottling and ringspotting, plant stunting, and flower distortion persist after the infection of *Dianthus caryophyllus* and *D. barbatus* by CRSV. CRSV probably occurs worldwide, wherever *Dianthus* species are grown. Virions are distributed in all parts of the infected plant, including in the cell cytoplasm and nuclei. Several species are susceptible to CRSV, showing systemic or local symptoms (Tremaine & Dodds, 1985). Its transmission by nematodes *Longidorus elongatus*, *L. macrosoma*, and *Xiphinema diversicaudatum* has been reported (Fritzsche, Kegler, Thiele, & Gruber, 1979) but is still contentious (Brunt, Crabtree, Dallwitz, Gibbs, & Watson, 1996).

2.2. RCNMV

Symptoms such as necrotic lesions accompanied with mosaic and leaf distortion persist in *Trifolium pratense*, *T. repens*, *Medicago sativa*, and *Melilotus officinalis* infected with RCNMV. The virus was first reported in the former Czechoslovakia (Musil, 1969; Musil & Matisová, 1967) but is known to occur in the United Kingdom, Australia, Canada, New Zealand, Poland, Sweden, and the United States. The virus was reported to be experimentally transmissible by the chytrid *Olpidium radicale* but is not transmitted by seed (Gerhardson & Insunza, 1979).

2.3. SCNMV

The main symptoms of SCNMV are mosaic, ringspots, and systemic veinal necrosis in the leaves of *M. officinalis* and *M. alba* (Hiruki, Rao, Chen,

Okuno, & Figueiredo, 1984). The most severe symptoms include stunting and distortion. SCNMV is widely spread in sweet clover growing in regions of dark-gray soil and black soil in Alberta, Canada (Hiruki, 1986a, 1986b, 1987). A different strain of the virus occurs in *M. sativa* (Pappu & Hiruki, 1989; Pappu, Hiruki, & Inouye, 1988). SCNMV is transmitted by contact and by drainage water but not by *Olpidium brassicae* (Hiruki, 1986b, 1994). Sixteen of 25 species in 6 dicotyledonous families were infected by inoculation with sap (Hiruki, 1986a, 1987).

2.4. FNSV

The major symptoms of FNSV are plant stunting and chlorotic streaks, which later coalesce to produce necrosis, drastically reducing the length and quality of the fiber of *Furcraea* (fique) spp., the most important fiber crops in Colombia. The virus is transmitted by mechanical inoculation and by grafting but not transmitted by *Planococcus citri*, *Saissetia coffeae*, *Xiphinema*, or *Trichodorus* spp. (Morales et al., 1992). The virus is consistently detected in the roots of infected fique plants (Morales et al., 1992).

2.5. Other suspected dianthoviruses

In 1996, Editors of *Viruses of Plant Viruses* included the following statement (Brunt et al., 1996, p. 1069) that was apparently collated by Dr. M. Hollings in 1980 under *Taxonomy and Relationships* of *Red clover necrotic mosaic dianthovirus*. Virus(es) with serologically unrelated virions: 60 strains of 45 isometric viruses. Some isolates induce different symptoms, for example, with those from England and Scotland are different. *Tetragonia expansa* is infected systemically only by RCNMV strain TpM-34.

An RNA with a gene organization very similar to that of dianthovirus RNA1 was reported in grassy stunt-diseased rice plants, although no virions were isolated nor any biological characters investigated (Miranda, Aliyar, & Shirako, 2001).

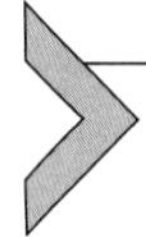

3. ECONOMIC IMPORTANCE OF DISEASES CAUSED BY DIANTHOVIRUSES

The economic importance of the diseases caused by the dianthoviruses varies according to the kind of crop affected. CRSV, the type virus of the dianthovirus group, is fairly common, infecting *Dianthus* spp. worldwide, spreading across the Central Asian region and India, the Eurasian region, and the North American region. It has been found in Australia but with no evidence of field spread (Brunt et al., 1996).

RCNMV has been reported to occur in the former Czechoslovakia (Musil, 1969; Musil & Matisová, 1967), Sweden (Gerhardson & Lindsten, 1973), and the United Kingdom (Bowen & Plumb, 1979). It has been found in Australia, Canada, Poland, and New Zealand but with no evidence of field spread (Brunt et al., 1996).

In Alberta, Canada, the production of forage seeds, as well as hay, is very important in supporting the $3 billion cattle industry in the province, which is a major source of income in agricultural industry. Sweet clover is also important to beekeeping as a source of nectar and pollen, and beekeeping is an important enterprise in Alberta's agriculture diversification strategy. It contributes almost $15 million to the agricultural economy. SCNMV infection of sweet clover not only hampers plant growth but also significantly diminishes the production of nectar and pollen.

The genus *Furcraea* (Agavaceae) contains several species found primarily in the American tropics. In Colombia, South America, four varieties of *Furcraea* spp., collectively known as "fique," constitute the major fiber crop with over 40,000 tons of "cabuya" fiber produced annually by small farmers. Among the various biotic constraints that affect commercial *Furcraea* production, necrotic streak caused by FNSV is the most difficult to control because no reliable control methods are available (Morales et al., 1992).

4. MOLECULAR BIOLOGY OF DIANTHOVIRUS

Dianthovirus belongs to the family *Tombusviridae*. The unifying feature of this family is that each member has a highly conserved RNA-dependent RNA polymerase (RdRP) motif. The family *Tombusviridae* is classified into the supergroup that includes the families *Flaviviridae* and *Luteoviridae* and the phage lineage (Koonin, 1991; Koonin & Dolja, 1993). The molecular mechanisms underlying the infection processes of *Dianthovirus*, including its translation, RNA replication, packaging, cell-to-cell movement, and suppression of RNA silencing, have been investigated predominantly in RCNMV (Basnayake, Sit, & Lommel, 2006; Giesman-Cookmeyer & Lommel, 1993; Mine & Okuno, 2012; Okuno, 2012; Takeda et al., 2005). Recent studies have increased our understanding of how the dianthoviruses replicate in host cells during successful infection. They use several unique strategies to regulate the expression of viral genes, to form viral replication factories by recruiting host proteins on cellular membranes, and to counteract host defense mechanisms. Here, we comprehensively review the molecular biology of *Dianthovirus*.

4.1. Genome organization

The genomes of the dianthoviruses consist of two positive-sense single-stranded RNAs, RNA1 and RNA2 (Gould, Francki, Hatta, & Hollings, 1981; Hiruki, 1987; Okuno, Hiruki, Rao, & Figueiredo, 1983). This bipartite nature of the genome is unique among the viruses of the family *Tombusviridae*, the genomes of which are usually monopartite. RNA1 (3.9 kb) encodes the RNA replicase components, an auxiliary 27-kDa protein (p27), and an N-terminally overlapping 88-kDa protein (p88) (Xiong, Kim, Kendall, & Lommel, 1993; Xiong & Lommel, 1989) that contains RdRP motif (Koonin & Dolja, 1993). RNA1 also encodes a coat protein (CP), which is translated from subgenomic RNA (sgRNA) (Sit, Vaewhongs, & Lommel, 1998; Tatsuta, Mizumoto, Kaido, Mise, & Okuno, 2005; Zavriev, Hickey, & Lommel, 1996). RNA1 can replicate in a single cell in the absence of RNA2 (Osman & Buck, 1987; Paje-Manalo & Lommel, 1989; Pappu & Hiruki, 1988). RNA2 encodes a 35-kDa movement protein (MP), which is essential for the cell-to-cell and systemic movement in plants (Osman & Buck, 1987; Xiong, Kim, Giesman-Cookmeyer, & Lommel, 1993). In addition to the genomic RNAs, a small noncoding RNA (0.4 kb) that consists of nearly the entire region of the 3′-untranslated region (UTR) of RNA1 accumulates in dianthovirus-infected plants and protoplasts (Iwakawa et al., 2008; H. Nagano, K. Omote, & T. Okuno, unpublished data). The genomic organization of RCNMV, several RNA elements of which are described in this review, is shown in Fig. 2.1.

4.2. Cap-independent translation

The genomic RNAs of RCNMV lack both a 5′-cap structure (Mizumoto, Tatsuta, Kaido, Mise, & Okuno, 2003) and a 3′ poly(A) tail (Lommel et al., 1988; Mizumoto, Hikichi, & Okuno, 2002; Mizumoto et al., 2003; Xiong & Lommel, 1989). Mutagenesis studies using a reporter mRNA that included the firefly luciferase (*Luc*) gene revealed that RNA1 contains an RNA element essential for the cap-independent translation in the 3′-UTR (Mizumoto et al., 2003). The element designated “3′TE-DR1” is predicted to have five stem-loop (SL) structures (Mizumoto et al., 2003;

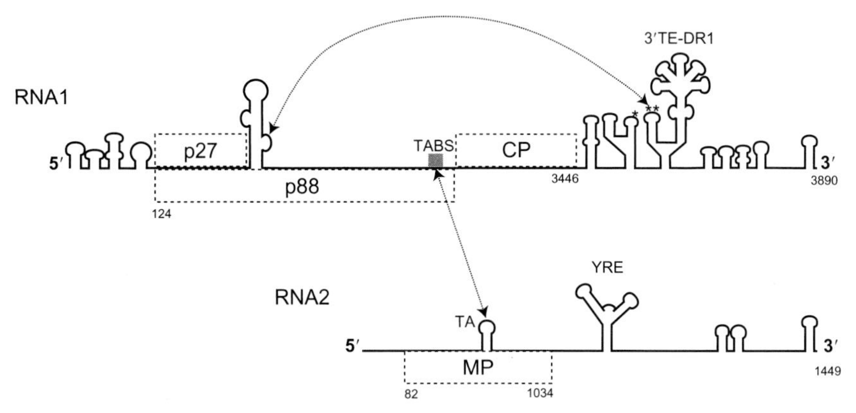

Figure 2.1 Schematic representation of the genome structure of RCNMV, with the predicted RNA structures. TABS, *trans*-activator-binding site; 3′TE-DR1, cap-independent translational enhancer; TA, *trans*-activator; YRE, Y-shaped RNA element. *A-rich sequence (ARS): the PABP-binding site. **Loop sequences required for the long-distance RNA–RNA interaction that facilitates −1PFS. Nucleotide numbers from the 5′ end are shown below the lines.

Sarawaneeyaruk et al., 2009), and this overall structure has been confirmed by structure probing (Wang, Kraft, Hui, & Miller, 2010; Y. Tajima & T. Okuno, unpublished data). Of these SLs, the 5′-proximal SL and two 3′-proximal SLs are required for its efficient translation (Mizumoto et al., 2003; Sarawaneeyaruk et al., 2009).

Cap-independent translation enhancers like 3′TE-DR1 have been identified in the 3′-UTRs of many viral RNA genomes (Kneller, Rakotondrafara, & Miller, 2006; Nicholson & White, 2011). These RNA elements replace the cap-structure and poly(A) tail in facilitating the translation of viral proteins and are termed as 3′ cap-independent translation elements (3′CITEs). The 3′CITEs of plant RNA viruses are classified into several structural classes (Miller, Wang, & Treder, 2007; Nicholson & White, 2011). The 5′-proximal SL of 3′TE-DR1 that is conserved in dianthoviruses, including SCNMV and CRSV, is very similar to an essential SL structure in the 3′CITE of *Barley yellow dwarf virus* (BYDV) (Guo, Allen, & Miller, 2000; Mizumoto et al., 2003; Shen & Miller, 2004).

3′CITEs interact with the eukaryotic initiation factors eIF4F and eIFiso4F (Gazo, Murphy, Gatchel, & Browning, 2004; Nicholson, Wu, Chevtchenko, & White, 2010; Treder et al., 2008; Wang, Treder, & Miller, 2009) or the 60S ribosomal subunits (Stupina et al., 2008). The 3′-UTR of RCNMV RNA1 interacted with the eIF4F/iso4F components in an RNA-aptamer (Strepto-Tag) affinity assay in a cell-free extract of evacuolated BY-2 protoplasts (BYL) (Iwakawa et al., 2012). BYL (Komoda, Naito, & Ishikawa, 2004) efficiently supports the 3′TE-DR1-mediated cap-independent translation and negative-strand RNA synthesis of RCNMV (Iwakawa, Kaido, Mise, & Okuno, 2007). Mutations in the 5′-proximal SL of 3′TE-DR1 abolish the association of these translation initiation factors with the 3′-UTR.

Poly(A)-binding protein (PABP) also coprecipitates with the Strepto-Tag-fused 3′-UTR (Iwakawa et al., 2012). The binding site of PABP resides in an adenine-rich sequence (ARS) (3518-AAACAGUAAAAUUGC AAAAAA-3538) located 60 nucleotides upstream from 3′TE-DR1 in the 3′-UTR. Interestingly, the mutations in ARS that compromise PABP binding also compromise the binding of the eIF4F/iso4F components and abolish 3′TE-DR1-mediated cap-independent translation, whereas the mutations in 3′TE-DR1 that abolish translation have no effect on PABP binding. Therefore, PABP binding seems to be required for the recruitment of the eIF4F/iso4F components to 3′TE-DR1. The binding of both the eIF4F/iso4F components and PABP to the 3′ RNA elements is required for the efficient recruitment of the 48S and 80S ribosome complexes to the viral RNA (Iwakawa et al., 2012).

Some form of communication must occur between the 5′-UTR and the 3′CITE of the viral RNAs to deliver translation factors for recruiting ribosomes to the 5′ translation initiation sites. Such communications could be mediated by long-distance RNA–RNA base pairing, as reported for BYDV (Guo, Allen, & Miller, 2001), *Black currant reversion nepovirus* (Karetnikov & Lehto, 2008), *Carnation Italian ringspot tombusvirus* (Nicholson & White, 2008), *Cucumber leaf spot aureusvirus* (Xu & White, 2009), *Saguaro cactus virus* (Chattopadhyay, Shi, Yuan, & Simon, 2011), and *Tomato bushy stunt virus* (TBSV) (Fabian & White, 2004, 2006). Alternatively and additionally, mechanisms other than RNA–RNA base pairing could mediate the 5′–3′ interaction of viral RNAs. This is the case for RCNMV RNA1.

The requirement for the RNA1 5′-UTR in 3′TE-DR1-mediated cap-independent translation differs greatly between plant species (Sarawaneeyaruk et al., 2009). The deletion of any one of the four SLs predicted in the 5′-UTR of RNA1 inhibited the 3′TE-DR1-mediated cap-independent activity of reporter mRNAs in BY-2 tobacco protoplasts, whereas their deletion had no effect in cowpea protoplasts. The RNA1 5′-UTR contributes to RNA stability in BY-2 protoplasts. A compensatory mutagenesis analysis used to identify possible interactions between the 5′- and 3′-UTRs of RNA1 in BY-2 protoplasts suggested that no long-distance RNA–RNA interaction is essential for 3′–5′ communication during

3′TE-DR1-mediated cap-independent translation (Sarawaneeyaruk et al., 2009; S. Sarawaneeyaruk & T. Okuno, unpublished data). Similarly, little or no involvement of RNA–RNA interactions between the 5′- and 3′-UTRs has been reported in the 3′CITE-mediated cap-independent translation of *Pea enation mosaic virus* RNA 2 (Wang et al., 2009), satellite *Tobacco necrosis virus* (Gazo et al., 2004), or *Turnip crinkle virus* (TCV) (Qu & Morris, 2000; Stupina et al., 2008). The mechanisms for delivering the translation initiation factors and translation initiation ribosome complexes to the 5′-UTR of the viral RNAs in these viruses, including RCNMV, remain to be resolved.

CP is translated from CPsgRNA, which is transcribed from RNA1 and coterminates with RNA1. This implies that CP is translated in a 3′TE-DR1-mediated cap-independent manner. Indeed, uncapped chimeric CPsgRNA, in which the CP open reading frame (ORF) is substituted with the Luc ORF, is translated as efficiently as RNA1 (Sarawaneeyaruk et al., 2009).

In contrast to RNA1 and CPsgRNA, RNA2 lacks RNA elements, such as ARS and 3′TE-DR1, that function efficiently in a reporter mRNA. Reporter Luc mRNAs with the 5′- and 3′-UTRs of RNA2 (R2–UTR–Luc) or with the Luc ORF inserted between the 5′-UTR and the MP ORF (RNA2–Luc) in RNA2 are not translated efficiently in protoplasts in the absence of the 5′-cap structure (Mizumoto, Iwakawa, Kaido, Mise, & Okuno, 2006). The cap-independent translational activities of R2–UTR–Luc and RNA2–Luc are less than 1% of 3′TE-DR1-mediated cap-independent translation. Instead, RNA2–Luc, which can replicate in the presence of RCNMV replicase proteins, functions as an efficient mRNA when replicated (Mizumoto et al., 2006). This suggests a strong link between the cap-independent translation of RNA2 and RNA replication and a difference in the translational mechanisms of RNA1 and RNA2. Host factors associated with the viral RNA replication process might facilitate the translation of RNA2. Alternatively, RNA2 may have a silencer element(s) that suppresses the translation of RNA2. The low translational activity of RNA2 may be comprehensible, if MP is only required in the late stage of replication to move the virus to neighboring cells. The difference in the translation strategy of RNA1 and RNA2 may be important for the temporal, spatial, and quantitative regulation of RCNMV gene expression during infection.

4.3. Translation via −1 programmed ribosome frameshifting

−1 Programmed ribosome frameshifting is one of the translation strategies used by many RNA viruses to regulate viral gene expression (Brierley, 1995). p88 is translated by −1PRF from RCNMV RNA1 (Xiong, Kim, Kendall, et al., 1993). The *cis*-acting RNA elements required for −1PRF have been mapped to a shifty heptanucleotide sequence GGAUUUU at the slippage site (Kim & Lommel, 1994) and a highly-structured bulged SL structure predicted just downstream from the slippage site in RNA1 (Kim & Lommel, 1998). In addition to these elements, a third *cis*-acting RNA element that facilitates −1PRF was identified using an *in vitro* translation/replication system (Tajima, Iwakawa, Kaido, Mise, & Okuno, 2011). This third RNA element is a small stable SL structure predicted between ARS and 3′TE-DR1 in the 3′-UTR of RNA1. The loop sequence of this element can base pair with the bulge of the SL adjacent to the shifty site. Such long-distance base pairing is possible in all dianthoviruses. Disruption and restoration mutagenesis analyses have demonstrated the importance of long-distance base pairing for efficient −1PRF in RCNMV RNA1 (Tajima et al., 2011). A similar requirement for long-distant RNA–RNA communication has been reported for the −1PRF of BYDV (Barry & Miller, 2002).

4.4. *cis*-Preferential requirement of replicase protein

An RNA1 mutant expressing p88 could be replicated efficiently in protoplasts when p27 was supplied in *trans*, but an RNA1 mutant expressing p27 alone could not be replicated when p88 was supplied in *trans* (Okamoto et al., 2008). Thus, only RCNMV RNA1, from which p88 is translated, is an effective template, on which viral RNA replicase can initiate RNA synthesis in the presence of p27. The *cis*-Preferential function of viral-encoded replication proteins or the coupling of translation and replication has been reported for several viruses, including *Alfalfa mosaic virus* (Neeleman & Bol, 1999; van Rossum, Garcia, & Bol, 1996), *Brome mosaic virus* (Yi & Kao, 2008), *Clover yellow mosaic virus* (White, Bancroft, & Mackie, 1992), Coronavirus (Chang, Hofmann, Sethna, & Brian, 1994), *Cowpea mosaic virus* (Van Bokhoven et al., 1993), Poliovirus (Hagino-Yamagishi & Nomoto, 1989; Johnson & Sarnow, 1991; Novak & Kirkegaard, 1994), TCV (White, Skuzeski, Li, Wei, & Morris, 1995), *Tobacco etch virus* (Mahajan, Dolja, & Carrington, 1996; Schaad, Haldeman-Cahill, Cronin, & Carrington, 1996), *Tobacco mosaic virus* (TMV) (Lewandowski & Dawson, 2000), TBSV (Oster, Wu, & White, 1998), *Turnip yellow mosaic virus* (Weiland & Dreher, 1993), and *Rubella virus* (Liang & Gillam, 2001).

4.5. RNA elements required for genome replication

The *cis*-acting RNA elements required for viral RNA replication have been identified for many viruses with mutagenesis studies, when the RNA can be replicated by a viral replicase supplied in *trans*. However, it is difficult to apply this method to viral RNAs that encode replication proteins that are required in *cis* for RNA replication. This is the case for RCNMV RNA1.

The RNA elements required for the negative-strand synthesis of RNA1 have been determined in its 3′-UTR using capped viral RNA transcripts in BYL and BY-2 protoplasts (Iwakawa et al., 2007). The use of capped viral RNA transcripts in BYL allowed the effects of introduced mutations on cap-independent translation and the negative-strand RNA synthesis of RNA1 to be distinguished. Two SL structures at the 3′ end of RNA1 and the intervening sequence between the two SLs are essential for negative-strand synthesis (Iwakawa et al., 2007) (see Fig. 2.1). These RNA elements are conserved between RNA1 and RNA2 and between SCNMV and CRSV. Interestingly, the core RNA element of 3′TE-DR1, which is essential for cap-independent translation of RNA1, is not essential for negative-strand RNA synthesis (Iwakawa et al., 2007). Thus, the RNA elements in the 3′-UTR of RCNMV RNA1 that are required for negative-strand RNA synthesis are separated from those required for cap-independent translation.

RCNMV-Australian (Aus) and RCNMV-Canadian (Can) strains have different temperature sensitivities during infection (Mizumoto et al., 2002). RCNMV-Can does not replicate in protoplasts at temperatures higher than 22 °C, indicating that replication processes are involved in the temperature-sensitive phenotype. Mutagenesis studies have shown that the temperature sensitivity of RCNMV is attributable to at least the 3′ terminal SL of RNA1, because a single-nucleotide substitution from U to C, which changes a U–G wobble pair to a stable C–G pair in the stem of the 3′ terminal SL in RNA1-Can allows the RNA to replicate and to support RNA2 accumulation at nonpermissive temperatures (Mizumoto et al., 2002). The lack of temperature sensitivity in RNA synthesis activity in the crude membrane-bound cellular fraction prepared from RCNMV-Can-infected plants or protoplasts suggests that the promoter elements are distorted in the initiation of minus-strand synthesis at the nonpermissive temperatures (K. Shimada & T. Okuno, unpublished data).

In contrast to RNA1, RNA2 is replicated efficiently by the RCNMV replicase proteins supplied in *trans*. The core promoter of negative-strand RNA synthesis is located in the 3′ proximal region of RNA2, which is homologous to that of RNA1, except for the presence of 66 extra nucleotides upstream from the 3′ terminal SL, which are not essential for replication (Iwakawa et al., 2007; Turner & Buck, 1999). Three discontinuous nucleotides in the loop of the 3′ terminal SL are thought to be involved in the interaction with the RCNMV RNA replicase (Weng & Xiong, 2009).

In addition to the core promoter, RNA2 has other unique replication elements. These elements include a *trans*-activator (TA) in the MP ORF (Tatsuta et al., 2005) and a Y-shaped RNA element (YRE) in the 5′ proximal region of the 3′-UTR (An et al., 2010; Iwakawa et al., 2011; Fig. 2.1). Interestingly, TA is the RNA element that interacts with RNA1 and enhances the RNA-mediated transcription of CPsgRNA (Sit et al., 1998) (see Section 4.10). Both TA and YRE play crucial roles in the negative-strand RNA synthesis of RNA2 (An et al., 2010).

YRE consists of two small SLs with a short intervening region between them on the basal stem structure (An et al., 2010). This structure is conserved among the dianthoviruses. The entire Y-shaped structure (84 nucleotides), including the two small SLs, is important for the negative-strand RNA synthesis of RNA2 (An et al., 2010). YRE is the only RNA element of RCNMV that interacted with p27 supplied in *trans*, when assessed with the RNA-aptamer (Strepto-Tag) affinity and immunoprecipitation assay in BYL (Iwakawa et al., 2011). YRE also interacts with the 480-kDa replicase complex via p27 (Iwakawa et al., 2011). The 480-kDa replicase complex contains p88 and host proteins, and is thought to be a key player in RCNMV RNA replication (Mine, Takeda, et al., 2010) (see Sections 4.7 and 4.8). The interaction between YRE and p27 is required for the recruitment of RNA2 to the endoplasmic reticulum (ER) membrane (Hyodo et al., 2011), which is the site of RCNMV RNA replication (Hyodo et al., 2013; Turner, Sit, Callaway, Allen, & Lommel, 2004).

4.6. Template recognition mechanisms of replicase proteins

RNA1 lacks the RNA elements that interact with p27 and p88 supplied in *trans*, as determined in the Strepto-Tag affinity and immunoprecipitation assay in BYL (Iwakawa et al., 2011). Interestingly, however, a protein-mediated coimmunoprecipitation analysis showed that both p27 and p88 can interact with their translating template RNA1, which are associated with ribosomes. The interaction between p27 or p88 and its translation templates is abolished or compromised, respectively, by puromycin treatment, which induces the dissociation of polyribosomes from mRNA (Blobel & Sabatini, 1971; Lehninger, Nelson, & Cox, 1993). These results imply that p27 binds to the template RNA1 via a polyribosome-dependent and puromycin-sensitive mechanism, whereas p88 mainly binds to the template RNA in a translation-coupled, polyribosome-independent, and puromycin-tolerant manner. The interaction mechanism unique to p88 is maintained after the dissociation of the polyribosomes. p88 seems to bind specifically to the 3′-UTR of RNA1 in this translation-coupled mechanism, because SR1f, a degradation product of the 3′-UTR of RNA1 (see Section 4.11), is coimmunoprecipitated with p88 but not with p27. The translation-coupled binding of p88 may partly explain the strong *cis*-preferential requirement for p88 in the replication of RNA1 (Okamoto et al., 2008).

4.7. Composition of RNA replicase complexes

A template-bound solubilized RNA polymerase has been isolated from RCNMV-infected *Nicotiana clevelandii* plants (Bates, Farjah, Osman, & Buck, 1995). The polymerase becomes template dependent after removing endogenous RNA templates with micrococcal nuclease. The RdRP contains p27 and p88 with several unknown host proteins and produces double-stranded RNA in a template-specific manner (Bates et al., 1995).

The *Agrobacterium*-mediated expression of RCNMV replicase proteins and RNAs in *N. benthamiana* leaves allowed large detergent-solubilized membrane-associated protein complexes to be isolated, with an apparent molecular mass of 480-kDa on blue-native polyacrylamide gel electrophoresis (BN–PAGE) (Mine, Takeda, et al., 2010). The 480-kDa integral membrane complex contains both p27 and p88, and is associated with possible host proteins. In sucrose gradient sedimentation, the 480-kDa complex cofractionates with both endogenous template-bound and exogenous template-dependent RdRP activities. The 480-kDa complex corresponds predominantly to the exogenous template-dependent RdRP activity, and specifically recognizes the 3′ core promoter sequences of the RCNMV genomic RNAs to produce viral RNA fragments. In contrast, the endogenous template-bound RdRP produces genome-sized RNAs without the addition of RNA templates (Mine, Takeda, et al., 2010).

Analysis of the affinity-purified solubilized membrane-bound RdRP complexes using two-dimensional BN/SDS–PAGE and mass spectrometry showed that the RdRP complexes contain many host proteins, including Hsp70, Hsp90, glyceraldehyde 3-phosphate dehydrogenase, ADP-ribosylation factor 1 (Arf1), histone deacetylase 1, ubiquitin, and several ribosomal proteins, in addition to viral replicase proteins (Hyodo et al., 2013; Mine, Takeda, et al., 2010).

In addition to the 480-kDa complex, a 380-kDa complex is formed in BYL and RCNMV-infected plant tissues expressing p27 alone or both p27 and p88

(Mine, Hyodo, et al., 2010; Mine, Takeda, et al., 2010). The 380-kDa complex could be a p27 oligomer with some host proteins.

4.8. Assembly of the replicase complex

The assembly of the viral replicase complexes of eukaryotic positive-strand RNA viruses is a regulated process: multiple viral and host proteins and template RNAs must be assembled on intracellular membranes and organized into quaternary complexes capable of synthesizing viral RNAs (Mine & Okuno, 2012; Nagy & Pogany, 2012).

p27 interacts with both p27 and p88 through direct protein–protein contacts. The C-terminal half of p27 is responsible for these interactions, whereas the nonoverlapping region unique to p88 is responsible for the p27–p88 interaction (Mine, Hyodo, et al., 2010). Both the p27–p27 and p27–p88 interactions are required for the formation of the 480-kDa complex *in vitro* and *in planta* (Mine, Hyodo, et al., 2010). A mutant p27 incapable of interacting with p27, but capable of interacting with p88, fails to form the 480-kDa complex. Another mutant p27 capable of interacting with p27, but incapable of interacting with p88, formed the 380-kDa complex but failed to form the 480-kDa complex. Thus, p27 oligomerization is a critical step in the formation of the 480-kDa complex. It appears that the formation of the 480-kDa complex is directed by the p27-oligomer (the 380-kDa complex), which interacts with the p88 protein(s).

The formation of the 480-kDa replicase complex is enhanced by the presence of the RCNMV genomic RNAs (Mine, Takeda, et al., 2010). The RNA-binding domain of p27 was identified using an RNA-aptamer (Strepto-Tag)-fused YRE that binds p27 in an affinity and immunoprecipitation assay in BYL (Hyodo et al., 2011). Deletion and alanine-scanning mutation analyses indicated that the main functional domains required for RNA binding differ from those required for protein binding, although they partially overlap.

There is a robust correlation between the RNA-binding activity of p27 and its RNA-recruiting activity to the ER membrane (Hyodo et al., 2011). Interestingly, several p27 mutants that retain the ability to bind to RNA2, to recruit RNA2 to the membrane, and to interact with both p27 and p88, fail to form the 480-kDa complex or to support RNA replication (Hyodo et al., 2011). These p27 mutants might lack the ability to interact, directly or indirectly, with the host proteins that are required for the proper assembly of the 480-kDa replicase complex.

The roles in viral RNA replication of Hsp70 and Hsp90 detected in the affinity-purified solubilized membrane-bound RdRP complexes of RCNMV (Mine, Hyodo, et al., 2010) were investigated (Mine et al., 2012). A bimolecular fluorescence complementation (BiFC) assay using confocal microscopy showed that p27, but not p88, interacts directly with both Hsp70 and Hsp90 within the p27-induced large aggregated structures on the ER in the perinuclear region. Gene silencing and the pharmacological inhibition of Hsp70 and Hsp90 compromised RCNMV RNA replication in plant cells. The inhibition of p27–Hsp70 interaction by 2-phenylethynesulfonamide (a specific inhibitor of Hsp70) inhibited the formation of the 480-kDa complex, but instead induced the formation of nonfunctional large complexes (~1024 kDa) in BYL. Hsp70 appears to control the proper assembly of the viral replicase complexes by preventing the aggregation of p27. Alternatively, the large complexes of p27 could be intermediates in the assembly of the RCNMV replicase complex, and Hsp70 might assist the assembly of these complexes into functional replicase complexes. In contrast, the inhibition of p27–Hsp90 interaction by geldanamycin (a specific inhibitor of Hsp90) inhibited the formation of the 480-kDa complex without inducing large complexes, and rendered p27 incapable of binding to a specific viral RNA element (YRE), which is a critical step in the assembly of the 480-kDa replicase complex. These findings suggest that Hsp70 and Hsp90 play essential roles in the assembly of the 480-kDa replicase complex mainly by regulating protein–protein interactions and protein–RNA interactions, respectively. Thus, Hsp70 and Hsp90 regulate different steps in the assembly of the RCNMV replicase complex.

4.9. Viral RNA replication factory

The replication of eukaryotic positive-strand RNA viruses occurs on membranes of selected subcellular organelles, such as the ER, chloroplasts, mitochondria, peroxisomes, or tonoplasts in infected cells (den Boon & Ahlquist, 2010; Mine & Okuno, 2012; Nagy & Pogany, 2012; Salonen, Ahola, & Kaariainen, 2005). Viral proteins play essential roles in targeting the viral replication complexes to these membranes and induce morphological changes in these membranes (Belov & van Kuppeveld, 2012; den Boon & Ahlquist, 2010; Miller & Krijnse-Locker, 2008).

Confocal microscopy using green fluorescent protein (GFP)-fused p27 and p88 showed that both proteins colocalize to the cortical and cytoplasmic

ER, and that p27 causes the restructuring and proliferation of the ER membrane (Kusumanegara et al., 2012; Mine et al., 2012; Turner et al., 2004). The domains and critical amino acids in p27 required for its association with and targeting of ER membranes were determined using a C-terminally GFP-fused p27 (p27–GFP), which supported viral RNA replication in the presence of p88. Membrane-flotation assays combined with microscopic observation revealed that the membrane association of p27 is mediated by a stretch of 20 amino acids located in its N-terminal region (amino acids 31–50) (Kusumanegara et al., 2012). These 20 amino acids are predicted to form an amphipathic α-helix, with one side having a cluster of hydrophilic, polar residues and the other side hydrophobic, nonpolar residues, and were sufficient to target the nonviral GFP to ER membranes. Mutations that impede the membrane association of p27 compromise the formation of the RCNMV RNA replication complexes and negative-strand RNA synthesis (Kusumanegara et al., 2012).

The role in viral RNA replication of the Arf1 identified in the affinity-purified RCNMV RdRP fraction was investigated (Hyodo et al., 2013). Arf1 is a ubiquitous, highly conserved, small GTPase that is implicated in the formation of the coat protein complex I (COPI) vesicles on the Golgi membranes and in the membrane transport from the Golgi to the ER (D'Souza-Schorey & Chavrier, 2006). GTP-bound Arf1 facilitates the formation of several types of vesicle coat complexes and activates lipid-modifying enzymes on the Golgi membranes (Donaldson & Jackson, 2011; Memon, 2004).

The pharmacological inhibition of the nucleotide-exchange activity of Arf1 using the inhibitor brefeldin A (BFA) and the expression of dominant-negative Arf1 mutants compromised RCNMV RNA replication in tobacco BY-2 protoplasts. *In vitro* pull-down and BiFC analyses showed that p27 interacts with Arf1 within the virus-induced large punctate structures of the ER membrane. Inhibition of Arf1 activity by BFA disrupted p27-mediated ER remodeling and the assembly of the 480-kDa viral replication complex. Thus, p27 interacts with Arf1, and recruits and relocalizes this protein to the aggregate structures on the ER membranes, where they colocalize (Hyodo et al., 2013).

BFA treatment or the expression of dominant-negative mutants of Arf1 in plant cells not only inhibited the COPI pathway but also compromised COPII vesicle trafficking (Stefano et al., 2006). Interestingly, the expression of a dominant-negative mutant of Sar1, a key regulator of the biogenesis of COPII vesicles at ER exit sites, also compromises RCNMV RNA replication (Hyodo et al., 2013). These results suggest that the replication of RCNMV depends on the host membrane trafficking machinery or that RCNMV rewires the cellular trafficking pathways to build a viral replication factory. A model for the formation of the viral replication factory in RCNMV-infected cells is presented (Fig. 2.2).

4.10. Subgenomic RNA

RCNMV uses a unique strategy to produce sgRNA. In addition to the core promoter that is predicted to exist within a stable SL (Zavriev et al., 1996), the transcription of CPsgRNA from RNA1 requires RNA2 *in trans* and is regulated by intermolecular interactions via base pairing between the eight-nucleotide sequence in the loop of TA in RNA2 and the complementary eight-nucleotide sequence in the TA-binding site (TABS) immediately upstream from the CPsgRNA transcription start site in RNA1 (Sit et al., 1998) (refer to Fig. 2.1). Biophysical analyses, using short synthetic oligonucleotides that mimic the TA of RNA2 and the TABS of RNA1, suggested the formation of a weak but stable bimolecular complex between these two mimics (Guenther et al., 2004).

Several lines of evidence have suggested that CPsgRNA is generated by a premature termination mechanism (Guenther et al., 2004; Sit et al., 1998; Tatsuta et al., 2005). The premature termination model is supported by the accumulation of negative-strand CPsgRNA in protoplasts (Iwakawa et al., 2008) because the production of negative-sense sgRNA templates is a fundamental step in premature termination (Eckerle, Albarino, & Ball, 2003; Lin & White, 2004; White, 2002). CPsgRNA is also unlikely to be an efficient RNA replicon because its 3′-UTR is identical to that of RNA1 whose replication depends predominantly on replication proteins translated from its own RNA in *cis* (Iwakawa et al., 2011; Okamoto et al., 2008).

4.11. A viral noncoding RNA

A small noncoding RNA (0.4 kb), designated "SR1f," which consists of nearly the entire 3′-UTR of RNA1, accumulates in RCNMV-infected plants and protoplasts, and is packaged into virions (Iwakawa et al., 2008). SR1f is neither a sgRNA nor a defective RNA replicon, but a stable degradation product generated by *cis*-RNA element-mediated protection against 5′ → 3′ decay. A 58 nucleotide sequence in the 5′ proximal region of the 3′-UTR of RNA1 is necessary and sufficient to protect against

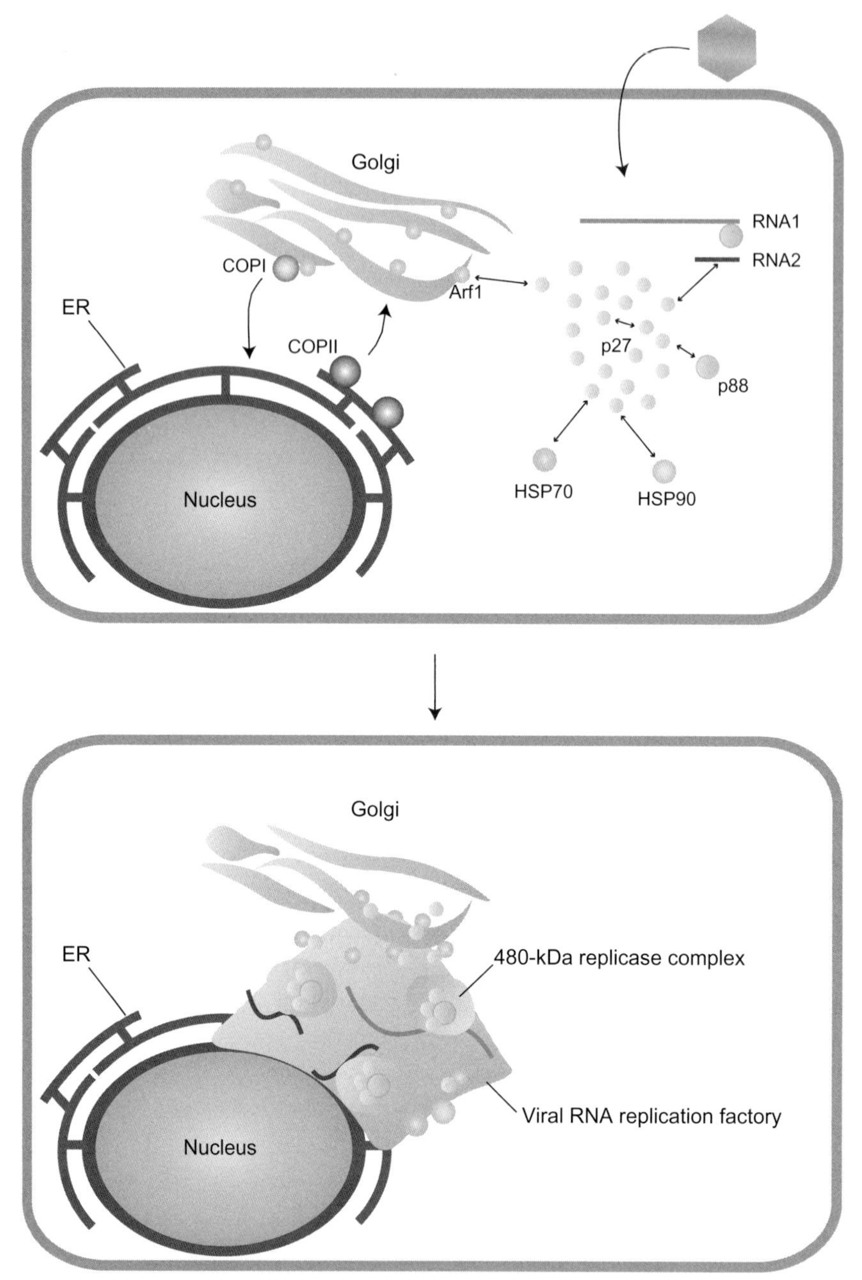

Figure 2.2 Schematic representation of the formation of the viral replication factory in RCNMV-infected cells. The abundant auxiliary replicase protein p27 interacts with p27 itself, p88, and viral RNAs, and also interacts with host proteins, such as Hsp70, Hsp90, and Arf1. These interactions are essential for the recruitment of viral genomic RNAs, viral proteins, and host factors to the ER membranes and for the reorganization of the membranes to form the viral RNA replication factory. (For color version of this figure, the reader is referred to the online version of this chapter.)

$5' \rightarrow 3'$ decay (Iwakawa et al., 2008). SR1f efficiently suppresses both cap-independent and cap-dependent translation, both *in vitro* and *in vivo*. SR1f *trans*-inhibits the negative-strand RNA synthesis of RCNMV genomic RNAs *via* the repression of replicase protein production, but not *via* competition with the replicase proteins *in vitro* (Iwakawa et al., 2008). SR1f seems to play important roles in virus infection and survival, although its precise roles in RCNMV infection in nature remain to be resolved.

The stable decay products of viral noncoding RNAs generated by a mechanism similar to that for SR1f have been reported in flaviviruses (Moon et al., 2012; Pijlman et al., 2008) and *Beet necrotic yellow vein virus* (Peltier et al., 2012). The *Flavivirus* noncoding RNA play important roles in viral pathogenicity by affecting host mRNA stability (Moon et al., 2012).

4.12. Cell-to-cell and systemic movement

A 35-kDa MP encoded by RNA2 is essential for the cell-to-cell and systemic movement of RCNMV (Osman & Buck, 1987; Paje-Manalo & Lommel, 1989; Xiong, Kim, Giesman-Cookmeyer, et al., 1993). RCNMV MP has the ability to bind single-stranded nucleic acids cooperatively (Giesman-Cookmeyer & Lommel, 1993; Osman, Hayes, & Buck, 1992; Osman, Thömmes, & Buck, 1993), localizes in the cell wall (Osman & Buck, 1991; Tremblay, Vaewhongs, Turner, Sit, & Lommel, 2005), and increases the size exclusion limit (SEL) of the plasmodesmata (PD) (Fujiwara, Giesman-Cookmeyer, Ding, Lommel, & Lucas, 1993). RCNMV MP fused with GFP (RCNMV MP:GFP) can form a homopolymer and is targeted to the cell wall, and this targeting is required for viral cell-to-cell movement (Tremblay et al., 2005).

Further analysis of the subcellular localization of RCNMV MP:GFP expressed from a recombinant virus in *N. benthamiana* epidermal cells and protoplasts showed that MP:GFP first appeared in the cell wall and was subsequently observed as punctate spots in the cortical ER (Kaido, Tsuno, Mise, & Okuno, 2009). The ER-localization of RCNMV MP:GFP was associated with the replication of RNA1, but not with that of RNA2 or the viral replicase component proteins (p27 and p88) *per se* (Kaido et al., 2009). This characteristic of MP:GFP is dependent on the 70 amino acids in the C-terminal region of MP and is required for the efficient cell-to-cell movement of the recombinant virus (Kaido, Funatsu, Tsuno, Mise, & Okuno, 2011). Interestingly, the deletion of the C-terminal 70 amino acids of MP had no deleterious effects on its localization to the cell wall, or its ability to increase the PD SEL, to bind to single-stranded RNA, or to interact with MP *in vivo* (Kaido et al., 2011). These findings show that the recruitment of MP to the viral replication complexes is required for viral cell-to-cell movement. The mechanism that targets RCNMV MP:GFP to the ER, dependent on the replication of RNA1, which does not encode MP, may reflect a strategy to retain MP close to RNA1, thereby achieving its intracellular movement, followed by the intercellular movement of the viral genome.

Transgenic *N. benthamiana* expressing RCNMV MP supported both the cell-to-cell and systemic movement of movement-defective TMV and CMV (Giesman-Cookmeyer et al., 1995; Rao et al., 1998). A chimeric TMV with the MP gene replaced with the RCNMV MP gene systemically infected *N. benthamiana* (Giesman-Cookmeyer et al., 1995). Therefore, the MPs of tobamoviruses and dianthoviruses are functionally homologous.

In contrast, transgenic *N. benthamiana* expressing RCNMV MP supported the cell-to-cell movement, but not the systemic movement, of movement-defective *Cowpea chlorotic mottle bromovirus* (Rao et al., 1988). A chimeric *Barley stripe mosaic virus*, in which the triple gene block was replaced with the RCNMV MP gene, accumulated in the inoculated leaves of *N. benthamiana*, but not in the upper uninoculated leaves (Solovyev et al., 1997). Immunocytochemical studies by Wang, Wang, Giesman-Cookmeyer, Lommel, and Lucas (1998) showed that point mutations in RCNMV MP that do not affect the cell-to-cell movement of the virus prevent systemic viral movement, presumably by inhibiting RCNMV loading into the companion cell–sieve element complex. These findings suggest that the roles of RCNMV MP in the two processes are genetically distinct.

RCNMV CP is not essential for the cell-to-cell movement of the virus but virion formation is important for its systemic movement in *N. benthamiana* (Vaewhongs & Lommel, 1995; Xiong, Kim, Giesman-Cookmeyer, et al., 1993). The N-terminal lysine-rich motif of CP is involved in the systemic viral movement, virion accumulation, and symptomatology in *N. benthamiana* (Park, Sit, Kim, & Lommel, 2012). A sobemovirus CP gene complemented the long-distance movement of a CP-null RCNMV (Callaway, George, & Lommel, 2004).

4.13. Virions

The virions of *Dianthovirus* contain 180 copies of the 37-kDa CP subunit, forming an icosahedral particle with $T=3$ symmetry and a diameter of about 36 nm (Hiruki, 1987; Sherman et al., 2006). Analysis of the RCNMV virions with heat treatment and UV cross-linking suggested that RCNMV is composed of two virion populations (Basnayake et al., 2006): one type contains RNA1 and RNA2, and the other type contains multiple copies (possibly four copies) of RNA2. The origin of assembly sequence (OAS) was identified in RNA2 using the TBSV-vector-based expression of RCNMV CP (Basnayake, Sit, & Lommel, 2009). RNA1 is copackaged with RNA2 via an OAS–RNA1 interaction. Interestingly, the OAS is the TA that is required for the transcription of CPsgRNA from RNA1 via an RNA–RNA interaction (Sit et al., 1998). The N-terminal lysine-rich motif of CP is involved in the formation of the virions in *N. benthamiana* (Park et al., 2012). The heterologous combinations of RNA1 and RNA2 between RCNMV and SCNMV were infectious and produced progeny virions in their host plants (Chen, Hiruki, & Okuno, 1984; Okuno et al., 1983). A mixture of RNA1 from CRSV and RNA2 from RCNMV was also infectious and produced progeny virions in the host plants (Callaway, Giesman-Cookmeyer, Gillock, Sit, & Lommel, 2001). However, unilateral compatibility of RNA1 and RNA2 was reported using the different strains of RCNMV (Rao & Hiruki, 1987).

The electrophoretic mobility of purified virions differs among the dianthoviruses, even in the different strains of the same virus (Pappu & Hiruki, 1989).

Analysis of the RCNMV virions by cryoelectron microscopy and three-dimensional image reconstruction showed that the structures of the CP subunits and the entire virion are similar to those observed in the other viruses of the *Tombusviridae*, such as TBSV and TCV (Sherman et al., 2006). Preliminary diffraction data are available for the X-ray structure of RCNMV virion crystals at higher than 4 Å resolution (Martin et al., 2010). Atomic absorption spectroscopic analysis of RCNMV virions showed that the virions contain significant amounts of Ca^{2+} and Mg^{2+} ions (Sherman et al., 2006). Removal of these ions from the virions with a chelator, followed by exposure to ribonucleases, reduces viral infectivity, suggesting a role for these ions in virion stabilization (Sherman et al., 2006).

Studies of the use of RCNMV virions as the viral nanoparticles that function as drug carriers in human cells are in progress (Lockney et al., 2011).

4.14. Suppression of RNA silencing

RNA silencing is an RNA-mediated plant defense mechanism against viral infection (Baulcombe, 2004). To counteract RNA silencing, viruses have developed a variety of strategies and evolved to encode a viral suppressor of RNA (VSR) silencing (Csorba, Pantaleo, & Burgyan, 2009; Ding & Voinnet, 2007; Li & Ding, 2006). Many plant viruses encode at least one VSR and some viruses encode more than one. Viral proteins that function as VSRs include CPs, replicase proteins, MPs, and other nonstructural proteins, and their targets in RNA silencing are diverse (Csorba et al., 2009; Ding & Voinnet, 2007; Li & Ding, 2006).

RCNMV has at least two strategies for suppressing RNA silencing. One strategy requires p27, p88, and viral RNAs (Takeda et al., 2005). In a sense-transgene-mediated posttranscriptional gene silencing (S-PTGS) assay involving the *Agrobacterium*-mediated transient expression of viral components and GFP in a GFP-expressing transgenic plant, the RNA-silencing-suppression activity of RCNMV was linked to the ability of viral factors to initiate negative-strand RNA synthesis (Takeda et al., 2005), and correlated with the formation of the 480-kDa replicase complex (Mine, Hyodo, et al., 2010). A close relationship between negative-strand RNA synthesis and the suppression of RNA silencing implies a possible scenario, in which RCNMV sequesters the host factors required for RNA silencing and reduces the antiviral silencing response. RCNMV replication also inhibits microRNA biogenesis, in which DCL1 plays an essential role, and dcl1 mutant plants show reduced susceptibility to RCNMV infection (Takeda et al., 2005). Therefore, it has been suggested that DCL1 or its homologues are recruited by the viral RNA replication complex (Takeda et al., 2005). Alternatively, the reduced susceptibility of a *dcl1* mutant to RCNMV infection can also be explained by the observations that the disruption of the DCL1 function leads to the higher expression of DCL4 and DCL3 in *Arabidopsis* leaves (Qu, Ye, & Morris, 2008) and that the destabilization of the miRNA pathway, including the disruption of DCL1, leads to the increased accumulation of AGO1 and positively influences S-PTGS (Martinez de Alba, Jauvion, Mallory, Bouteiller, & Vaucheret, 2011). Given that RCNMV infection inhibits miRNA biogenesis, reduced accumulations of miRNAs, including miR168 that is a regulator of AGO1 mRNA, could increase the accumulation levels of AGO1, which is the core component of the RNA-induced silencing complex.

MP has been identified as the second VSR of RCNMV in another assay system (Powers, Sit, Heinsohn, et al., 2008; Powers, Sit, Qu, et al., 2008). This assay relies on TCV that contains a reporter GFP gene in place of the CP ORF (TCV-sGFP) (Powers, Sit, Qu, et al., 2008). The TCV CP is a VSR (Qu, Ren, & Morris, 2003), and TCV requires CP for its cell-to-cell movement in *N. benthamiana* (Cohen, Gisel, & Zambryski, 2000). TCV-sGFP moved from cell to cell in leaves infiltrated with *Agrobacterium* expressing RCNMV MP, but did not in leaves infiltrated with *Agrobacterium* expressing RCNMV p27, p88, and CP (Powers, Sit, Heinsohn, et al., 2008). The functional domains in RCNMV MP required for RNA silencing (Powers, Sit, Heinsohn, et al., 2008) differ from those required for the cell-to-cell movement of the virus and SEL modification (Fujiwara et al., 1993; Tremblay et al., 2005). Thus, RCNMV has evolved two strategies for suppressing RNA silencing and counteracting the host defenses.

5. TRANSMISSION STUDIES AND EPIDEMIOLOGY

Dianthoviruses multiply in infected leaves to high concentrations and are very stable *in vitro* (Hiruki, 1987). Soil transmission has been reported to be a general feature of dianthoviruses (Brunt et al., 1996).

The transmission of CRSV by nematodes such as *L. macrosoma* and *X. diversicaudatum* (Fritzsche & Schmelzer, 1967; Kegler et al, 1977) and *L. elongates* (Fritzsche et al., 1979) has been reported but not confirmed (Brunt et al., 1996).

The nematode transmission of FNSV is suspected but not demonstrated (Morales et al., 1992). No other vectors have been investigated in terms of the soil transmission of this virus.

The possible transmission of RCNMV by soil-inhabiting fungi was suggested (Bowen & Plumb, 1979). A Swedish isolate of RCNMV was transmitted to *N. clevelandii* when the seedlings were planted into the soil that had been infested by growing RCNMV-infected plants, or by adding virus suspension to the soil prior to planting. The concomitant presence of *O. brassicae* only increased the rate of transmission, but not essential as an intrinsic vector for the soil transmission of the virus (Gerhardson & Insunza, 1979). Independent tests by different groups did not demonstrate that *O. brassicae* serves as a vector of RCNMV (Lynes, Teakle, & Smith, 1981). Hiruki (1986b) reported that the presence of a tobacco strain of *O. brassicae* was not required for the transmission of SCNMV in sand culture.

Compared with leguminous forage crops such as alfalfa, alsike clover, red clover, crown vetch, and white clover, sweet clover is highly susceptible to SCNMV, while the others remain practically resistant to the virus. It is very interesting to note that the most prevalent area of SCNMV occurrence in Alberta coincides with the area where beekeeping industry is most active and where sweet clover is commonly used for seed production. Hiruki, Kudo, and Figueiredo (1989) reported that the pollen from SCNMV-infected sweet clover is highly contaminated with the virus and that the virus, detectable by enzyme-linked immunosorbent assays, could be recovered by washing.

In 2007, an area of approximately a quarter section of land, consisting of 50% arable land and 50% popular forest, was cleared for residential development in a suburb of Edmonton. The topsoil, about 10 cm in depth, was completely removed and piled aside. However, due to unexpected economic downturn, the construction was halted for 3 years. To prevent erosion, sweet clover was seeded over the whole area (Fig. 2.3), offering a rare opportunity to observe the natural development of plant virus epidemic.

Regular visual inspections and samplings were conducted, and followed by infectivity tests on the indicator plants in the greenhouse. The results indicated that SCNMV infection occurred at a rate of about 6%. The source of the natural infection remains unknown. In the past, the natural infection of a solitary sweet clover plant in the forest, roadside, or riverbanks was observed frequently and shown to test positive (C. Hiruki, unpublished data). When infected sweet clover is not disturbed, abundant seeds are shed around the plant, and the young seedlings are subsequently infected by the soil transmission of SCNMV.

Figure 2.3 A general view of the site where sweet clover plants were grown to prevent soil erosion. This site was used to observe the natural spread of SCNMV. (For color version of this figure, the reader is referred to the online version of this chapter.)

6. EFFECTS ON ENVIRONMENT

Although considerable attention has been directed toward the pollution of environmental waters with human and animal viruses, there have been only sporadic attempts to shed light on the contamination of rivers, lakes, and sea with plant viruses (Buttner, Jacobi, & Koenig, 1987; Koenig & Lesemann, 1985; Tomlinson, Faithfull, Flewett, & Beards, 1982; Tomlinson, Faithfull, & Fraser, 1983; Tomlinson, Faithfull, & Seeley, 1982).

6.1. Release of dianthoviruses from intact roots, decaying plant materials, and pollen grains

Generally, dianthoviruses occur at high concentrations in infected cells and are very stable *in vitro*, thus serve as potentially highly infective inoculum sources. The release of CRSV from infected roots to drainage water was reported in Germany (Kegler and Kegler, 1981). Actively growing plants of *N. clevelandii* released much larger amounts of RCNMV into drainage water than did plants from which the aerial parts had been removed or plants which were senescent or dead (Gerhardson & Insunza, 1979). CRSV was isolated from the water samples collected from a canal near a sewage plant in Germany (Koenig, An, Lesemann, & Burgermeister, 1988).

While sweet clover is a biennial crop, successive infections with SCNMV can occur from the soil around an infected plant to young germinating plants in the second year, and from the soil around the dead infected plants to young plants emerging from fallen seeds in the third year (Hiruki, unpublished data).

Another strong possibility is pollen transmission in the case of SCVMV, although its direct experimental evidence is lacking at present. However, significantly high degrees of SCNMV occurrence and release of the virus particles from pollen grains were detected serologically (Hiruki et al., 1989). Moreover, in systematic field surveys over many years, high-specific incidences of SCNMV coincide with the zones 3 (black soil) and 4 (gray-wooded soil) in the Agriculture map of Alberta (Hiruki, 1987) where legume seed industry and bee keeping industries are concentrated (Hiruki, unpublished data). There are three main species of bees to pollinate legume crops such as honey bees (*Apis mellifera*), leafcutting bees (*Megachile rotundata*), and American bumble bees (*Bombus fervidus*). Honey bees are effective pollinators of a number of legume crops including sweet clover. However, honey bees are of little value as pollinator of alfalfa (Fairey,

2003). This fact may explain a minor incidence of SCNMV in alfalfa while frequent incidence of the same virus is observed in sweet clover in the same area (Hiruki, 1986a, 1986b; Hiruki et al., 1984). The occurrence of SCNMV in sweet clover in the forests, river banks, roadsides, and abandoned field may be explained by pollen transmission mediated mostly by honey bees and to lesser degrees by other bees.

6.2. Sewage

CRSV was isolated from river water that was fed by the outlets of a city's sewage plant in Germany (Koenig et al., 1988). The virus was identified by means of nucleic acid hybridization analysis and serology. CRSV and other viruses can also be released into streams or rivers. It is known that several plant viruses are capable of surviving a passage through the alimentary tracts of animals (Kegler et al., 1987; Tomlinson, Faithfull, Flewett, et al., 1982; Tomlinson, Faithfull, & Seeley, 1982). SCNMV was not transmitted by *Sitona cylindricollus* from sweet clover to sweet clover but remained infectious in feces after a passage through the alimentary tract (Hiruki, 1986a, 1986b).

At any rate, these observations on the occurrence of plant viruses in streams, rivers, and lakes illustrate the potential danger of using polluted water in horticultural and agricultural practices (Buttner et al., 1987; Koenig, 1986; Koenig & Lesemann, 1985; Koenig et al., 1988).

7. CONCLUDING REMARKS AND PERSPECTIVES

The dianthoviruses have moderately broad natural host ranges, restricted to dicot plant species, including legume plants (Hiruki, 1986a, 1986b; Descriptions lists from the VIDE Database; Virus Taxonomy, Ninth Report of the International Committee on Taxonomy of Viruses, 2012), and the viruses and many of their strains have been reported frequently worldwide except for the tropical and subtropical areas of the world. However, a natural host of FNSV, an apparent member of the dianthoviruses, is *Furcraea* spp., which are xerophytic monocots native to the tropical regions of Mexico, the Caribbean, Central America, and northern South America. Dianthovirus RNA1-like RNA has also been reported in grassy stunt-diseased rice plants coinfected with *Rice grassy stunt virus* in the Philippines (Miranda et al., 2001). These reports imply that other unidentified *Dianthovirus* species or viruses related to the dianthoviruses might exist in tropical and subtropical areas. Therefore, a systematic search for dianthoviruses worldwide, including in wild plants in fields, forests, and areas of water such as river, ponds, and seas, should open the way to finding additional members of *Dianthovirus*.

Dianthoviruses are known to release large numbers of infective viral particles from infected roots into soil and drainage waters in the surrounding area (Gerhardson & Insunza, 1979; Hiruki, 1986a, 1986b; Kegler and Kegler, 1981). FNSV is detected consistently in the roots of infected fique plants (Morales et al., 1992). In nature, soil is a complex medium and acts as a heterogeneous environment. Its properties can be altered by seasonal climatic changes and agricultural practices in a particular area. This implies that any infective agents, such as viruses, introduced into the soil and drainage water may be affected by a variety of environmental factors or combination of several factors that are influenced by the interplay between the climate, vegetation, and soil type. It is also expected that the application of sewage and sludge to agricultural land may alter some of the soil's physical and chemical properties that affect viral survival.

Our understanding of the molecular mechanisms underlying the infection processes of the dianthoviruses has advanced greatly in recent years, as described in this review. These advances have also given rise to many interesting questions that are yet to be addressed. These questions include the differences in the gene expression strategies of RNA1 and RNA2. For example, RNA1 contains strong translation enhancer elements (3′TE-DR1 and ARS), whereas RNA2 has no such elements: instead, the translation of MP is linked to the replication of RNA2 (Mizumoto et al., 2006). What do these differences in translation strategies mean? RNA1 and RNA2 might use distinct translation factors. The recruitment of MP to viral RNAs, a process essential for viral cell-to-cell movement, also differs between RNA1 and RNA2. The replication of RNA1, but not RNA2, is associated with MP recruitment to the ER-associated viral replication complexes (Kaido et al., 2009). This strategy might assist the cell-to-cell movement of RNA1, which does not encode MP. These findings also suggest that the replication complexes formed on RNA1 might differ from those on RNA2, and may localize to different sites on the ER. Indeed, RNA2 has an RNA element (YRE) that recruits p27 and the replicase complex (An et al., 2010; Hyodo et al., 2011), whereas RNA1 has no such recruiting element. The replication of RNA1 is linked to the translation of the replicase proteins (Iwakawa et al., 2011; Okamoto et al., 2008).

The TA of RNA2 is a key regulator of translation, transcription, replication, virion formation, and cell-to-cell movement. The structural requirements of TA seem to differ depending on its role, because nucleotide substitutions in RNA2 that disrupt TA-mediated base pairing have no effect on the copackaging of RNA1 and RNA2 (Basnayake et al., 2009). Further analyses of the temporal changes in the TA structure and the host and viral factors that interact with TA are necessary to clarify the complex regulatory mechanisms of the *Dianthovirus* infection processes.

The auxiliary replication protein p27 plays multiple roles in RNA replication. These roles include the recruitment of the viral RNA to the ER membranes, the assembly of the replication complexes, and the formation of the viral replication factory. p27 performs these roles by interacting with p27 itself, p88, and the viral RNA, and also by interacting with host proteins, such as Hsp70, Hsp90, and Arf1. How can p27 perform these roles properly? A model proposed by Nagy, Barajas, & Pogany (2012) for the *Tombusvirus* auxiliary replication protein p33 might help to answer this question. These researchers propose that "depending on the given interacting partners, the avandant p33 molecules are divided into many groups that perform different functions at different subcellular locations." Further identification and analysis of the host proteins associated with the different functions of RCNMV p27 are required to answer this question.

Another interesting question to be addressed is the mechanism that delivers the translation initiation factors from the 3′-UTR of RCNMV RNA1 to the 5′-UTR, where the ribosome starts to scan the template (Sarawaneeyaruk et al., 2009). In 3′TE-DR1-mediated cap-independent translation, PABP plays an essential role in the binding of eIF4F/eIFiso4F to 3′TE-DR1 (Iwakawa et al., 2012). It is generally believed that the binding of eIF4F and eIFiso4F to viral 3′CITEs facilitates ribosome recruitment to the 5′ end of the RNA via 5′–3′ communication, which is mediated by a long-distance RNA–RNA interaction or a protein factor (Miller & White, 2006; Nicholson & White, 2011). However, the factors other than eIF4F/eIFiso4F or PABP that are required for 3′CITE-mediated translation remain unclear. How the ribosome recognizes the uncapped 5′ end, even when eIF4F/iso4F is delivered to the vicinity of the 5′ end of RNA1 by 5′–3′ communication is also unclear. Although we do not know how RCNMV RNA1 achieves the communication between its 5′- and 3′-ends in 3′TE-DR1-mediated cap-independent translation, our recent data suggest that the binding of eIF4F/iso4F and PABP to the 3′-UTR of RCNMV RNA1 is important for delivering cap-binding factors other than eIF4F/iso4F and PABP to its nonfunctional G-capped 5′ end, and in enhancing the recruitment of the 40S ribosomal subunit to the viral mRNA (H. Iwakawa & T. Okuno, unpublished data). Identification of the proteins associated with the 5′ end of RNA1 will help to clarify these unknown mechanisms.

As described in this review, recent advances in our knowledge of the dianthoviruses, especially of their molecular biology, have relied on the development and availability of several experimental systems, such as an *in vitro* translation/replication system (BYL) (Komoda et al., 2004), and the *Agrobacterium*-mediated expression of viral and host components in combination with a protein-mediated or RNA-aptamer-mediated pull-down assay and mass spectrometric analysis. Genome-wide screens of the host factors that affect viral replication have been performed using yeast as the model host. The lists of host proteins obtained from these studies in yeast (Kushner et al., 2003; Panavas, Serviene, Brasher, & Nagy, 2005; Serviene et al., 2005), together with those obtained with other approaches, including those involving RCNMV, will enhance our understanding of the molecular mechanisms underlying viral infection. This knowledge will contribute to the development of antiviral measures that are more efficient, which will reduce the disease-mediated damage caused by the virus infections. However, we must consider that both the viral and host factors that affect viral infection might differ in different combinations of viruses and hosts (Li & Wong, 2007; Sarawaneeyaruk et al., 2009).

ACKNOWLEDGMENTS

The authors thank all their colleagues and associates who have contributed to studies of the dianthoviruses. This work was supported in part by funds from the Ministry of Education, Culture, Sports, Science, and Technology, Japan to Tetsuro Okuno, and from Natural Sciences and Engineering Research Council of Canada (Grant No. A3843, STR G1450, IC0145) and University of Alberta Distinguished University Professor Research Grant to Chuji Hiruki.

REFERENCES

An, M., Iwakawa, H. O., Mine, A., Kaido, M., Mise, K., & Okuno, T. (2010). A Y-shaped RNA structure in the 3′ untranslated region together with the trans-activator and core promoter of Red clover necrotic mosaic virus RNA2 is required for its negative-strand RNA synthesis. *Virology*, *405*, 100–109.

Barry, J. K., & Miller, W. A. (2002). A −1 ribosomal frameshift element that requires base pairing across four kilobases suggests a mechanism of regulating ribosome and replicase traffic on a viral RNA. *Proceedings of the National Academy of Sciences of the United States of America*, *99*, 11133–11138.

Basnayake, V. R., Sit, T. L., & Lommel, S. A. (2006). The genomic RNA packaging scheme of *Red clover necrotic mosaic virus*. *Virology*, *345*, 532–539.

Basnayake, V. R., Sit, T. L., & Lommel, S. A. (2009). The *Red clover necrotic mosaic virus* origin of assembly is delimited to the RNA-2 *trans*-activator. *Virology*, *384*, 169–178.

Bates, H. J., Farjah, M., Osman, T. A., & Buck, K. W. (1995). Isolation and characterization of an RNA-dependent RNA polymerase from *Nicotiana clevelandii* plants infected with red clover necrotic mosaic dianthovirus. *Journal of General Virology*, *76*, 1483–1491.
Baulcombe, D. (2004). RNA silencing in plants. *Nature*, *431*, 356–363.
Belov, A. G., & van Kuppeveld, J. M. (2012). (+)RNA viruses rewire cellular pathways to build replication organelles. *Current Opinion in Virology*, *2*, 734–741.
Blobel, G., & Sabatini, D. (1971). Dissociation of mammalian polyribosomes into subunits by puromycin. *Proceedings of the National Academy of Sciences of the United States of America*, *68*, 390–394.
Bowen, R., & Plumb, R. T. (1979). The occurrence and effects of red clover necrotic mosaic virus in red clover (*Trifolium pretense*). *Annals of Applied Biology*, *91*, 227–236.
Brierley, I. (1995). Ribosomal frameshifting viral RNAs. *Journal of General Virology*, *76*, 1885–1892.
Brunt, A. A., Crabtree, K., Dallwitz, M. J., Gibbs, A. J., & Watson, L. (1996). *Viruses of plants*. Wallingford, UK: CAB International.
Buttner, C., Jacobi, V., & Koenig, R. (1987). Isolation of carnation Italian ringspot virus from a creek in a forested area Southwest of Bonn. *Journal of Phytopathology*, *118*, 131–134.
Callaway, A. S., George, C. G., & Lommel, S. A. (2004). A Sobemovirus coat protein gene complements long-distance movement of a coat protein-null Dianthovirus. *Virology*, *330*, 186–195.
Callaway, A., Giesman-Cookmeyer, D., Gillock, E. T., Sit, T. L., & Lommel, S. A. (2001). The multifunctional capsid proteins of plant RNA viruses. *Annual Review of Phytopathology*, *39*, 419–460.
Chang, R. Y., Hofmann, M. A., Sethna, P. B., & Brian, D. A. (1994). A cis-acting function for the coronavirus leader in defective interfering RNA replication. *Journal of Virology*, *68*, 8223–8231.
Chattopadhyay, M., Shi, K., Yuan, X., & Simon, A. E. (2011). Long-distance kissing loop interactions between a 3′ proximal Y-shaped structureand apical loops of 5′ hairpins enhance translation of *Saguaro cactus virus*. *Virology*, *417*, 113–125.
Chen, M. H., Hiruki, C., & Okuno, T. (1984). Immunosorbent electron microscopy of dianthoviruses and their pseudorecombinants. *Canadian Journal of Plant Pathology*, *6*, 191–195.
Cohen, Y., Gisel, A., & Zambryski, P. C. (2000). Cell-to-cell and systemic movement of recombinant green fluorescent protein-tagged turnip crinkle viruses. *Virology*, *273*, 258–266.
Csorba, T., Pantaleo, V., & Burgyan, J. (2009). RNA silencing: An antiviral mechanism. *Advances in Virus Research*, *75*, 35–71.
den Boon, J. A., & Ahlquist, P. (2010). Organelle-like membrane compartmentalization of positive-strand RNA virus replication factories. *Nature Reviews. Microbiology*, *64*, 241–256.
Ding, S. W., & Voinnet, O. (2007). Antiviral immunity directed by small RNAs. *Cell*, *130*, 413–426.
Donaldson, J. G., & Jackson, C. L. (2011). ARF family G proteins and their regulators: Roles in membrane transport, development and disease. *Nature Reviews. Molecular Cell Biology*, *12*, 362–375.
D'Souza-Schorey, C., & Chavrier, P. (2006). ARF proteins: Roles in membrane traffic and beyond. *Nature Reviews. Molecular Cell Biology*, *7*, 347–358.
Eckerle, L. D., Albarino, C. G., & Ball, L. A. (2003). Flock House virus subgenomic RNA3 is replicated and its replication correlates with transactivation of RNA2. *Virology*, *317*, 95–108.
Fabian, M. R., & White, K. A. (2004). 5′-3′ RNA–RNA interaction facilitates cap- and poly(A) tail-independent translation of tomato bushy stunt virus mRNA: A potential common mechanism for tombusviridae. *Journal of Biological Chemistry*, *279*, 28862–28872.

Fabian, M. R., & White, K. A. (2006). Analysis of a 39-translation enhancer in a tombusvirus: A dynamic model for RNA-RNA interactions of mRNA termini. *RNA*, *12*, 1304–1314.

Fairey, D. T. (2003). Growing forage legume for seed. Government of Alberta, Canada, Agdex 120/15-1.

Fritzsche, R., Kegler, H., Thiele, S., & Gruber, G. (1979). Contribution to epidemiology and transmission of carnation ringspot virus in fruit plantations. *Archiv fuer Phytopathologie und Pflanzenschutz*, *15*, 177–180.

Fritzsche, R., & Schmelzer, K. (1967). Nematode-transmissibility of carnation ringspot virus. *Naturwissenschaften*, *54*, 498–499.

Fujiwara, T., Giesman-Cookmeyer, D., Ding, B., Lommel, S. A., & Lucas, W. J. (1993). Cell-to-cell trafficking of macromolecules through plasmodesmata potentiated by the red clover necrotic mosaic virus movement protein. *The Plant Cell*, *5*, 1783–1794.

Gazo, B. M., Murphy, P., Gatchel, J. R., & Browning, K. S. (2004). A novel interaction of cap-binding protein complexes eukaryotic initiation factor (eIF)4F and eIF(iso)4F with a region in the 3′-untranslated region of satellite tobacco necrosis virus. *Journal of Biological Chemistry*, *279*(4), 13584–13592.

Ge, Z., Hiruki, C., & Roy, K. L. (1992). A comparative study of the RNA-2 nucleotide sequences of two sweet clover necrotic mosaic virus strains. *Journal of General Virology*, *73*, 2483–2486.

Ge, Z., Hiruki, C., & Roy, K. L. (1993). Nucleotide sequence of sweet clover necrotic mosaic dianthovirus RNA-1. *Virus Research*, *28*, 113–124.

Gerhardson, B., & Insunza, V. (1979). Soil transmission of red clover necrotic mosaic virus. *Phytopathologische Zeitschrift*, *94*, 67–71.

Gerhardson, B., & Lindsten, K. (1973). Red clover mottle virus and red clover necrotic mosaic virus in Sweden. *Phytopathologische Zeitschrift*, *76*, 67–79.

Giesman-Cookmeyer, D., & Lommel, S. A. (1993). Alanine scanning mutagenesis of a plant virus movement protein identifies three functional domains. *The Plant Cell*, *5*, 973–982.

Giesman-Cookmeyer, D., Silver, S., Vaewhong, A. A., Lommel, S. A., & Deom, C. M. (1995). Tobamovirus and dianthovirus movement proteins are functionally homologous. *Virology*, *213*, 38–45.

Gould, A. R., Francki, R. I., Hatta, T., & Hollings, M. (1981). The bipartite genome of red clover necrotic mosaic virus. *Virology*, *108*, 499–506.

Guenther, R. H., Sit, T. L., Gracz, H. S., Dolan, M. A., Townsend, H. L., Liu, G., et al. (2004). Structural characterization of an intermolecular RNA-RNA interaction involved in the transcription regulation element of a bipartite plant virus. *Nucleic Acids Research*, *32*, 2819–2828.

Guo, L., Allen, E., & Miller, W. A. (2000). Structure and function of a cap-independent translation element that functions in either the 3' or the 5' untranslated region. *RNA*, *6*, 1808–1820.

Guo, L., Allen, E., & Miller, W. A. (2001). Base-pairing between untranslated regions facilitates translation of uncapped, nonpolyadenylated viral RNA. *Molecular Cell*, *7*, 1103–1109.

Hagino-Yamagishi, K., & Nomoto, A. (1989). *In vitro* construction of poliovirus defective interfering particles. *Journal of Virology*, *63*, 5386–5392.

Hiruki, C. (1986a). Sweet clover necrotic mosaic virus. *AAB Descriptions of Plant Viruses*, *321*, 1–4, Association of Applied Biologists, U.K.

Hiruki, C. (1986b). Incidence and geographic distribution of sweet clover necrotic mosaic virus in Alberta. *Plant Disease*, *70*, 1129–1131.

Hiruki, C. (1987). The dianthoviruses: A distinct group of isometric plant viruses with bipartite genome. *Advances in Virus Research*, *33*, 257–300.

Hiruki, C. (1994). Transmission of dianthoviruses. *Acta Horticulturae*, *377*, 341–347.

Hiruki, C., Kudo, K., & Figueiredo, G. (1989). Transmission of sweet clover necrotic mosaic virus. *Proceedings of the Japan Academy, 65B*, 234–237.
Hiruki, C., Rao, D. V., Chen, M. H., Okuno, T., & Figueiredo, G. C. (1984). Characterization of sweet clover necrotic mosaic virus. *Phytopathology, 74*, 482–486.
Hyodo, K., Mine, A., Iwakawa, H. O., Kaido, M., Mise, K., & Okuno, T. (2011). Identification of amino acids in auxiliary replicase protein p27 critical for its RNA-binding activity and the assembly of the replicase complex in Red clover necrotic mosaic virus. *Virology, 413*, 300–309.
Hyodo, K., Mine, A., Iwakawa, H. O., Kaido, M., Mise, K., & Okuno, T. (2013). The ADP-ribosylation factor 1 plays an essential role in the replication of a plant RNA virus. *Journal of Virology, 87*, 163–176.
Iwakawa, H. O., Kaido, M., Mise, K., & Okuno, T. (2007). *cis*-Acting core RNA elements required for negative-strand RNA synthesis and cap-independent translation are separated in the 3′-untranslated region of *Red clover necrotic mosaic virus* RNA1. *Virology, 369*, 168–181.
Iwakawa, H. O., Mine, A., Hyodo, K., An, M., Kaido, M., Mise, K., et al. (2011). Template recognition mechanisms by replicase proteins differ between bipartite positive-strand genomic RNAs of a plant virus. *Journal of Virology, 85*, 497–509.
Iwakawa, H. O., Mizumoto, H., Nagano, H., Imoto, Y., Takigawa, K., Sarawaneeyaruk, S., et al. (2008). A viral noncoding RNA generated by cis-element-mediated protection against 5′ → 3′ RNA decay represses both cap-independent and cap-dependent translation. *Journal of Virology, 82*, 10162–10174.
Iwakawa, H. O., Tajima, Y., Taniguchi, T., Kaido, M., Mise, K., Tomari, Y., et al. (2012). Poly(A)-binding protein facilitates translation of an uncapped/nonpolyadenylated viral RNA by binding to the 3′ untranslated region. *Journal of Virology, 86*(15), 7836–7849.
Johnson, K. L., & Sarnow, P. (1991). Three poliovirus 2B mutants exhibit noncomplementable defects in viral RNA amplification and display dosage-dependent dominance over wild-type poliovirus. *Journal of Virology, 65*, 4341–4349.
Kaido, M., Funatsu, N., Tsuno, Y., Mise, K., & Okuno, T. (2011). Viral cell-to-cell movement requires formation of cortical punctate structures containing Red clover necrotic mosaic virus movement protein. *Virology, 413*, 205–215.
Kaido, M., Tsuno, Y., Mise, K., & Okuno, T. (2009). Endoplasmic reticulum targeting of the Red clover necrotic mosaic virus movement protein is associated with the replication of viral RNA1 but not that of RNA2. *Virology, 395*, 232–242.
Karetnikov, A., & Lehto, K. (2008). Translation mechanisms involving long-distance base pairing interactions between the 5′ and 3′ non-translated regions and internal ribosomal entry are conserved for both genomic RNAs of *Blackcurrant reversion nepovirus*. *Virology, 371*, 292–308.
Kegler, G., & Kegler, H. (1981). On vectorless transmission of plant pathogenic viruses. *Archiv fur Phytopathologie und Pflanzenschutz, 17*, 307–323.
Kegler, H., Verderevskaja, T. D., Proll, E., Fritzsche, R., Schmidt, H. B., Kalasjan, J. A., et al. (1977). Isolation and characterization of a virus from pears with pear stony pit. *Archiv fur Phytopathologie und Pflanzenschutz, 13*, 297–310.
Kendall, T. L., & Lommel, S. A. (1992). Nucleotide sequence of carnation ringspot dianthovirus RNA-2. *Journal of General Virology, 73*, 2479–2482.
Kim, K. H., & Lommel, S. A. (1994). Identification and analysis of the site of −1 ribosomal frameshifting in red clover necrotic mosaic virus. *Virology, 200*, 574–582.
Kim, K. H., & Lommel, S. A. (1998). Sequence element required for efficient −1 ribosomal frameshifting in red clover necrotic mosaic dianthovirus. *Virology, 250*, 50–59.
Kneller, E. L., Rakotondrafara, A. M., & Miller, W. A. (2006). Cap-independent translation of plant viral RNAs. *Virus Research, 119*, 63–75.
Koening, R. (1986). Plant viruses in rivers and lakes. *Advances in Virus Research, 31*, 321–333.

Koenig, R., An, D., Lesemann, D. E., & Burgermeister, W. (1988). Isolation of carnation ringspot virus from a canal near a sewage plant: cDNA hybridization analysis, serology, and cytopathology. *Journal of Phytopathology, 121*, 346–356.

Koenig, R., & Lesemann, D. E. (1985). Plant viruses in German rivers and lakes. I. Tombusviruses, a potexvirus and carnation mottle virus. *Phytopathologische Zeitschrift, 112*, 105–116.

Komoda, K., Naito, S., & Ishikawa, M. (2004). Replication of plant RNA virus genomes in a cell-free extract of evacuolated plant protoplasts. *Proceedings of the National Academy of Sciences of the United States of America, 101*, 1863–1867.

Koonin, E. V. (1991). The phylogeny of RNA-dependent RNA polymerases of positive-strand RNA viruses. *Journal of General Virology, 72*, 2197–2206.

Koonin, E. V., & Dolja, V. V. (1993). Evolution and taxonomy of positive-strand RNA viruses: Implications of comparative analysis of amino acid sequences. *Critical Reviews in Biochemistry and Molecular Biology, 28*, 375–430.

Kushner, D. B., Lindenbach, B. D., Grdzelishvili, V. Z., Noueiry, A. O., Paul, S. M., & Ahlquist, P. (2003). Systematic, genome-wide identification of host genes affecting replication of a positive-strand RNA virus. *Proceedings of the National Academy of Sciences of the United States of America, 100*, 15764–15769.

Kusumanegara, K., Mine, A., Hyodo, K., Kaido, M., Mise, K., & Okuno, T. (2012). Identification of domains in p27 auxiliary replicase protein essential for its association with the endoplasmic reticulum membranes in *Red clover necrotic mosaic virus*. *Virology, 433*, 131–141.

Lehninger, A. L., Nelson, D. L., & Cox, M. M. (1993). *Principles of biochemistry*. New York, NY: Worth Publishers, pp. 927–928.

Lewandowski, D. J., & Dawson, W. O. (2000). Functions of the 126- and 183-kDa proteins of tobacco mosaic virus. *Virology, 271*, 90–98.

Li, F., & Ding, S. W. (2006). Virus counterdefense: Diverse strategies for evading the RNA-silencing immunity. *Annual Review of Microbiology, 60*, 503–531.

Li, W., & Wong, S. M. (2007). Host-dependent effects of the 3′ untranslated region of turnip crinkle virus RNA on accumulation in Hibiscus and Arabidopsis. *Journal of General Virology, 88*, 680–687.

Liang, Y., & Gillam, S. (2001). Rubella virus RNA replication is *cis*-preferential and synthesis of negative- and positive-strand RNAs is regulated by the processing of nonstructural protein. *Virology, 282*, 307–319.

Lin, H. X., & White, K. A. (2004). A complex network of RNA–RNA interactions controls subgenomic mRNA transcription in a tombusvirus. *EMBO Journal, 23*, 3365–3374.

Lockney, D. M., Guenther, R. N., Loo, L., Overton, W., Antonelli, R., Clark, J., et al. (2011). The Red clover necrotic mosaic virus capsid as a multifunctional cell targeting plant viral nanoparticle. *Bioconjugate Chemistry, 22*, 67–73.

Lommel, S. A., Weston-Fina, M., Xiong, Z., & Lomonossoff, G. P. (1988). The nucleotide sequence and gene organization of red clover necrotic mosaic virus RNA-2. *Nucleic Acids Research, 16*, 8587–8602.

Lynes, E. W., Teakle, D. S., & Smith, P. R. (1981). Red clover necrotic mosaic virus isolated from *Trifolium repens* and *Medicago sativa* in Victoria. *Australasian Plant Pathology, 10*, 6–7.

Mahajan, S., Dolja, V. V., & Carrington, J. C. (1996). Roles of the sequence encoding tobacco etch virus capsid protein I genome amplification: Requirements for the translation process and a cis-active element. *Journal of Virology, 70*, 4370–4379.

Martin, S. L., Guenther, R. H., Sit, T. L., Swartz, P. D., Meilleur, F., Lommel, S. A., et al. (2010). Crystallization and preliminary X-ray diffraction analysis of red clover necrotic mosaic virus. *Acta Crystallographica, F66*, 1458–1462.

Martinez de Alba, A. E., Jauvion, V., Mallory, A. C., Bouteiller, N., & Vaucheret, H. (2011). The miRNA pathway limits AGO1 availability during siRNA-mediated PTGS defense against exogenous RNA. *Nucleic Acids Research, 39*, 9339–9344.

Matthews, R. E. F. (1982). Classification and nomenclature of viruses. In: *4th Report of the international committee for taxonomy of viruses. Karger, Basel.*
Memon, A. R. (2004). The role of ADP-ribosylation factor and SAR1 in vesicular trafficking in plants. *Biochimica et Biophysica Acta, 1664*, 9–30.
Miller, S., & Krijnse-Locker, J. (2008). Modification of intracellular membrane structure for virus replication. *Nature Reviews. Microbiology, 6*, 363–374.
Miller, W. A., Wang, Z., & Treder, K. (2007). The amazing diversity of cap-independent translation elements in the 3′-untranslated regions of plant viral RNAs. *Biochemical Society Transactions, 35*, 1629–1633.
Miller, W. A., & White, K. A. (2006). Long-distance RNA-RNA interactions in plant virus gene expression and replication. *Annual Review of Phytopathology, 44*, 447–467.
Mine, A., Hyodo, K., Tajima, Y., Taniguchi, T., Taniguchi, H., Kaido, M., et al. (2012). Differential roles of Hsp70 and Hsp90 in the assembly of the replicase complex of a positive-strand RNA plant virus. *Journal of Virology, 86*, 12091–12104.
Mine, A., Hyodo, K., Takeda, A., Kaido, M., Mise, K., & Okuno, T. (2010). Interactions between p27 and p88 replicase proteins of *Red clover necrotic mosaic virus* play an essential role in viral RNA replication and suppression of RNA silencing via the 480-kDa viral replicase complex assembly. *Virology, 407*, 213–224.
Mine, A., & Okuno, T. (2012). Composition of plant virus RNA replicase complexes. *Current Opinion in Virology, 2*, 663–669.
Mine, A., Takeda, A., Taniguchi, T., Taniguchi, H., Kaido, M., Mise, K., et al. (2010). Identification and characterization of the 480-kilodalton template-specific RNA-dependent RNA polymerase complex of *Red clover necrotic mosaic virus. Journal of Virology, 84*, 6070–6081.
Miranda, G. J., Aliyar, R., & Shirako, Y. (2001). Nucleotide sequence of a Dianthovirus RNA1-like RNA found in grassy stunt-diseased rice plants. *Archives of Virology, 146*, 225–238.
Mizumoto, H., Hikichi, Y., & Okuno, T. (2002). The 3′-untranslated region of RNA1 as a primary determinant of temperature sensitivity of *Red clover necrotic mosaic virus* Canadian strain. *Virology, 293*, 320–327.
Mizumoto, H., Iwakawa, H. O., Kaido, M., Mise, K., & Okuno, T. (2006). Cap-independent translation mechanism of *Red clover necrotic mosaic virus* RNA2 differs from that of RNA1 and is linked to RNA replication. *Journal of Virology, 80*, 3781–3791.
Mizumoto, H., Tatsuta, M., Kaido, M., Mise, K., & Okuno, T. (2003). Cap-independent translational enhancement by the 3′ untranslated region of *Red clover necrotic mosaic virus* RNA1. *Journal of Virology, 77*, 12113–12121.
Moon, S. L., Anderson, J. R., Kumagai, Y., Wilusz, C. J., Akira, S., Khromykh, A. A., et al. (2012). A noncoding RNA produced by arthropod-borne flaviviruses inhibits the cellular exoribonuclease XRN1 and alters host mRNA stability. *RNA, 18*, 2029–2040.
Morales, F., Castaño, M., Calvert, L., & Arroyave, J. A. (1992). *Furcraea necrotic streak virus*: An apparent new member of the dianthovirus group. *Journal of Phytopathology, 134*, 247–254.
Musil, M. (1969). Red clover necrotic mosaic virus, a new virus infecting red clover (*Trifolium pratense*) in Czechoslovakia. *Biologia (Bratislava), 24*, 33–45.
Musil, M., & Matisová, J. (1967). Contribution to the knowledge of mosaic viruses of red clover in Slovakia. *Ochrana Rostlin, 3*, 225–234.
Nagy, P. D., Barajas, D., & Pogany, J. (2012). Host factors with regulatory roles in tombusvirus replication. *Current Opinion in Virology, 2*, 691–698.
Neeleman, L., & Bol, J. F. (1999). *Cis*-acting functions of alfalfa mosaic virus proteins involved in replication and encapsidation of viral RNA. *Virology, 254*, 324–333.
Nicholson, B. L., & White, K. A. (2008). Context-influenced cap-independent translation of Tombusvirus mRNAs in vitro. *Virology, 380*, 203–212.

Nicholson, B. L., & White, K. A. (2011). 3′ Cap-independent translation enhancers of positive-strand RNA plant viruses. *Current Opinion in Virology*, *1*, 373–380.

Nicholson, B. L., Wu, B., Chevtchenko, I., & White, K. A. (2010). Tombusvirus recruitment of host translational machinery via the 3′ UTR. *RNA*, *16*, 1402–1419.

Novak, J. E., & Kirkegaard, K. (1994). Coupling between genome translation and replication in an RNA virus. *Genes & Development*, *8*, 1726–1737.

Okamoto, K., Nagano, H., Iwakawa, H., Mizumoto, H., Takeda, A., Kaido, M., et al. (2008). cis-Preferential requirement of a −1 frameshift product p88 for the replication of *Red clover necrotic mosaic virus* RNA1. *Virology*, *375*, 205–212.

Okuno, T. (2012). Replication mechanisms of plant RNA viruses: Current understanding and perspectives. *Journal of General Plant Pathology*, *78*, 404–408.

Okuno, T., Hiruki, C., Rao, D. V., & Figueiredo, G. C. (1983). Genetic determinants distributed in two genomic RNAs of sweet clover necrotic mosaic, red clover necrotic mosaic and clover primary leaf necrosis viruses. *Journal of General Virology*, *64*, 1907–1914.

Osman, T. A., & Buck, K. W. (1987). Replication of red clover necrotic mosaic virus RNA in cowpea protoplasts: RNA 1 replicates independently of RNA 2. *Journal of General Virology*, *68*, 289–296.

Osman, T. A., & Buck, K. W. (1991). Detection of the movement protein of red clover necrotic mosaic virus in a cell wall fraction from infected Nicotiana clevelandii plants. *Journal of General Virology*, *72*, 2853–2856.

Osman, T. A., Hayes, R. J., & Buck, K. W. (1992). Cooperative binding of the red clover necrotic mosaic virus movement protein to single-stranded nucleic acids. *Journal of General Virology*, *73*, 223–227.

Osman, T. A., Thömmes, P., & Buck, K. W. (1993). Localization of a single-stranded RNA-binding domain in the movement protein of red clover necrotic mosaic dianthovirus. *Journal of General Virology*, *74*, 2453.

Oster, S. K., Wu, B., & White, K. A. (1998). Uncoupled expression of p33 and p92 permits amplification of Tomato bushy stunt virus RNAs. *Journal of Virology*, *72*, 5845–5851.

Paje-Manalo, L. L., & Lommel, S. A. (1989). Independent replication of red clover necrotic mosaic virus RNA-1 in electroporated host and nonhost Nicotiana species protoplasts. *Phytopathology*, *79*, 457–461.

Panavas, T., Serviene, E., Brasher, J., & Nagy, P. D. (2005). Yeast genome-wide screen reveals dissimilar sets of host genes affecting replication of RNA viruses. *Proceedings of the National Academy of Sciences of the United States of America*, *102*, 7326–7331.

Pappu, H. R., & Hiruki, C. (1988). Replication of sweet clover necrotic mosaic virus in cowpea protoplasts. *Canadian Journal of Plant Pathology*, *10*, 110–115.

Pappu, H. R., & Hiruki, C. (1989). Electrophoretic variability among dianthoviruses. *Phytopathology*, *79*, 1253–1257.

Pappu, H. R., Hiruki, C., & Inouye, N. (1988). A new serotype of sweet clover necrotic mosaic virus. *Phytopathology*, *78*, 1343–1348.

Park, S.-H., Sit, T. L., Kim, K.-H., & Lommel, S. A. (2012). The *Red clover necrotic mosaic virus* capsid protein N-terminal lysine-rich motief is a determinant of symptpmatology and virion accumulation. *Molecular Plant Pathology*, *13*, 744–753.

Peltier, C., Klein, E., Hleibieh, K., D'Alonzo, M., Hammann, P., Bouzoubaa, S., et al. (2012). Beet necrotic yellow vein virus subgenomic RNA3 is a cleavage product leading to stable non-coding RNA required for long-distance movement. *Journal of General Virology*, *93*, 1093–1102.

Pijlman, G. P., Funk, A., Kondratieva, N., Leung, J., Torres, S., van der Aa, L., et al. (2008). A highly structured, nuclease-resistant, noncoding RNA produced by flaviviruses is required for pathogenicity. *Cell Host & Microbe*, *4*, 579–591.

Powers, J. G., Sit, T. L., Heinsohn, C., George, C. G., Kim, K. H., & Lommel, S. A. (2008). The *Red clover necrotic mosaic virus* RNA-2 encoded movement protein is a second suppressor of RNA silencing. *Virology*, *381*, 277–286.

Powers, J. G., Sit, T. L., Qu, F., Morris, T. J., Kim, K. H., & Lommel, S. A. (2008). A versatile assay for the identification of RNA silencing suppressors based on complementation of viral movement. *Molecular Plant-Microbe Interactions*, *21*, 879–890.

Qu, F., & Morris, T. J. (2000). Cap-independent translational enhancement of turnip crinkle virus genomic and subgenomic RNAs. *Journal of Virology*, *74*, 1085–1093.

Qu, F., Ren, T., & Morris, T. J. (2003). The coat protein of turnip crinkle virus suppresses posttranscriptional gene silencing at an early initiation step. *Journal of Virology*, *77*, 511–522.

Qu, F., Ye, X., & Morris, T. J. (2008). *Arabidopsis* DRB4, AGO1, AGO7, and RDR6 participate in a DCL4-initiated antiviral RNA silencing pathway negatively regulated by DCL1. *Proceedings of the National Academy of Sciences of the United States of America*, *105*, 14732–14737.

Rao, A. L. N., Cooper, B., & Deom, C. M. (1998). Defective movement of viruses in the family *Bromoviridae* is differentially complemented in *Nicotiana benthamiana* expressing tobamovirus or dianthovirus movement proteins. *Phytopthology*, *88*, 666–672.

Rao, A. L. N., & Hiruki, C. (1987). Unilateral compatibility of genome segments from two distinct strains of red clover necrotic mosaic virus. *Journal of General Virology*, *36*, 191–194.

Ryabov, E. V., Generozov, E. V., Kendall, T. L., Lommel, S. A., & Zavriev, S. K. (1994). Nucleotide sequence of carnation ringspot dianthovirus RNA-1. *Journal of General Virology*, *75*, 243–247.

Salonen, A., Ahola, T., & Kaariainen, L. (2005). Viral RNA replication in association with cellular membranes. *Current Topics in Microbiology and Immunology*, *285*, 139–173.

Sarawaneeyaruk, S., Iwakawa, H.-O., Mizumoto, H., Murakami, H., Kaido, M., Mise, K., et al. (2009). Host-dependent roles of the viral 5′ untranslated region (UTR) in RNA stabilization and cap-independent translational enhancement mediated by the 3′ UTR of *Red clover necrotic mosaic virus* RNA1. *Virology*, *391*, 107–118.

Schaad, M. C., Haldeman-Cahill, R., Cronin, S., & Carrington, J. C. (1996). Analysis of the VPg-proteinase (NIa) encoded by tobacco etch potyvirus: Effects of mutations on subcellular transport, proteolytic processing, and genome amplification. *Journal of Virology*, *70*, 7039–7048.

Serviene, E., Shapka, N., Cheng, C. P., Panavas, T., Phuangrat, B., Baker, J., et al. (2005). Genome-wide screen identifies host genes affecting viral RNA recombination. *Proceedings of the National Academy of Sciences of the United States of America*, *102*, 10545–10550.

Shen, R., & Miller, W. A. (2004). The 3′ untranslated region of tobacco necrosis virus RNA contains a barley yellow dwarf virus-like cap-independent translation element. *Journal of Virology*, *78*, 4655–4664.

Sherman, M. B., Guenther, R. H., Tama, F., Sit, T. L., Brooks, C. L., Mikhailov, A. M., et al. (2006). Removal of divalent cations induces structural transitions in *Red clover necrotic mosaic virus*, revealing a potential mechanism for RNA release. *Journal of Virology*, *80*, 10395–10406.

Sit, T. L., Haikal, P. R., Callaway, A. S., & Lommel, S. A. (2001). A single amino acid mutation in the carnation ringspot virus capsid protein allows virion formation but prevents systemic infection. *Journal of Virology*, *75*, 9538–9542.

Sit, T. L., Vaewhongs, A. A., & Lommel, S. A. (1998). RNA-mediated *trans*-activation of transcription from a viral RNA. *Science*, *281*, 829–832.

Solovyev, A. G., Zelenina, D. A., Savenkov, E. I., Grdzelishvili, V. Z., Morozov, S. Yu., Maiss, E., et al. (1997). Host-controlled cell-to-cell movement of a hybrid barley stripe mosaic virus expressing a dianthovirus movement protein. *Intervirology*, *40*, 1–6.

Stefano, G., Renna, L., Chatre, L., Hanton, S. L., Moreau, P., Hawes, C., et al. (2006). In tobacco leaf epidermal cells, the integrity of protein export from the endoplasmic reticulum and of ER export sites depends on active COPI machinery. *The Plant Journal*, *46*, 95–110.

Stupina, V. A., Meskauskas, A., McCormack, J. C., Yingling, Y. G., Kasprzak, W., Shapiro, B. A., et al. (2008). The 3′ proximal translational enhancer of Turnip crinkle virus binds to 60S ribosomal subunits. *RNA*, *14*, 2379–2393.

Tajima, Y., Iwakawa, H. O., Kaido, M., Mise, K., & Okuno, T. (2011). A long-distance RNA-RNA interaction plays an important role in programmed −1 ribosomal frameshifting in the translation of p88 replicase protein of Red clover necrotic mosaic virus. *Virology*, *417*, 169–178.

Takeda, A., Tsukuda, M., Mizumoto, H., Okamoto, K., Kaido, M., Mise, K., et al. (2005). A plant RNA virus suppresses RNA silencing through viral RNA replication. *EMBO Journal*, *24*, 3147–3157.

Tatsuta, M., Mizumoto, H., Kaido, M., Mise, K., & Okuno, T. (2005). The *Red clover necrotic mosaic virus* RNA2 *trans*-activator is also a cis-acting RNA2 replication element. *Journal of Virology*, *79*, 978–986.

Tomlinson, J. A., Faithfull, E., Flewett, T. H., & Beards, G. (1982). Isolation of infective tomato bushy stunt virus after passage through the human alimentary tract. *Nature*, *300*, 637–638.

Tomlinson, J. A., Faithfull, S. M., & Fraser, R. S. S. (1983). Plant viruses in river water. In: *33rd Annual report for 1982 national vegetable research station Wellesbourne, Warwick, UK* (pp. 81–82).

Tomlinson, J. A., Faithfull, E. M., & Seeley, N. D. (1982). Plant viruses in rivers. *Rothamsted experimental station report for 1981* (pp. 86–87).

Treder, K., Kneller, E. L., Allen, E. M., Wang, Z., Browning, K. S., & Miller, W. A. (2008). The 3′ cap-independent translation element of *Barley yellow dwarf virus* binds eIF4F via the eIF4G subunit to initiate translation. *RNA*, *14*, 134–147.

Tremaine, J. H., & Dodds, J. A. (1985). Carnation ringspot virus. *CMI/AAB Descriptions of Plant Viruses. No. 308.*

Tremblay, D., Vaewhongs, A. A., Turner, K. A., Sit, T. L., & Lommel, S. A. (2005). Cell wall localization of *Red clover necrotic mosaic virus* movement protein is required for cell-to-cell movement. *Virology*, *333*, 10–21.

Turner, R. L., & Buck, K. W. (1999). Mutational analysis of *cis*-acting sequences in the 3′- and 5′-untranslated regions of RNA2 of *Red clover necrotic mosaic virus*. *Virology*, *253*, 115–124.

Turner, K. A., Sit, T. L., Callaway, A. S., Allen, N. S., & Lommel, S. A. (2004). Red clover necrotic mosaic virus replication proteins accumulate at the endoplasmic reticulum. *Virology*, *320*, 276–290.

Vaewhongs, A. A., & Lommel, S. A. (1995). Virion formation is required for the long-distance movement of red clover necrotic mosaic virus in movement protein transgenic plants. *Virology*, *212*, 607–613.

Van Bokhoven, H., Le Gall, O., Kasteel, D., Verver, J., Wellink, J., & Van Kammen, A. B. (1993). *Cis*- and *trans*-acting elements in cowpea mosaic virus RNA replication. *Virology*, *195*, 377–386.

van Rossum, C. M., Garcia, M. L., & Bol, J. F. (1996). Accumulation of alfalfa mosaic virus RNAs 1 and 2 requires the encoded proteins in cis. *Journal of Virology*, *70*, 5100–5105.

Wang, Z., Kraft, J. J., Hui, A. Y., & Miller, W. A. (2010). Structural plasticity of Barley yellow dwarf virus-like cap-independent translation elements in four genera of plant viral RNAs. *Virology*, *402*, 177–186.

Wang, Z., Treder, K., & Miller, W. A. (2009). Structure of a viral cap-independent translation element that functions via high affinity binding to the eIF4E subunit of eIF4F. *Journal of Biological Chemistry*, *284*, 14189–14202.

Wang, H.-L., Wang, Y., Giesman-Cookmeyer, D., Lommel, S. A., & Lucas, W. A. (1998). Mutations in viral movement protein alter systemic infection and identify an intercellular barrier to entry into the phloem long-distance transport system. *Virology*, *245*, 75–89.

Weiland, J. J., & Dreher, T. W. (1993). Cis-preferential replication of the turnip yellow mosaic virus RNA genome. *Proceedings of the National Academy of Sciences of the United States of America, 90*, 6095–6099.
Weng, Z., & Xiong, Z. (2009). Three discontinuous loop nucleotides in the 3' terminal stem-loop are required for *Red clover necrotic mosaic virus* RNA-2 replication. *Virology, 393*, 346–354.
White, K. A. (2002). The premature termination model: A possible third mechanism for subgenomic mRNA transcription in (+)-strand RNA viruses. *Virology, 304*, 47–154.
White, K. A., Bancroft, J. B., & Mackie, G. A. (1992). Coding capacity determinants in vivo accumulation of a defective RNA of clover yellow mosaic virus. *Journal of Virology, 66*, 3069–3076.
White, K. A., Skuzeski, J. M., Li, W., Wei, N., & Morris, T. J. (1995). Immunodetection, expression strategy and complementation of turnip crinkle virus p28 and p88 replication components. *Virology, 211*, 525–534.
Xiong, Z., Kim, K. H., Giesman-Cookmeyer, D., & Lommel, S. A. (1993). The roles of the red clover necrotic mosaic virus capsid and cell-to-cell movement proteins in systemic infection. *Virology, 192*, 27–32.
Xiong, Z., Kim, K. H., Kendall, T. L., & Lommel, S. A. (1993). Synthesis of the putative red clover necrotic mosaic virus RNA polymerase by ribosomal frameshifting *in vitro*. *Virology, 193*, 213–221.
Xiong, Z., & Lommel, S. A. (1989). The complete nucleotide sequence and genome organization of red clover necrotic mosaic virus RNA-1. *Virology, 171*, 543–554.
Xiong, Z. G., & Lommel, S. A. (1991). Red clover necrotic mosaic virus infectious transcripts synthesized in vitro. *Virology, 182*, 388–392.
Xu, W., & White, K. A. (2009). RNA-based regulation of transcription and translation of aureusvirus subgenomic mRNA1. *Journal of Virology, 83*, 10096–10105.
Yi, G., & Kao, C. C. (2008). *cis-* and *trans-*Acting functions of brome mosaic virus protein 1a in genomic RNA1 replication. *Journal of Virology, 82*, 3045–3053.
Zavriev, S. K., Hickey, C. M., & Lommel, S. A. (1996). Mapping of the red clover necrotic mosaic virus subgenomic RNA. *Virology, 216*, 407–410.

CHAPTER THREE

Viral and Nonviral Elements in Potexvirus Replication and Movement and in Antiviral Responses☆

Mi-Ri Park[*,†,1], **Jang-Kyun Seo**[*,†,‡,1], **Kook-Hyung Kim**[*,†,2]
[*]Department of Agricultural Biotechnology, Plant Genomics and Breeding Institute, Seoul National University, Seoul, Republic of Korea
[†]Research Institute for Agriculture and Life Sciences, Seoul National University, Seoul, Republic of Korea
[‡]Crop Protection Division, National Academy of Agricultural Science, Rural Development Administration, Suwon, Republic of Korea
[1]These authors contributed equally to this work
[2]Corresponding author: e-mail address: kookkim@snu.ac.kr

Contents

☆The English in this document has been checked by at least two professional editors, both native speakers of English. For a certificate, please see: http://www.textcheck.com/certificate/t9pReV.

Advances in Virus Research, Volume 87
ISSN 0065-3527
http://dx.doi.org/10.1016/B978-0-12-407698-3.00003-X

Abstract

In *Potato virus X*, a member of the genus *Potexvirus*, special sequences and structures at the 5′ and 3′ ends of the nontranslated region function as *cis*-acting elements for viral replication. These elements greatly affect interactions between viral RNAs and those between viral RNAs and host factors. The potexvirus genome encodes five open-reading frames. Viral replicase, which is required for the synthesis of viral RNA, binds viral RNA elements and host factors to form a viral replication complex at the host cellular membrane. The coat protein (CP) and three viral movement proteins (TGB1, TGB2, and TGB3) have critical roles in mediating cell-to-cell viral movement through plasmodesmata by virion formation or by nonvirion ribonucleoprotein (RNP) complex formation with viral movement proteins (TGBs). The RNP complex, like TGB1–CP–viral RNA, is associated with viral replicase and used for immediate reinitiation of viral replication in newly invaded cells. Higher plants have defense mechanisms against potexviruses such as *Rx*-mediated resistance and RNA silencing. The CP acts as an avirulence effector for plant defense mechanisms, while TGB1 functions as a viral suppressor of RNA silencing, which is the mechanism of innate immune resistance. Here, we describe recent findings concerning the involvement of viral and host factors in potexvirus replication and in antiviral responses to potexvirus infection.

1. INTRODUCTION

The interaction between plant viral elements and host cellular factors in host plants is important for both viral replication and movement because most plant viruses encode only a few viral proteins and because host plants have defense mechanisms. Host plant cellular factors are used by plant RNA viruses for viral replication and movement, but because plants lack an RNA-dependent RNA polymerase (RdRp, replicase), plant RNA viruses require their own RNA polymerase for the synthesis of viral RNA in plants. Viral replicase associates with and rearranges the host cellular membrane to create a suitable environment for viral RNA replication (Mackenzie, 2005; Salonen, Ahola, & Kaariainen, 2005; Sanfacon & Zhang, 2008; Schwartz, Chen, Lee, Janda, & Ahlquist, 2004). Replication of the viral genome progresses with the formation of a viral replication complex (VRC), which comprises the viral RNA genome, the viral replicase, and host factor(s) at the host cellular membrane. After viral RNA is synthesized by the VRC, cell-to-cell movement and virion assembly are subsequently required for viral systemic infection in plants. The movement and assembly require the association of viral movement proteins and coat protein (CP) with specific host cellular factors. These viral proteins may act as avirulence (Avr) effectors by inducing host defense-related cellular proteins or molecules during the initiation stage of a viral infection. Therefore, the interaction between plant viral elements and host cellular factors in host plants is a critical aspect of the virus life cycle and the host response to viral infection. The interaction between plant viral elements and host cellular factors in relation to viral infection and host defense mechanisms has been well studied for the plant RNA viruses of the potexvirus group.

The genus *Potexvirus*, which is one of the eight genera belonging to the new family *Flexiviridae*, contains 28 plant viruses with plus single-stranded RNA (ssRNA; Tables 3.1 and 3.2). Members of the *Potexvirus* are flexuous filamentous monopartite viruses about 13 nm in diameter and 470–580 nm in length. The viral genome, which has a 5′-cap structure and a 3′-poly(A) tail, consists of 5.9–7.0 kb of plus ssRNA that encodes five open-reading frames (ORFs). ORF1 encodes replicase, ORF2–4 encode triple-gene

Table 3.1 Eight genera in the family *Flexiviridae* (Adams et al., 2004)

Genus	Species	Virion length (nm)	ORFs	MP(s)	CP (kDa)
Potexvirus	*Potato virus X* (PVX)	470–580	5	TGB	22–27
Mandarivirus	*Indian citrus ringspot virus* (ICRSV)	650	6	TGB	34
Allexivirus	*Shallot virus X* (ShVX)	~800	6	TGB	26–29
Carlavirus	*Carnation latent virus* (CLV)	610–700	6	TGB	32–36
Foveavirus	*Apple stem pitting virus* (ASPV)	~800	5	TGB	28–44
Capillovirus	*Apple stem growing virus* (ASGV)	640–700	2 or 3	30 K[a]	25–27
Vitivirus	*Grapevine virus A* (GVA)	725–785	5	30 K	18–22
Trichovirus	*Apple chlorotic leafspot virus* (ACLSV)	640–760	3 or 4	30 K	21–24

[a]30 K, single protein of the 30-kDa superfamily (Melcher, 2000).

Table 3.2 Species in the genus *Potexvirus* (Adams et al., 2004)

Species	**Abbreviation**	**Accession numbers**
Alternanthera mosaic virus	AltMV	[AF080448]
Asparagus virus 3	AV-3	[D26017, L77962, AF018156]
Bamboo mosaic virus	BaMV	[AF308158, AY241392]
Cactus virus X	CVX	[U23414]
Cassava common mosaic virus	CsCMV	[D29630]
Cassava virus X	CsVX	[U62963, AF016914]
Clover yellow mosaic virus	ClYMV	[AY121833, M62730]
Commelina virus X	ComVX	[AY181252]
Cymbidium mosaic virus	CymMV	[AJ270987, AJ550524]
Daphne virus X	DVX	[X15342]
Foxtail mosaic virus	FoMV	[D13747]
Hosta virus X	HVX	[D13957]
Hydrangea ringspot virus	HdRSV	[AF484251, AJ438767]
Lily virus X	LVX	[Z21647]
Narcissus mosaic virus	NMV	[S73580]
Nerine virus X	NVX	[D00344, M38480, X05198, X55802, M95516, Z23256, X72214, AF111193, AF172259, AF373782, AB056718]
Papaya mosaic virus	PapMV	[AJ316085]
Pepino mosaic virus	PepMV	[D12517]
Plantago asiatica mosaic virus	PlAMV	[AB066288]
Plantago severe mottle virus	PlSMoV	[X06728, X16636]
Plantain virus X	PlVX	
Potato aucuba mosaic virus	PAMV	
Potato virus X	PVX	
Scallion virus X	ScaVX	
Strawberry mild yellow edge virus	SMYEV	
Tamus red mosaic virus	TRMV	
Tulip virus X	TVX	
White clover mosaic virus	WClMV	

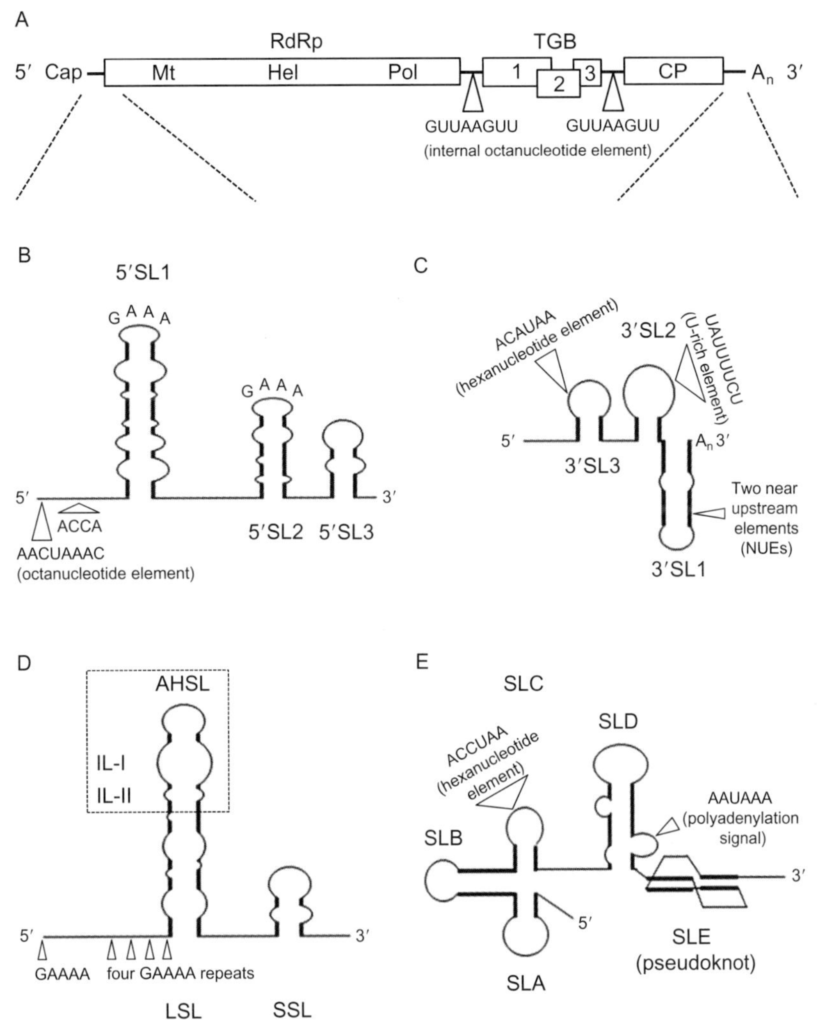

Figure 3.1 The genome organization of *Potexvirus* and the *cis*-acting elements in the 5′ and 3′NTR regions of the PVX and BaMV genomes. (A) The genome organization of *Potexvirus*. Mt, methyltransferase domain; Hel, helicase domain; Pol, RNA polymerase domain; RdRp, RNA-dependent RNA polymerase; TGB, triple-gene block; CP, coat protein. (B) The secondary structure of the 5′NTR region of PVX. (C) The secondary structure of the 3′NTR region of PVX. (D) The secondary structure of the 5′NTR region of BaMV. (E) The secondary structure of the 3′NTR region of BaMV. The thin and bold lines indicate single- and double-strand RNA forms, respectively.

blocks (TGBs) that function as viral movement proteins, and ORF5 encodes CPs from 22 to 27 kDa in size (Adams et al., 2004; Fig. 3.1A).

Viral replication, movement, and host–virus interactions have been well studied for many potexviruses including *Potato virus X* (PVX). Here, we review the current understanding of the viral elements and nonviral host factors involved in potexvirus replication and movement and in host resistance and/or host defense to potexviruses.

2. ELEMENTS THAT REGULATE POTEXVIRUS REPLICATION

2.1. Viral RNA elements

The viral RNA genomes of both PVX and *Bamboo mosaic virus* (BaMV) include special sequences and structures in the nontranslated region (NTR) that function as *cis*-acting elements for viral RNA replication. As summarized in a previous review article by Verchot-Lubicz, Ye, and Bamunusinghe (2007), three RNA stem-loop (5′SL1, 5′SL2, and 5′SL3) structures, the first of five repeated ACCA sequence elements, and a GAAA sequence in the 5′SL1 and 5′SL2 structures in the 5′NTR of PVX genomic RNA (gRNA) are all required for viral RNA replication (Kim & Hemenway, 1996; Kwon & Kim, 2006; Miller, Kim, & Hemenway, 1999; Miller, Plante, Kim, Brown, & Hemenway, 1998; Park, Kwon, Choi, Hemenway, & Kim, 2008; Fig. 3.1B). The 5′SL1, in particular, is important for PVX translation, viral RNA replication, and initiation of PVX virion assembly (Kim & Hemenway, 1997; Kwon & Kim, 2006). Interactions between a conserved octanucleotide sequence (5′-AACUAAAC-3′) and the complementary, conserved octanucleotide sequence (5′-GUUAAGUU-3′) located in the upstream subgenomic (sg) promoter regions of TGB and CP are required for sgRNA and plus-strand gRNA accumulation (Hu, Pillai-Nair, & Hemenway, 2007; Kim & Hemenway, 1997, 1999; Fig. 3.1A and B). According to a recent report (Park et al., 2008), the 5′NTR region of PVX gRNA contains five repeated ACCA sequences. Deletion, addition, and site-directed mutation of the 5′NTR including the five repeated ACCA sequences revealed that the first ACCA element (nt 10–13) affects the synthesis of plus-strand gRNA and sgRNA but not minus-strand RNA, the fourth ACCA element (nt 29–32) weakly affects virus replication, and mutations in the fifth ACCA element (nt 38–41) reduce virus replication by disrupting the SL1 structure (Park et al., 2008).

The 3′NTR of PVX genome contains three stem-loop (3′SL1, 3′SL2, and 3′SL3) structures and the poly(A) tail (Pillai-Nair, Kim, & Hemenway, 2003; Fig. 3.1C). Two overlapping sequences (i.e., 5′-UAUAAA-3′ and 5′-AAUAAA-3′ named NUE1 and NUE2, respectively) present in the 3′SL1 are important for the synthesis of plus-strand gRNA and polyadenylation (Pillai-Nair et al., 2003). The U-rich sequence (5′-UAUUUUCU-3′) in the 3′SL2 affects minus-strand RNA accumulation and is required for host protein binding (Pillai-Nair et al., 2003; Sriskanda, Pruss, Ge, & Vance, 1996). The hexanucleotide (5′-ACAUAA-3′ in the 3′SL3) and internal octanucleotide sequences (5-GUUAAGUU-3′ located in the sg promoter regions of TGB and CP) are important for the synthesis of minus-strand RNA (Hu et al., 2007).

Like PVX, BaMV contains RNA elements in the 5′ and 3′NTRs of the viral gRNA that are required for viral replication. Some BaMV strains are associated with satellite RNAs (satRNAs) which are linear RNA molecules 836 nt in length. Thus far, two satRNAs of BaMV (satBaMV) have been well studied. One is BSL6, which reduces the accumulation of BaMV RNA and attenuates BaMV-induced symptoms in coinfected plants. Another is the noninfecting satRNA, BSF4 (Hsu, Lee, Liu, & Liu, 1998). Both BaMV and its satRNAs have the repeated 5′-GAAA(A)-3′ sequence in the 5′NTR and the conserved apical hairpin stem-loop (AHSL) secondary structure in the long SL (LSL) and small SL (SSL; Chen, Desprez, & Olsthoorn, 2010; Chen, Lin, et al., 2010). These two elements are important in that they provide a specific spacing for the initiation of plus-strand RNA synthesis from minus-strand templates (Chen, Desprez, et al., 2010; Chen et al., 2012; Chen, Lin, et al., 2010; Lin et al., 2005; Lin & Hsu, 1994). The AHSL comprises an apical loop followed by two internal loops (IL-I and IL-II) located at the top of the LSL in the 5′NTR region. IL-I and IL-II are critical elements for the replication of satBaMVs (Annamalai, Hsu, Liu, Tsai, & Lin, 2003; Chen, Hsu, & Lin, 2007; Hsu et al., 2006). These elements are also key determinants for interference with BaMV replication by satBaMV (Hsu et al., 2006; Figs. 3.1D and 3.2A).

The 3′NTR of BaMV gRNA contains three secondary structures (SLA, SLB, and SLC) in the cloverleaf-like ABC domain, a D domain (SLD), and a tertiary pseudoknot domain (SLE; Cheng & Tsai, 1999; Tsai et al., 1999). The SLB, SLC, SLD, and SLE domains are important for BaMV replication, whereas the SLA element plays a role in the long-distance movement of BaMV (Chen, Meng, Hsu, & Tsai, 2003; Huang, Huang, Meng, Hsu, & Tsai, 2001). Like the secondary structures of BaMV in the 3′NTR, satBaMVs have similar structural domains, excluding SLA and SLD, in the 3′NTR. However, these satRNA structures do not function in the same manner as the similar structures in BaMV (Huang et al., 2009).

2.2. Viral protein: RdRp

Potexviruses have plus-strand RNA genomes that contain five ORFs. One of five ORFs, ORF1, encodes a viral replicase that contains a capping enzyme

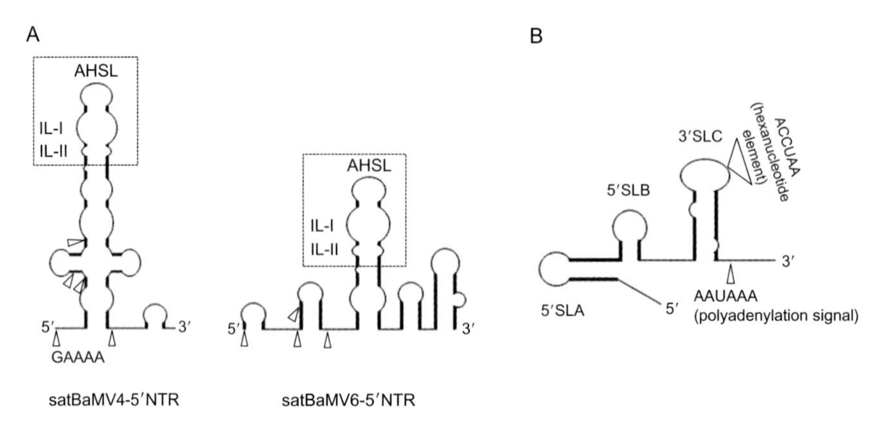

Figure 3.2 The *cis*-acting elements of the 5′ and 3′NTR regions of satBaMV. (A) The elements of the 5′NTR region in two types of satBaMV (BSF4-5′NTR and BSL6-5′NTR). Open triangles (Δ) indicate GAAA(A) repeat sequences in the 5′NTR regions. Conserved AHSLs containing two internal loops (IL-I and IL-II) in the 5′NRT regions of BaMV and satBaMVs are indicated by boxes. Three or four small stem loops (SSL) and one long stem loop (LSL) are present in the 5′NTR regions. (B) The elements in the 3′NTR region of satBaMV. The open triangle (Δ) indicates a polyadenylation signal (5′-AAUAAA-3′). The 5′-ACCUAA-3′ sequence located in the SLC in the 3′NTR is an hexanucleotide element.

domain, a helicase-like domain, which has RNA 5′-triphosphatase (RTPase) and nucleoside triphosphatase (NTPase) activities, and a polymerase domain (Huang et al., 2009; Li, Chen, Hsu, & Meng, 2001; Li et al., 1998; Li, Shih, et al., 2001). Helicases are defined as a class of enzymes that catalyze the separation of duplex nucleic acids into single strands in an ATP-dependent reaction and function in DNA modification processing, including DNA replication, DNA repair, recombination, transcription, translation, and many other nucleic acid-related processes (Jankowsky & Fairman, 2007). Helicase-like domains encoded by RNA viruses can be placed into one of three superfamilies (Kadare & Haenni, 1997). NTPase activity of the helicase-like domain is a common feature of the viral helicases, and RTPase activity is responsible for the removal of 5′-γ-phosphates from nascent viral RNAs (Han, Hsu, Lo, & Meng, 2009). The BaMV genome encodes an 115-kDa replicase consisting of an *S*-adenosyl-L-methionine (AdoMet)-dependent guanylyltransferase domain (a capping enzyme domain), a helicase-like domain, and a polymerase domain (Li, Chen, et al., 2001; Li, Shih, et al., 2001). The BaMV helicase-like domain has NTPase and RTPase activities (Li, Shih, et al., 2001). After removal of 5′-γ-phosphates from newly synthesized viral RNAs by the NTPase and RTPase activities of the BaMV helicase-like domain, the AdoMet-dependent guanylyltransferase activity catalyzes cap structure formation at the 5′-diphosphate end of the viral RNAs during viral RNA replication (Huang, Hsu, Han, & Meng, 2005). The BaMV helicase-like domain can be classified as a member of the RNA helicase superfamily 1 and be expressed as a fusion protein with a thioredoxin/hexahistidine/S-tag fused at the N-terminus (Han et al., 2007; Li, Shih, et al., 2001). The BaMV helicase-like domain has five functional motifs based on formaldehyde cross-linking and liquid chromatography coupled with tandem mass spectrometry analyses (Han et al., 2009). Two conserved motifs I and II play essential roles in removing the 5′-γ-phosphate from RNA substrates and three other nonconserved motifs are involved in RNA binding (Han et al., 2009). Results from a yeast two-hybrid screening assay demonstrated that the BaMV helicase-like domain interacts with CP (Lee et al., 2011). Mutational analyses revealed that interaction between the BaMV helicase-like domain and CP is critical for cell-to-cell movement of BaMV in plants (Lee et al., 2011). The specific amino acid motifs of BaMV CP that are required to interact with the helicase-like domain were well conserved among many potexviruses including BaMV, *Foxtail mosaic virus* (FoMV), PVX, *Cassava common mosaic virus* (CsCMV), and *Hydrangea ringspot virus* (HdRSV; Lee et al., 2011). The functional significance of the interaction between the helicase-like domain and CP in potexvirus cell-to-cell movement was further confirmed in FoMV (Lee et al., 2011).

2.3. Host cellular proteins involved in potexvirus replication

Most plant plus-strand RNA viruses encode their own proteins with RdRp activity for their replication. In general, viral replicases associate with host cellular membranes and rearrange the membranes to create suitable environments for viral replication (Mackenzie, 2005; Salonen et al., 2005; Sanfacon & Zhang, 2008; Schwartz et al., 2004). Viral replicases are thought to interact with host cellular proteins and recruit them to replication sites to form functional replication complexes.

In BaMV, the VRC associates with *cis*-acting elements in the 3′NTR of BaMV gRNAs for the synthesis of minus-strand viral RNAs. In particular, binding of the VRC to the pseudoknot poly(A) tail of BaMV gRNAs is required to initiate the synthesis of minus-strand viral RNA (Cheng, Peng, Hsu, & Tsai, 2002; Huang et al., 2001). Host cellular proteins such as heat-shock proteins (Hsps) are important factors in the VRC. Hsp90 from *Nicotiana benthamiana* (NbHsp90) associates with the BaMV VRC and binds to the SLE in the BaMV 3′NTR for BaMV RNA replication (Huang et al.,

2012; Pillai-Nair et al., 2003). NbHsp90 is involved as an RNA chaperone in the folding of the BaMV pseudoknot structure in the BaMV 3′NTR (Huang et al., 2012). NbHsp90 functions as a protein chaperone by interacting with the ORF1 protein and converting it to an active form (Huang et al., 2012). However, the replication of satBaMV and PVX was shown to be independent of NbHsp90 (Huang et al., 2012). A cellular glycolytic enzyme, glyceraldehyde 3-phosphate dehydrogenase (GAPDH), also binds to the pseudoknot poly(A) tail of BaMV and the SLC domain of satBaMV, resulting in a decrease in minus-strand synthesis (Prasanth et al., 2011). These studies have shown that host cellular factors are implicated in the regulation of viral RNA replication.

Many molecular techniques have been recently developed allowing many candidate host cellular proteins to be screened for physical interactions with viral proteins or RNAs. A methyltransferase (PNbMTS1) that interacts with BaMV replicase was identified from a yeast two-hybrid screen of an *N. benthamiana* cDNA library (Cheng et al., 2009). Overexpression of PNbMTS1 decreased the accumulation of the viral CP in protoplasts and transient gene silencing of PNbMTS1 increased the accumulation of viral CP as well as viral RNA in plants, suggesting that PNbMTS1 may have a role in a plant defense mechanism that counteracts viral infection by targeting BaMV replicase (Cheng et al., 2009). Proteomic analyses showed that many cellular proteins from *N. benthamiana* including *Tobacco mosaic virus*-movement protein 30-binding protein 2C (NbMP2Cb), putative multiprotein-bridging factor 1 (NbMPB2C), DnaJ-like protein (NbDnaJ), and potyviral CP-interacting proteins 2a and 2b (NbCPIP2s) bind to the 5′SL1 *cis*-acting element in PVX (Table 3.3; Cho, Cho, & Kim, 2012). NbMP2Cb binds to the 5′SL1 of plus-strand gRNA and the corresponding region of the minus-strand RNA (Cho, Cho, Choi, & Kim, 2012). Virus challenge inoculation assays performed on plants in which NbMP2Cb was overexpressed or silenced demonstrated that NbMP2Cb negatively regulates PVX movement in *N. benthamiana* plants (Cho, Cho, Choi, et al., 2012). Binding of NbDnaJ to the 5′SL1 region of minus-strand PVX RNA inhibits PVX systemic movement and viral RNA accumulation in *N. benthamiana* plants (Cho, Cho, Sohn, & Kim, 2012). Moreover, the above cellular proteins that interact with the PVX 5′SL1 have been reported to interact with other viral proteins as well (Kim, Huh, Ham, & Paek, 2008; Kragler, Curin, Trutnyeva, Gansch, & Waigmann, 2003; Matsushita, Miyakawa, Deguchi, Nishiguchi, & Nyunoya, 2002; Shimizu et al., 2009; Yamanaka et al., 2000), suggesting that some host cellular proteins may be commonly associated with the viral replication system.

Table 3.3 Tobacco cellular proteins that bind to PVX 5′SL1 based on proteomics analyses (Cho, Cho, & Kim, 2012)

Protein name	Function	PVX 5′SL1 strand	Accession numbers
Putative multiprotein bridge factor 1	Transcription factor	(−)	[20086364]
Elongation factor 2	Signaling protein	(−)	[1841462]
DnaJ heat-shock protein	Heat-shock protein	(−)	[90654964]
LIM domain protein WLIM1	Transcription factor	(+)/(−)	[5932413]
LIM domain protein WLIM2	Transcription factor	(+)/(−)	[5932420]
Heat-shock protein 26 (Type I)	Heat-shock protein	(+)/(−)	[3256375]
14-3-3 Protein	Signaling protein	(+)/(−)	[15778156]
Floral homeotic protein GLOBOSA	Transcription factor	(+)	[417063]
Myb-related transcription factor LBM1	Transcription factor	(−)	[6552359]
Transcription factor Myb1	Transcription factor	(−)	[1732247]
Heat-shock factor	Heat-shock protein	(+)	[5821136]
Potyviral CP-interacting protein 2a	Heat-shock protein	(+)	[34811738]
Potyviral CP-interacting protein 2b	Heat-shock protein	(+)	[34811740]
Tobamovirus multiplication 3	Transcription factor	(−)	[74038607]
DNA-binding protein NtWRKY3	Transcription factor	(+)	[4760596]
TMV-MP30-binding protein 2C	Cytoskeletal protein	(+)/(−)	[19716176]

Continued

Table 3.3 Tobacco cellular proteins that bind to PVX 5′SL1 based on proteomics analyses (Cho, Cho, & Kim, 2012)—cont'd

Protein name	Function	PVX 5′SL1 strand	Accession numbers
Ribosomal protein L2	rRNA-binding protein	(+)	[57014014]
DnaJ-like protein	Heat-shock protein	(−)	[6179940]
CMV 1a-interacting protein 1	Signaling protein	(+)	[39598908]
Mitogen-activated protein kinase	Signaling protein	(−)	[78096654]
Wound-induced protein kinase	Signaling protein	(−)	[18143321]
Trehalose-phosphate phosphatase	Signaling protein	(−)	[51458330]
Shaggy-related protein kinase NtK-1	Signaling protein	(−)	[11133532]
NSK6; Shaggy-like kinase 6	Signaling protein	(−)	[1617200]
PVX CP	Viral protein	(+)/(−)	[70994254]

3. ELEMENTS THAT REGULATE POTEXVIRUS MOVEMENT

3.1. CP: Virion structure and assembly

The potexviruses were early subjects of virion structure analyses (Bernal & Fankuchen, 1941). Various physical and biochemical methods such as *in vitro* assembly assays, spectroscopy, microscopy, and diffraction studies were used to characterize flexible filamentous virions of potexviruses including PVX, *Narcissus mosaic virus* (NMV), and *Papaya mosaic virus* (PapMV; Kendall et al., 2007, 2008; Parker, Kendall, & Stubbs, 2002; Tollin, Bancroft, Richardson, Payne, & Beveridge, 1979; Tollin, Wilson, & Bancroft, 1980; Tollin, Wilson, & Mowat, 1975). While all members of the genus *Potexvirus* have been generally considered to have similar virion architectures with high α-helical contents, some controversial observations regarding potexvirus symmetry have been reported (Kendall et al., 2007; Radwan, Wilson, & Duncan, 1981; Wilson, Al-Mukhtar, Tollin, & Hutcheson, 1978).

While no high-resolution structural information for the flexible filamentous viruses including potexviruses is yet available despite decades of intensive efforts, recent studies have reported low-resolution structures of potexviruses including PVX, PapMV, and NMV based on fiber diffraction, cryo-electron microscopy (cryo-EM), and scanning transmission EM (Kendall et al., 2013, 2008; Fig. 3.3A–I). Potexviruses share a common helical symmetry with slightly less than nine subunits per helical turn (Kendall et al., 2008, 2013). The three-dimensional structure of PVX CP has been proposed to consist of two domains: an inner four-helix bundle and an outer βαββα-type ribonucleoprotein (RNP) fold (Nemykh et al., 2008). PVX CP can exist in two different functional states, so this two-domain model explains the high plasticity of the PVX CP structure (Atabekov, Rodionova, Karpova, Kozlovsky, & Poljakov, 2000; Nemykh et al., 2008). Indeed, the PVX CP structure can be converted to a more compact structure referred to as a "sandwich" version after interaction with TGB1 (Lukashina et al., 2009).

Another line of research has characterized the crystal structure of a C-terminal truncated PapMV CP (Yang et al., 2012; Fig. 3.3J). The predicted structure of PapMV CP consists of a novel all-helix fold with seven α-helices, which is different from the four-helix-bundle fold of TMV CP (Namba, Pattanayek, & Stubbs, 1989; Yang et al., 2012; Fig. 3.3K). Note that PapMV CP was crystallized as an asymmetrical dimer in which one CP subunit interacted with its neighboring subunit via N-terminal peptide contacts. Based on the crystal structure and the previous findings that mutations at the N-terminus of PapMV CP entirely abolished the interactions between CP subunits (Laliberte Gagne, Lecours, Gagne, & Leclerc, 2008; Lecours, Tremblay, Gagne, Gagne, & Leclerc, 2006), hydrophobic interaction between the N-terminal peptides of one CP subunit and the deep hydrophobic groove of a neighboring subunit has been suggested to be critical for virion assembly for the potexviruses (Yang et al., 2012). Docking the crystal structure of the PapMV CP with the cryo-EM reconstruction of the PapMV virion particle indicated that the N-terminus of the CP is exposed to the surface and that the C-terminal domain is positioned in the interior hollow of the filamentous particle (Yang et al., 2012). The cryo-EM study indicated a helical symmetry for PapMV with 10.25 subunits per turn (Yang et al., 2012). However, this result is somewhat different from the symmetry determined by Kendall et al. (2012). This contradiction may be

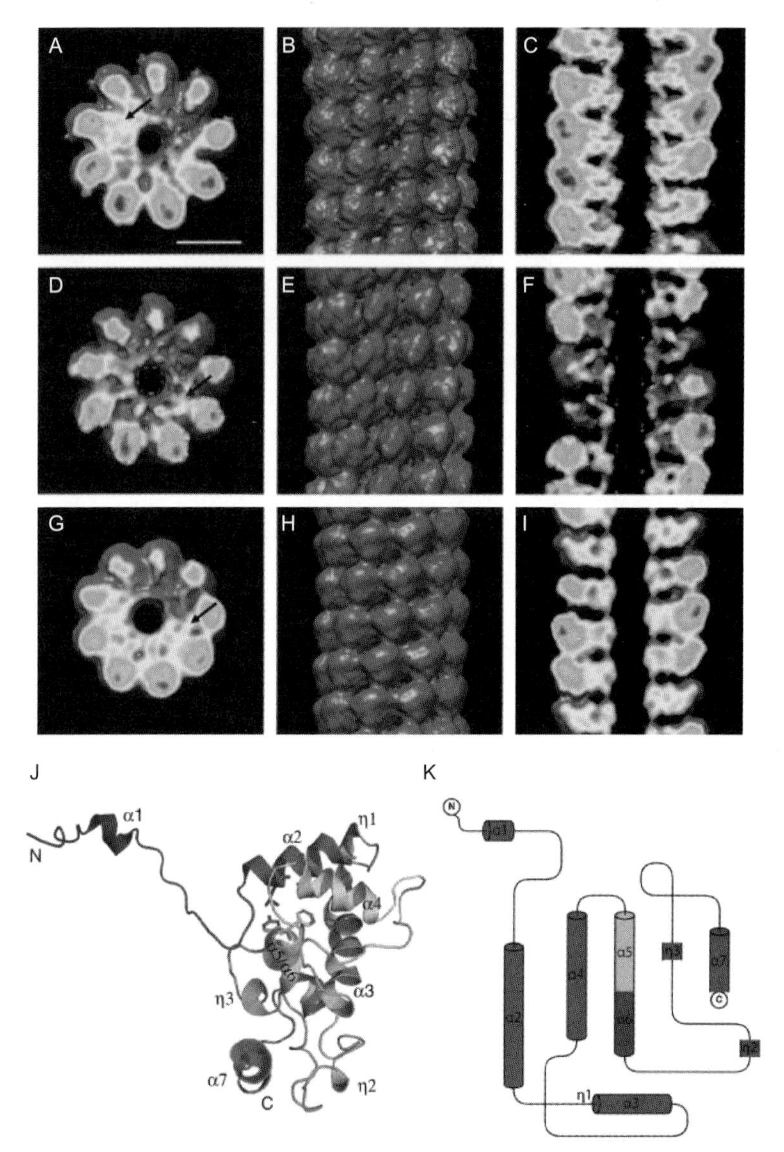

Figure 3.3 The predicted structures of potexviruses. (A–I) The iterative helical real space reconstruction (IHRSR) models of PVX (A–C), NMV (D–F), and PapMV (G–I); a section normal to the viral axis, outside surface view, and a section through the viral axis (scale bar = 50 Å). (J and K) Crystal structure of the PapMV CP monomer. A predicted three-dimensional structure in cartoon view (J) and topology of the seven α-helices (K) of the PapMV CP. *Panels (A–I): Reprinted from Kendall et al. (2013), Copyright (2013), with permission from Elsevier and panels (J and K) Reprinted from Yang et al. (2012), Copyright (2012), with permission from Elsevier.* (For color version of this figure, the reader is referred to the online version of this chapter.)

due to the low resolution of cryo-EM mapping of flexible filamentous viruses. Improving the resolution of cryo-EM analyses may be difficult. Thus, data from other methods such as fiber diffraction, crystallography, and scanning transmission EM may be required to add more detailed refinements to the structure of potexviruses.

As noted above, CP subunits in PVX virions can exist in two different functional states. This possibility was evident from changes in the translatability of PVX virions after interaction with TGB1 (Atabekov et al., 2000). While PVX virions incubated with wheat germ extracts were not translatable, virions incubated with TGB1 acquired the ability to be translated (Atabekov et al., 2000). Interaction of TGB1 with the CP subunits at the end of a virion particle containing a 5′-terminus of viral RNA resulted in a rapid destabilization of the whole virion particle (Atabekov et al., 2000; Karpova et al., 2006; Rodionova et al., 2003). Destabilization of PVX particles can also be triggered by phosphorylation of CP by host serine/threonine (Ser/Thr) protein kinases (Atabekov et al., 2001; Rodionova et al., 2003). Immunochemical analyses revealed that phosphorylation of the N-terminus of the PVX CP, which is exposed at the virion surface, induced conformational changes in the PVX CP (Atabekov et al., 2001). Further mutational analyses suggested that the N-terminus of the PVX CP is implicated in the structural alteration of intravirus CP subunits and that this alteration affects the viral RNA-binding activity of the CP subunits (Lukashina et al., 2012). Phosphorylation of PVX CP subunits by host protein kinases has been hypothesized to trigger virion disassembly, thereby rendering PVX RNA translatable during the early infection stage, whereas interactions between TGB1 and intravirus CP subunits primarily trigger translational activation of the progeny virions as the virus replication cycle continues (Atabekov et al., 2001; Verchot-Lubicz et al., 2007).

In potexviruses, the CP is critical for cell-to-cell viral movement through the plasmodesmata (PD). Two major models have been suggested to explain the cell-to-cell movement of potexviruses. The basic difference between these two models lies in whether a virion or a nonvirion RNP complex (i.e., TGB1–CP–RNA) is transported through the PD. The virion model was suggested based on the observation by EM of fibrillar particles resembling PVX virions embedded in the PD in PVX-infected leaves (Santa Cruz, Roberts, Prior, Chapman, & Oparka, 1998). The RNP complex model was postulated based on microinjection experiments and mutational analyses of CP. A key experiment demonstrated that a C-terminal truncation mutant of the CP of *White clover mosaic virus* (WClMV) was capable of forming a virion but incapable of supporting cell-to-cell viral movement, suggesting that virion formation *per se* is not sufficient for the cell-to-cell movement of WClMV (Lough et al., 2000). A similar finding was reported for PVX: a C-terminal-truncated PVX CP, which was able to form a virion, was deficient for cell-to-cell movement (Fedorkin et al., 2000, 2001). Moreover, the movement deficiency of the C-terminal-truncated PVX CP mutant could be complemented by coexpression of the potyvirus CP, although PVX RNA was not encapsidated by the potyvirus CP (Fedorkin et al., 2000, 2001). These findings suggest that the C-terminal region of the potexvirus CP harbors a distinct function required for cell-to-cell movement, although it is dispensable for virion formation.

A direct interaction between BaMV replicase and CP was demonstrated using yeast two-hybrid assays (Lee et al., 2011). Mutational analyses identified two amino acids (alanine at position 209 and asparagine at position 210) in the BaMV CP that are critical for interaction with replicase (Lee et al., 2011). To determine the significance of the interaction between the replicase and CP for viral replication, movement, and encapsidation, the mutations were introduced into the identified positions in a BaMV infectious clone. The introduction of the mutations resulted in inhibition of cell-to-cell viral movement but had no effect on viral RNA accumulation or encapsidation (Lee et al., 2011). Based on these findings, Lee et al. (2011) speculated that the BaMV replicase may be associated with the infectious RNP movement complex. A similar possibility has been suggested for TMV viral movement complex based on the observation that TMV requires a significantly longer time for movement from primary inoculated cells to secondary cells than is required for movement from secondary to tertiary cells (Kawakami, Watanabe, & Beachy, 2004). The association of the viral replicase with the RNP complex would be beneficial for immediate reinitiation of viral replication in the newly invaded cells.

3.2. TGB proteins: Virus cell-to-cell movement

Potexviruses require three viral movement proteins, named TGB1, TGB2, and TGB3, for cell-to-cell movement. The role of TGB proteins in the cell-to-cell movement of potexviruses has been extensively studied and reviewed (Morozov & Solovyev, 2003; Verchot-Lubicz et al., 2007, 2010). This review will focus on recent advances in the functional analysis of TGB-mediated cell-to-cell movement of potexviruses.

TGB1 is a multifunctional protein that moves cell to cell through PD and has RNA binding, RNA helicase, and ATPase activities. TGB1 is known to trigger an increased size-exclusion limit (SEL) in the PD. The movement of potexvirus through PD has been suggested to be dependent on the ATPase activity of TGB1, since TGB1 mutants defective for ATPase activity inhibited virus movement and reduced cell-to-cell transport of TGB1 fused to GFP (Angell, Davies, & Baulcombe, 1996; Howard et al., 2004). However, quantitative analyses revealed that the cell-to-cell movement of TGB1 through the PD is a diffusion-driven, not an ATP-driven, active transport process (Obermeyer & Tyerman, 2005; Schonknecht, Brown, & Verchot-Lubicz, 2008), suggesting that the ATPase activity of TGB1 is not directly required for movement from cell to cell. Although additional studies are required to examine the exact role of TGB1 ATPase activity in potexvirus movement, TGB1 ATPase activity may be required to increase the SEL of the PD after targeting to the PD.

As discussed above, TGB1 is a component of the potexvirus RNP complex (i.e., TGB1–CP–RNA), which is thought to be transported through the PD. Several ectopic expression strategies using fluorescent protein–TGB1 fusions have been employed to visualize the subcellular localization of TGB1. When GFP–TGB1 was expressed alone, the fluorescence signal was diffused throughout the cytoplasm and nucleoplasm (Howard et al., 2004; Samuels et al., 2007; Zamyatnin et al., 2004). However, GFP–TGB1 expressed from viral genomic replicons accompanying viral replication was observed in the PD and in aggregates (Samuels et al., 2007). TGB1 was recently shown to be responsible for remodeling host actin and endomembranes and recruiting TGB2 and TGB3 to form the X-body, which is a virus-induced inclusion structure (Tilsner et al., 2012). This TGB1-mediated X-body organization was also shown to function in compartmentalization of the viral gene products during viral infection (Tilsner et al., 2012). A study by Yan, Lu, Lin, Zheng, and Chen (2012) also demonstrated the functional significance of this X-body (referred to as rod-like structures in Yan et al.'s study). The TGB1 mutant that lost the movement function could not form rod-like structures, whereas the mutants that still supported cell-to-cell movement formed the structure, indicating that the rod-like structure formation by TGB1 is necessary for the movement function of TGB1 (Yan et al., 2012). TGB1 is not required for PVX replication and virion assembly (Morozov & Solovyev, 2003; Tilsner et al., 2012). Potexvirus CP mutants that were defective for interaction with TGB1 still formed virus particles *in vitro*, but they were unable to move cell to cell (Ozeki et al., 2009; Tilsner et al., 2012; Zayakina et al., 2008). Thus, interactions between TGB1 and CP are thought to mediate targeting of the potexvirus movement complex (i.e., virions and/or RNA complexes) to the PD.

Although the majority of TGB1 proteins were shown to accumulate in the cytoplasm, PVX TGB1 was also found to be targeted partly to the nucleus (Samuels et al., 2007). The TGB1 proteins of *Alternanthera mosaic virus* (AltMV) and NMV were localized to the nucleolus as well as the nucleus (Lim, Vaira, Bae, et al., 2010). The functional importance of the nuclear/nucleolar localization of potexvirus TGB1 proteins in relation to their movement functions remains unclear. In some viruses, nucleolar localization of viral proteins is critical for virus movement (Taliansky, Brown, Rajamaki, Valkonen, & Kalinina, 2010). One of the best-studied plant viruses encoding a nucleolar-targeting protein is *Groundnut rosette virus* (GRV), an umbravirus. GRV encodes the ORF3 protein, which acts as a long-distance movement factor. GRV ORF3 protein was found in the nucleus and predominantly in the nucleolus (Ryabov, Robinson, & Taliansky, 1999). Nuclear/nucleolar trafficking of GRV ORF3 protein requires a mechanism involving reorganization of Cajal bodies and interaction with fibrillarin (Kim, Macfarlane, et al., 2007; Kim, Ryabov, et al., 2007). Mutagenic analyses showed that mutations in GRV ORF3 protein that blocked its nuclear/nucleolar trafficking capability prevented the formation of cytoplasmic viral RNPs and the long-distance movement of viruses (Kim, Macfarlane, et al., 2007; Kim, Ryabov, et al., 2007). Hordeiviruses and pomoviruses are other TGB-containing viruses. Unlike potexvirus TGB1 proteins, the TGB1 proteins of hordeiviruses and pomoviruses are predicted to contain at least one nucleolar localization signal, indicating that nucleolar targeting of the TGB1 proteins of hordeiviruses and pomoviruses is directed by their own localization signals (Semashko, Gonzalez, et al., 2012; Semashko, Rakitina, et al., 2012; Solovyev, Kalinina, & Morozov, 2012). Moreover, the hordeivirus TGB1 protein is able to interact with both fibrillarin (the major nucleolar protein) and coilin (the major structural component of Cajal bodies; Semashko, Gonzalez, et al., 2012; Semashko, Rakitina, et al., 2012). As for GRV ORF3, nuclear/nucleolar localization of hordeivirus TGB1 is probably required for long-distance viral movement. However, no evidence exists that nuclear/nucleolar targeting of potexvirus TGB1 is associated with virus movement.

Potexvirus TGB1 is also a virus-encoded suppressor of RNA silencing. Some PVX TGB1 mutants defective in the suppression of RNA silencing inhibited viral cell-to-cell movement, and this inhibition could be restored by providing heterologous viral RNA silencing suppressors such as P19 and HC-Pro *in trans* (Bayne, Rakitina, Morozov, & Baulcombe, 2005). Thus, the processes of cell-to-cell movement and RNA silencing suppression were proposed to possibly be coordinated. However, some TGB1 mutants that are fully functional for RNA silencing suppression did not support viral cell-to-cell movement, indicating that the other functions of TGB1 might be necessary for viral movement (Bayne et al., 2005). Another line of research has shown that a specific mutation in TGB1 from AltMV that caused a significant decrease in silencing suppression activity still maintained full cell-to-cell movement, indicating that the silencing suppression and movement functions of TGB1 may be uncoupled (Lim, Vaira, Reinsel, et al., 2010). This discrepancy may be due to different experimental approaches with different viral species. Also, potexvirus TGB1s may have different modes of action for silencing suppression and cell-to-cell movement in different viral species. Additional studies are required to investigate the relationship between potexvirus TGB1-mediated silencing suppression and cell-to-cell movement.

TGB1 trafficking to the PD requires the TGB proteins TGB2 and TGB3, which are endoplasmic reticulum (ER)-localized proteins (Verchot-Lubicz et al., 2010). Mutational analyses have shown that integration of TGB2 and TGB3 into the ER is critical for viral cell-to-cell movement (Krishnamurthy et al., 2003; Mitra et al., 2003). In addition, potexvirus TGB2 was shown to have RNA-binding activity in a sequence-independent manner (Cowan, Lioliopoulou, Ziegler, & Torrance, 2002; Hsu et al., 2009). Both TGB2 and TGB3 enhanced cell-to-cell diffusion of free GFP or GFP–sporamin fusion proteins, suggesting that TGB2 and TGB3 are capable of gating the PD (Haupt et al., 2005; Tamai & Meshi, 2001). TGB2 was predicted to contain two transmembrane domains and TGB3 to have at least one transmembrane domain located at the N-terminus based on protein sequence analyses (Krishnamurthy et al., 2003; Mitra et al., 2003; Wu, Lee, & Wang, 2011). Recent topological analyses predicted that the N- and C-termini of TGB2 are exposed to the cytoplasm and that the central loop region is located in the ER lumen (Hsu et al., 2009; Lee, Wu, & Wang, 2010; Mitra et al., 2003). Moreover, the cytoplasmic tail of TGB3 was shown to contain a sorting signal that is necessary and sufficient for TGB3 oligomerization and for targeting of integral membrane proteins to cortical ER tubules (Wu et al., 2011). The TGB3 sorting signal is a highly conserved feature of potexviruses, and TGB3 mutants with sorting signal defects had impaired cell-to-cell viral movement (Wu et al., 2011).

PVX TGB2 expressed alone or from a viral replicon induced small granular vesicles that were aligned on actin filaments (Ju et al., 2005). Mutational analyses showed that these TGB2-induced granular vesicles are necessary for PVX cell to cell (Ju, Brown, Ye, & Verchot-Lubicz, 2007), and subcellular localization studies showed that PVX TGB3 proteins are localized to the ER network when expressed alone in tobacco leaves (Ju et al., 2005). However, TGB3 was colocalized in the TGB2-induced granular vesicles in PVX-infected cells, indicating that TGB2 functions to induce vesicle formation from the ER membrane associated with TGB3 (Schepetilnikov et al., 2005). Recent studies have shown that TGB3 directs movement of the ER-derived vesicles induced by TGB2 from the perinuclear ER to the cortical ER in both yeast and plant systems (Lee et al., 2010; Wu et al., 2011), indicating that TGB3 serves as a driving factor for TGB-mediated intra- and intercellular trafficking. Stoichiometric analyses revealed that TGB2 and TGB3 may constitute a protein complex at a ratio of approximately 1:1 in punctuate structures required for trafficking (Lee et al., 2010), indicating a physical interaction between potexvirus TGB2 and TGB3 proteins. Indeed, based on a bimolecular fluorescence complementation assay, TGB2 was shown to physically interact with TGB3 in a membrane-associated form (Wu et al., 2011). This same study also showed an interaction between TGB2 and either CP or TGB1. The interactions of TGB1–TGB2 and TGB2–CP were associated with the ER network, which is consistent with the localization of TGB2, and these interaction complexes were translocated to the TGB3-containing puncta within ER tubules by coexpression of TGB3 (Wu et al., 2011). The findings by Wu et al. (2011) and Lee et al. (2011) suggest a model for an ER-localized potexvirus movement complex that consists of viral RNA, CP, TGB1, TGB2, TGB3, and a replicase (Fig. 3.4).

Unlike PVX TGB3, AltMV TGB3 was shown to be localized to the chloroplast membrane and to accumulate preferentially in mesophyll cells (Lim, Vaira, Bae, et al., 2010). Comparative sequence analyses revealed that AltMV TGB3 has limited similarity to other potexvirus TGB3 homologs, and mutational analyses demonstrated that AltMV TGB3 contains a novel signal for chloroplast membrane localization. In addition, the AltMV mutants carrying mutations in the TGB3 chloroplast localization signal showed very limited cell-to-cell movement and only in epidermal cells, indicating that chloroplast targeting of TGB3 is required for the systemic movement of AltMV (Lim, Vaira, Bae, et al., 2010). These findings for AltMV further support the general conclusion that different potexviruses may employ somewhat different mechanisms for viral movement (Lin, Hu, Lin, Chang, & Hsu, 2006).

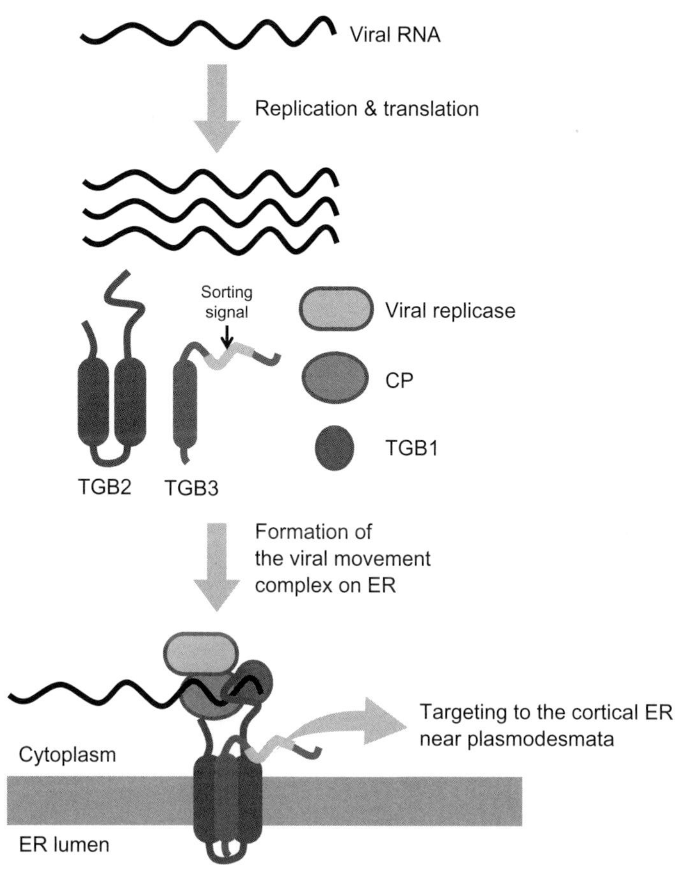

Figure 3.4 A model for the potexvirus movement complex. Viral replication and translation produce progeny viruses and viral protein components. TGB2 and TGB3 are targeted to the ER and form a stoichiometric complex. This complex recruits the viral RNA–TGB1–CP–relpicase complex by interaction with TGB2–TGB1 and/or TGB2–CP. The sorting signal in TGB3 targets the resulting movement complex to the cortical ER near the plasmodesmata for cell-to-cell viral movement. (For color version of this figure, the reader is referred to the online version of this chapter.)

3.3. Host cellular proteins involved in potexvirus movement

In addition to the involvement of the TGB proteins and CP in potexvirus movement, recent advances in identifying and characterizing host cellular proteins associated with potexvirus movement have shed light on how viruses utilize host intra/intercellular systems for their movement.

Membrane rafts, which act as docking sites for cellular import/export processes, including secretion, signal transduction, and perception of pathogens (Rajendran & Simons, 2005), may have important functions in the steps of a viral infection cycle, including virus entry, assembly, and budding (Chazal & Gerlier, 2003; Rajendran & Simons, 2005). Remorins (REMs) are located both at the PD and in 70-nm membrane domains, similar to lipid rafts in the cytosolic leaflet of the plasma membrane (Raffaele et al., 2009). REMs can serve as a counteracting membrane platform for viral RNP docking, thereby inhibiting cell-to-cell viral movement by physical interaction between REMs and PVX TGB1 (Raffaele et al., 2009).

TGB2-interacting host cellular proteins (TIP1, TIP2, and TIP3) were identified using a yeast two-hybrid screen of an *N. tabacum* cDNA library (Fridborg et al., 2003). All three TIPs were ankyrin repeat-containing proteins (AKRs) and specifically interacted with TGB2 but not with TGB1 or TGB3 (Fridborg et al., 2003). AKRs have been identified in a diverse group of proteins such as channels, toxins, enzymes, and proteins involved in the regulation of cytoskeletal organization, signal transduction, and transcription (Rubtsov & Lopina, 2000; Sedgwick & Smerdon, 1999). β-1,3-Glucanase is a TIP1-interacting host cellular protein that directly interacts with all three TIPs, but not with TGB2 (Fridborg et al., 2003). During PVX infection, TGB2 is required for an increased SEL in the PD and is associated with peripheral bodies in the presence of TGB3 (Tamai & Meshi, 2001) that colocalize with callose (Erhardt et al., 2000). Moreover, β-1,3-glucanase plays a role as a regulator of the PD SEL by mediating callose degradation (Iglesias & Meins, 2000). Subcellular localization studies demonstrated that GFP-fused TIP1 accumulated in the cytoplasm and translocated to deposits of additional cytoplasm in the presence of TGB2, suggesting that TGB2 mediates the targeting of TIP and β-1,3-glucanase to the PD and thereby alters callose deposition in the neck region of PD with a resulting increase in the PD SEL (Fridborg et al., 2003).

Some host cellular Hsps, acting as chaperones and folding enzymes, are also involved in viral movement (Verchot, 2012). Hsps are conserved molecular polypeptide chaperones that are involved in various processes in the cytoplasm, ER, mitochondria, and plastids of all living cellular organisms (Derocher, Helm, Lauzon, & Vierling, 1991; Wang, Vinocur, Shoseyov, & Altman, 2004). Hsp70 is upregulated in response to infection by viruses, including PVX, and may contribute to pathogenesis in some host plants (Aparicio et al., 2005). Suppressor of G2 allele of skp1 (SGT1) is a cochaperone required for stabilization of protein complexes. The SGT1–Hsp70 complex is important for plant responses to environmental stresses such as pathogen invasion (Cazale et al., 2009; Noel et al., 2007; Zabka, Lesniak, Prus, Kuznicki, & Filipek, 2008). PVX TGB3 induces the expression of SGT1 and the induced SGT increases the systemic accumulation of PVX in *N. benthamiana* (Ye, Kelly, Payton, Dickman, & Verchot, 2012). SGT1 is also required for the steady-state accumulation of the resistance protein *Rx* by forming a complex with Hsp90 and another cellular protein, *Rar1* (Boter et al., 2007). Interaction between *Pepino mosaic virus* (PepMV) CP and Hsc70 may facilitate viral cell-to-cell movement through the PD via the ATPase and translocation activities of Hsc70 (Mathioudakis et al., 2012).

PVX CP was shown to be phosphorylated by plant Ser/Thr protein kinases that are a mixture of casein kinase I and II (CKI and CKII) and a soluble *Nicotiana glutinosa* kinase (Atabekov et al., 2001). The phosphorylation of PVX CP was suggested to render PVX RNA translatable in primary inoculated cells, but that the translational activation of the progeny virions destined for PD trafficking was triggered by TGB1 (Atabekov et al., 2001). PVX TGB1 is phosphorylated by the *N. tabacum* CKII-like kinase, and this phosphorylation is important for the movement and replication of PVX (Modena, Zelada, Conte, & Mentaberry, 2008).

A host protein that interacts with PVX CP (NbPCIP1) was identified by screening an *N. benthamiana* cDNA library using the yeast two-hybrid system (Park, Park, Cho, & Kim, 2009). NbPCIP1 increased PVX replication and movement in *N. benthamiana* plants (Park et al., 2009). NbDnaJ (or Hsp40) was identified using yeast three-hybrid assays and reported to interact with PVX CP and viral RNA elements (Cho, Cho, Sohn, et al., 2012).

4. ELEMENTS INVOLVED IN ANTIVIRAL RESPONSES

4.1. Interactions between host and viral elements

Higher plants have evolved a multilayered pathogen surveillance system (Chisholm, Coaker, Day, & Staskawicz, 2006; Jones & Dangl, 2006). One of the layers in this process is referred to as pathogen-associated molecular pattern (PAMP)-triggered immunity (PTI). PTI involves the recognition of conserved PAMPs and the subsequent induction of basal defense responses. The second layer of defense is referred to as effector-triggered immunity (ETI). ETI is triggered by the recognition of pathogen Avr effectors by plant disease resistance (R) proteins. ETI is often associated with the hypersensitive response (HR), which includes the induction of defense-related genes and the oxidative burst resulting in localized cell death (Chisholm et al., 2006; Jones & Dangl, 2006). Many plant R proteins contain nucleotide-binding site (NBS) and leucine-rich repeat (LRR) domains

(Chisholm et al., 2006; Jones & Dangl, 2006). R proteins interact directly or indirectly with Avr proteins from diverse pathogens to confer disease resistance. In this regard, any protein component of a plant virus can be recognized as a specific Avr effector protein by a specific R protein.

The potato *Rx* gene, which encodes a member of the NBS–LRR family of R proteins, is well characterized and involved in conferring PVX resistance by recognizing PVX CP as an Avr effector (Bendahmane, Kanyuka, & Baulcombe, 1999; Bendahmane, Kohn, Dedi, & Baulcombe, 1995). Rx is known to confer strong resistance and to suppress virus accumulation without HR (Bendahmane et al., 1999). However, overexpression of CP, which is the elicitor for *Rx*-mediated resistance, resulted in HR (Bendahmane et al., 1999). Thus, Rx was suggested to have a mode of action similar to that of other NBS–LRR proteins that mediate the HR. Indeed, *Rx*-mediated resistance requires SGT1 and Hsp90 proteins, which are associated with HR-mediated resistance (Boter et al., 2007; Mayor, Martinon, De Smedt, Petrilli, & Tschopp, 2007). A recent study identified a specific *Rx*-associated protein, RanGAP2 (*N. benthamiana* Ran GTPase-Activating Protein 2) and demonstrated that the downregulation of RanGAP2 resulted in suppression of *Rx*-mediated resistance (Tameling & Baulcombe, 2007). This finding suggests that the activation of *Rx*-mediated resistance involves nucleocytoplasmic trafficking since eukaryotic RanGAP proteins have an important regulatory role in nucleocytoplasmic trafficking through nuclear pores (Meier, 2007).

The Rx protein maintains a folded resting state conformation in uninfected cells (Moffett, Farnham, Peart, & Baulcombe, 2002; Rairdan & Moffett, 2006). A systematic structure–function analysis demonstrated that the coiled-coil (CC) domain, which is located at the N-terminus of the NBS–LRR domain of the Rx protein, mediates intramolecular interactions between domains within the Rx protein and is also required for the interaction with RanGAP2 (Rairdan et al., 2008). The folded conformation of the Rx protein was altered by interaction with PVX CP, resulting in exposure of the NBS–LRR domain and activation of a signaling cascade (Moffett et al., 2002; Rairdan et al., 2008; Rairdan & Moffett, 2006). The Rx gene confers resistance against PVX and other potexviruses including NMV, WClMV, *Cymbidium mosaic virus* (CymMV), and PepMV by recognition of the CPs (Baures, Candresse, Leveau, Bendahmane, & Sturbois, 2008; Candresse et al., 2010). These potexviruses are distantly related to PVX, and the amino acid sequences of their CPs are only weakly conserved (approximately

40% identity; Baures et al., 2008; Candresse et al., 2010). Thus, the Rx-mediated complex has been suggested to recognize a conserved structural element among the potexvirus CPs rather than a conserved amino acid sequence motif (Baures et al., 2008).

A recent study showed that the PVX CP is associated with symptom severity in *N. benthamiana* through an interaction with the precursor of plastocyanin, a protein involved in photosynthesis (Qiao, Li, Wong, & Fan, 2009). The interaction between PVX CP and the plastocyanin precursor occurred in chloroplasts, and downregulation of plastocyanin precursor expression decreased symptom severity, suggesting that targeting of PVX CP into chloroplasts is associated with symptom severity (Qiao et al., 2009).

Overexpression of PNbMTS1 decreased virus accumulation in protoplasts, and downregulation of the protein resulted in increased virus accumulation in plants (Cheng et al., 2009). Although the authors did not show if the BaMV replicase could be methylated by PNbMTS1, PNbMTS1 may be involved in the plant innate defense response and act to suppress virus accumulation during the early phase of an infection (Cheng et al., 2009). The replicase of AltMV was shown to contain determinants of symptom severity by analyzing multiple biologically distinct infectious clones of AltMV and chimeric derivatives (Lim, Vaira, Reinsel, et al., 2010).

PVX TGB3 triggers the unfolded protein response (UPR) when overexpressed by agroinfiltration (Ye, Dickman, Whitham, Payton, & Verchot, 2011). The UPR, which is a conserved phenomenon across kingdoms, is caused by various ER stresses that can result in the accumulation of misfolded proteins in the ER (Walter & Ron, 2011). Under conditions of unmitigated ER stress, the function of the UPR is to maintain homeostasis in the ER or to induce programmed cell death by regulating the expression of numerous genes such as the basic region leucine zipper 60 (*bZIP60*), ER luminal-binding protein (*BiP*), protein disulfide isomerase (*PDI*), calreticulin (*CRT*), and calmodulin (*CAM*) (Walter & Ron, 2011; Ye et al., 2011). Both PVX infection and ectopic overexpression of TGB3 caused upregulation of the above UPR-related genes (Ye et al., 2011). The overexpression of TGB3 resulted in localized necrosis, but the necrosis could be abrogated by coexpression of TGB3 with BiP (Ye et al., 2011). This indicated that TGB3, which is a virus-encoded ER-targeting protein, is capable of inducing ER stress-related cell death in plant cells, but that the UPR attenuates cell death by helping to maintain proper ER protein folding.

4.2. RNA-based antiviral defense

RNA silencing is generally recognized as an important mechanism of innate immune resistance to viruses. Plants have evolved RNA silencing pathways to defend against virus infections (Baulcombe, 2004; Voinnet, 2005), while viruses have evolved to encode viral suppressors of RNA silencing (VSRs) to counteract host RNA-based antiviral activities at various levels (Baulcombe, 2004; Voinnet, 2005). More than 50 individual VSRs have been identified in almost all plant viruses (Burgyan & Havelda, 2011; Ding & Voinnet, 2007), and recent studies have shown that VSRs target various steps/components of the RNA silencing pathway through multiple mechanisms (Burgyan & Havelda, 2011; Qu & Morris, 2005).

In potexviruses, the TGB1 protein functions as a VSR, and PVX TGB1 was one of the first VSRs to be identified. When TGB1 was removed from the PVX genome, accumulation of a 25-nt short-interfering RNA (siRNA) associated with systemic silencing increased, indicating that PVX TGB1 suppresses systemic RNA silencing by regulating the accumulation of 25-nt siRNA (Voinnet, Lederer, & Baulcombe, 2000). At the same time, it was reported that transgene-mediated RNA silencing involves accumulation of the 25-nt siRNA and requires RDR6 (an RdRp homolog also known as SDE1; Dalmay, Hamilton, Rudd, Angell, & Baulcombe, 2000). Thus, PVX TGB1 was suggested to suppress RNA silencing at a step that is dependent on RDR6 functions (Voinnet et al., 2000). However, a later study from the same group showed that PVX TGB1 could suppress RNA silencing that was induced by an inverted repeat (IR) RNAi construct (Bayne et al., 2005). Because RNA silencing induced by IR constructs is not dependent on RDR6 (Beclin, Boutet, Waterhouse, & Vaucheret, 2002; Schwach, Vaistij, Jones, & Baulcombe, 2005), it was concluded that RDR6 might not be the primary target of TGB1 (Bayne et al., 2005). Alternatively, it was suggested that TGB1 may block the assembly or function of the effector complexes of RNA silencing because PVX TGB1 was able to block IR-induced RNA silencing, although it did not prevent siRNA synthesis from IRs (Bayne et al., 2005). Indeed, PVX TGB1 was shown to interact with Argonaute (AGO) proteins (the major components of the RNA silencing effector complex), and the level of AGO1 decreased in the presence of TGB1 but was restored by treatment with the proteasome inhibitor MG132 (Chiu, Chen, Baulcombe, & Tsai, 2010). Thus, PVX TGB1 has been proposed to target AGO1 and mediate the degradation of AGO1 through a proteasome degradation pathway (Chiu et al., 2010).

Recently, RNA silencing was demonstrated to play an important role in restricting virus–host ranges (Jaubert, Bhattacharjee, Mello, Perry, & Moffett, 2011). PVX can infect *Arabidopsis thaliana* plants, which is normally a nonhost for PVX when coinfected with *Pepper ringspot virus* (PepRSV), and the infection is enabled by a VSR encoded by PepRSV (Jaubert et al., 2011). Based on this finding, the roles of RNA-silencing pathway components including AGO and Dicer-like proteins in *Arabidopsis* resistance to PVX were examined by genetic screening. PVX could infect the *Arabidopsis* triple dicer mutant (*dcl2/dcl3/dcl4*) and the *ago2* mutant, indicating that RNA silencing is responsible for *Arabidopsis* nonhost resistance against PVX. These findings further indicate that PVX successfully suppresses AGO1-mediated RNA silencing activities in *Arabidopsis* and that TBG1 may be responsible for this suppression of AGO1 activity as shown in *N. benthamiana* (Chiu et al., 2010). AGO proteins may contribute to plant antiviral immunity in a hierarchical and synergistic manner (Chiu et al., 2010; Ding, 2010; Ding & Voinnet, 2007). In this regard, the findings by Jaubert et al. (2011) also suggest that AGO proteins have evolved in different host plants to target specific viruses.

5. CONCLUSIONS

Recently, host cellular factors have been a focus of plant virus studies owing to their essential role in viral propagation in plants. In the best-studied case of host cellular proteins interacting with potexviruses, the BaMV VRC associates with the NbHsp90 in the ER and binds to viral RNA elements in the BaMV 3′NTR for the replication of BaMV and satBaMV RNA (Huang et al., 2012; Pillai-Nair et al., 2003). NbHsp90, which is involved in the folding of the BaMV pseudoknot structure in the BaMV 3′NTR, interacts with the folded BaMV replicase (active form) and specifically leads to correct folding of the replicase in the assembly of the VRC for the replication of BaMV RNA (Huang et al., 2012). Meanwhile, GAPDH binds to the pseudoknot poly(A) tail of the BaMV 3′NTR and the SLC domain of the satBaMV 3′NTR to reduce the synthesis of minus-strand viral RNA in *N. benthamiana* (Prasanth et al., 2011). Thus, these studies provide evidence that cellular host factors are indispensable for the regulation of the mechanism of viral RNA replication.

TGB3 functions to direct movement of ER-derived vesicles induced by TGB2 from the perinuclear ER to the cortical ER in the host during PVX infection (Lee et al., 2010; Wu et al., 2011), indicating that TGB3 serves as a driving factor of TGB-mediated intra- and intercellular trafficking. TGB2 interacts with either CP or TGB1 in the ER network (Wu et al., 2011). In conclusion, all viral proteins are involved in the VRC, which associates with the host ER network leading to viral replication and movement in plants. The possibility exists that some ER-associated host cellular proteins are involved in the VRC.

In the plant immune response against PVX, PVX TGB3 triggers the UPR, which is known to be elicited by various ER stresses that cause accumulation of misfolded proteins in the ER (Walter & Ron, 2011). The UPR either maintains homeostasis in the ER or induces programmed cell death if the ER stress remains unmitigated and does so by regulating the expression of numerous genes, such as the *bZIP60*, the ER luminal *BiP*, *PDI*, *CRT*, and *CAM* (Walter & Ron, 2011; Ye et al., 2011).

Plants have evolved RNA silencing pathways to defend against viral infections (Baulcombe, 2004; Voinnet, 2005) and viruses have evolved to encode VSRs that counteract host RNA-based antiviral activities at various levels (Baulcombe, 2004; Voinnet, 2005). More than 50 individual VSRs have been identified in almost all plant viruses (Burgyan & Havelda, 2011; Ding & Voinnet, 2007). VSRs target various steps/components of the RNA silencing pathway through multiple mechanisms (Burgyan & Havelda, 2011; Qu & Morris, 2005). PVX TGB1 was one of the first VSRs identified, and it suppresses systemic RNA silencing by regulating accumulation of the RDR6-dependent 25-nt siRNA (Voinnet et al., 2000) or by an RDR6-independent mechanism in transgenic plants expressing an IR RNAi construct (Bayne et al., 2005; Beclin et al., 2002; Schwach et al., 2005). These results suggest that RDR6 might not be the primary target of TGB1 (Bayne et al., 2005). Indeed, a recent study demonstrated that PVX TGB1 interacted with AGO proteins and reduced the level of AGO1. Thus, PVX TGB1 was proposed to target AGO1 and mediate the degradation of AGO1 through a proteasome pathway (Chiu et al., 2010).

The finding that PVX can infect nonhost *Arabidopsis* plants when coinfected with PepRSV demonstrated that RNA silencing plays an important role in restricting virus–host ranges (Jaubert et al., 2011). A VSR encoded by PepRSV enabled PVX to infect *Arabidopsis*, demonstrating that RNA-silencing pathway components, including AGO and Dicer-like proteins, are responsible for *Arabidopsis* nonhost resistance against PVX (Jaubert et al., 2011). These findings suggest that AGO proteins evolve with different virus specificities in different host plants.

Potexvirus studies to date suggest that many unknown host cellular factors may be involved in viral propagation and resistance, and in host antiviral defense mechanisms, and that host cellular factors are indispensable for the regulation of viral RNA replication. The development of molecular biotechnology methods for the screening of specific interacting proteins has allowed the identification of many host proteins that interact with potexvirus proteins and/or RNA elements, but much more remains to be investigated about host–virus interactions. In this chapter, we summarized the results of potexvirus studies and anticipate that this knowledge will be useful for the further study of host–virus interactions.

ACKNOWLEDGMENTS

This research was supported in part by grants from the National Research Foundation (Grant No. 20110012328) funded by the Ministry of Education, Science, and Technology (MEST); the Vegetable Breeding Research Center (No. 710001-03) through the Agriculture Research Center program from the Ministry for Food, Agriculture, Forestry and Fisheries; and the Next-Generation BioGreen 21 Program of Rural Development Administration (No. PJ00819801).

REFERENCES

Adams, M. J., Antoniw, J. F., Bar-Joseph, M., Brunt, A. A., Candresse, T., Foster, G. D., et al. (2004). The new plant virus family Flexiviridae and assessment of molecular criteria for species demarcation. *Archives of Virology*, *149*, 1045–1060.

Angell, S. M., Davies, C., & Baulcombe, D. C. (1996). Cell-to-cell movement of potato virus X is associated with a change in the size-exclusion limit of plasmodesmata in trichome cells of Nicotiana clevelandii. *Virology*, *216*, 197–201.

Annamalai, P., Hsu, Y. H., Liu, Y. P., Tsai, C. H., & Lin, N. S. (2003). Structural and mutational analyses of cis-acting sequences in the 5'-untranslated region of satellite RNA of bamboo mosaic potexvirus. *Virology*, *311*, 229–239.

Aparicio, F., Thomas, C. L., Lederer, C., Niu, Y., Wang, D., & Maule, A. J. (2005). Virus induction of heat shock protein 70 reflects a general response to protein accumulation in the plant cytosol. *Plant Physiology*, *138*, 529–536.

Atabekov, J. G., Rodionova, N. P., Karpova, O. V., Kozlovsky, S. V., Novikov, V. K., & Arkhipenko, M. V. (2001). Translational activation of encapsidated *Potato virus X* RNA by coat protein phosphorylation. *Virology*, *286*, 466–474.

Atabekov, J. G., Rodionova, N. P., Karpova, O. V., Kozlovsky, S. V., & Poljakov, V. Y. (2000). The movement protein-triggered in situ conversion of *Potato virus X* virion RNA from a nontranslatable into a translatable form. *Virology*, *271*, 259–263.

Baulcombe, D. (2004). RNA silencing in plants. *Nature*, *431*, 356–363.

Baures, I., Candresse, T., Leveau, A., Bendahmane, A., & Sturbois, B. (2008). The Rx gene confers resistance to a range of potexviruses in transgenic *Nicotiana* plants. *Molecular Plant-Microbe Interactions*, *21*, 1154–1164.

Bayne, E. H., Rakitina, D. V., Morozov, S. Y., & Baulcombe, D. C. (2005). Cell-to-cell movement of potato potexvirus X is dependent on suppression of RNA silencing. *The Plant Journal*, *44*, 471–482.

Beclin, C., Boutet, S., Waterhouse, P., & Vaucheret, H. (2002). A branched pathway for transgene-induced RNA silencing in plants. *Current Biology, 12*, 684–688.

Bendahmane, A., Kanyuka, K., & Baulcombe, D. C. (1999). The Rx gene from potato controls separate virus resistance and cell death responses. *The Plant Cell, 11*, 781–792.

Bendahmane, A., Kohn, B. A., Dedi, C., & Baulcombe, D. C. (1995). The coat protein of potato virus X is a strain-specific elicitor of Rx1-mediated virus resistance in potato. *The Plant Journal, 8*, 933–941.

Bernal, J. D., & Fankuchen, I. (1941). X-ray and crystallographic studies of plant virus preparations: I. Introduction and preparation of specimens II. Modes of aggregation of the virus particles. *The Journal of General Physiology, 25*, 111–146.

Boter, M., Amigues, B., Peart, J., Breuer, C., Kadota, Y., Casais, C., et al. (2007). Structural and functional analysis of SGT1 reveals that its interaction with HSP90 is required for the accumulation of Rx, an R protein involved in plant immunity. *The Plant Cell, 19*, 3791–3804.

Burgyan, J., & Havelda, Z. (2011). Viral suppressors of RNA silencing. *Trends in Plant Science, 16*, 265–272.

Candresse, T., Marais, A., Faure, C., Dubrana, M. P., Gombert, J., & Bendahmane, A. (2010). Multiple coat protein mutations abolish recognition of *Pepino mosaic potexvirus* (PepMV) by the potato rx resistance gene in transgenic tomatoes. *Molecular Plant-Microbe Interactions, 23*, 376–383.

Cazale, A. C., Clement, M., Chiarenza, S., Roncato, M. A., Pochon, N., Creff, A., et al. (2009). Altered expression of cytosolic/nuclear HSC70-1 molecular chaperone affects development and abiotic stress tolerance in *Arabidopsis thaliana*. *Journal of Experimental Botany, 60*, 2653–2664.

Chazal, N., & Gerlier, D. (2003). Virus entry, assembly, budding, and membrane rafts. *Microbiology and Molecular Biology Reviews, 67*, 226–237, table of contents.

Chen, S. C., Desprez, A., & Olsthoorn, R. C. (2010). Structural homology between *Bamboo mosaic virus* and its satellite RNAs in the 5'untranslated region. *The Journal of General Virology, 91*, 782–787.

Chen, H. C., Hsu, Y. H., & Lin, N. S. (2007). Downregulation of *Bamboo mosaic virus* replication requires the 5' apical hairpin stem loop structure and sequence of satellite RNA. *Virology, 365*, 271–284.

Chen, H. C., Kong, L. R., Yeh, T. Y., Cheng, C. P., Hsu, Y. H., & Lin, N. S. (2012). The conserved 5' apical hairpin stem loops of *Bamboo mosaic virus* and its satellite RNA contribute to replication competence. *Nucleic Acids Research, 40*, 4641–4652.

Chen, I. H., Lin, J. W., Chen, Y. J., Wang, Z. C., Liang, L. F., Meng, M., et al. (2010). The 3'-terminal sequence of *Bamboo mosaic virus* minus-strand RNA interacts with RNA-dependent RNA polymerase and initiates plus-strand RNA synthesis. *Molecular Plant Pathology, 11*, 203–212.

Chen, I. H., Meng, M., Hsu, Y. H., & Tsai, C. H. (2003). Functional analysis of the cloverleaf-like structure in the 3' untranslated region of bamboo mosaic potexvirus RNA revealed dual roles in viral RNA replication and long distance movement. *Virology, 315*, 415–424.

Cheng, C. W., Hsiao, Y. Y., Wu, H. C., Chuang, C. M., Chen, J. S., Tsai, C. H., et al. (2009). Suppression of *Bamboo mosaic virus* accumulation by a putative methyltransferase in *Nicotiana benthamiana*. *Journal of Virology, 83*, 5796–5805.

Cheng, J. H., Peng, C. W., Hsu, Y. H., & Tsai, C. H. (2002). The synthesis of minus-strand RNA of bamboo mosaic potexvirus initiates from multiple sites within the poly(A) tail. *Journal of Virology, 76*, 6114–6120.

Cheng, C. P., & Tsai, C. H. (1999). Structural and functional analysis of the 3' untranslated region of bamboo mosaic potexvirus genomic RNA. *Journal of Molecular Biology, 288*, 555–565.

Chisholm, S. T., Coaker, G., Day, B., & Staskawicz, B. J. (2006). Host-microbe interactions: Shaping the evolution of the plant immune response. *Cell, 124*, 803–814.
Chiu, M. H., Chen, I. H., Baulcombe, D. C., & Tsai, C. H. (2010). The silencing suppressor P25 of *Potato virus X* interacts with Argonaute1 and mediates its degradation through the proteasome pathway. *Molecular Plant Pathology, 11*, 641–649.
Cho, S. Y., Cho, W. K., Choi, H. S., & Kim, K. H. (2012). Cis-acting element (SL1) of *Potato virus X* controls viral movement by interacting with the NbMPB2Cb and viral proteins. *Virology, 427*, 166–176.
Cho, S. Y., Cho, W. K., & Kim, K. H. (2012). Identification of tobacco proteins associated with the stem-loop 1 RNAs of *Potato virus X*. *Molecules and Cells, 33*, 379–384.
Cho, S. Y., Cho, W. K., Sohn, S. H., & Kim, K. H. (2012). Interaction of the host protein NbDnaJ with *Potato virus X* minus-strand stem-loop 1 RNA and capsid protein affects viral replication and movement. *Biochemical and Biophysical Research Communications, 417*, 451–456.
Cowan, G. H., Lioliopoulou, F., Ziegler, A., & Torrance, L. (2002). Subcellular localisation, protein interactions, and RNA binding of *Potato mop-top virus* triple gene block proteins. *Virology, 298*, 106–115.
Dalmay, T., Hamilton, A., Rudd, S., Angell, S., & Baulcombe, D. C. (2000). An RNA-dependent RNA polymerase gene in *Arabidopsis* is required for posttranscriptional gene silencing mediated by a transgene but not by a virus. *Cell, 101*, 543–553.
Derocher, A. E., Helm, K. W., Lauzon, L. M., & Vierling, E. (1991). Expression of a conserved family of cytoplasmic low molecular weight heat shock proteins during heat stress and recovery. *Plant Physiology, 96*, 1038–1047.
Ding, S. W. (2010). RNA-based antiviral immunity. *Nature Reviews. Immunology, 10*, 632–644.
Ding, S. W., & Voinnet, O. (2007). Antiviral immunity directed by small RNAs. *Cell, 130*, 413–426.
Erhardt, M., Morant, M., Ritzenthaler, C., Stussi-Garaud, C., Guilley, H., Richards, K., et al. (2000). P42 movement protein of *Beet necrotic yellow vein virus* is targeted by the movement proteins P13 and P15 to punctate bodies associated with plasmodesmata. *Molecular Plant-Microbe Interactions, 13*, 520–528.
Fedorkin, O. N., Merits, A., Lucchesi, J., Solovyev, A. G., Saarma, M., Morozov, S. Y., et al. (2000). Complementation of the movement-deficient mutations in *Potato virus X*: Potyvirus coat protein mediates cell-to-cell trafficking of C-terminal truncation but not deletion mutant of potexvirus coat protein. *Virology, 270*, 31–42.
Fedorkin, O., Solovyev, A., Yelina, N., Zamyatnin, A., Jr., Zinovkin, R., Makinen, K., et al. (2001). Cell-to-cell movement of *Potato virus X* involves distinct functions of the coat protein. *The Journal of General Virology, 82*, 449–458.
Fridborg, I., Grainger, J., Page, A., Coleman, M., Findlay, K., & Angell, S. (2003). TIP, a novel host factor linking callose degradation with the cell-to-cell movement of *Potato virus X*. *Molecular Plant-Microbe Interactions, 16*, 132–140.
Han, Y. T., Hsu, Y. H., Lo, C. W., & Meng, M. (2009). Identification and functional characterization of regions that can be crosslinked to RNA in the helicase-like domain of BaMV replicase. *Virology, 389*, 34–44.
Han, Y. T., Tsai, C. S., Chen, Y. C., Lin, M. K., Hsu, Y. H., & Meng, M. (2007). Mutational analysis of a helicase motif-based RNA 5'-triphosphatase/NTPase from *Bamboo mosaic virus*. *Virology, 367*, 41–50.
Haupt, S., Cowan, G. H., Ziegler, A., Roberts, A. G., Oparka, K. J., & Torrance, L. (2005). Two plant-viral movement proteins traffic in the endocytic recycling pathway. *The Plant Cell, 17*, 164–181.
Howard, A. R., Heppler, M. L., Ju, H. J., Krishnamurthy, K., Payton, M. E., & Verchot-Lubicz, J. (2004). *Potato virus X* TGBp1 induces plasmodesmata gating and moves

between cells in several host species whereas CP moves only in *N. benthamiana* leaves. *Virology*, *328*, 185–197.

Hsu, Y. H., Chen, H. C., Cheng, J., Annamalai, P., Lin, B. Y., Wu, C. T., et al. (2006). Crucial role of the 5' conserved structure of *Bamboo mosaic virus* satellite RNA in down-regulation of helper viral RNA replication. *Journal of Virology*, *80*, 2566–2574.

Hsu, Y. H., Lee, Y. S., Liu, J. S., & Liu, N. S. (1998). Differential interactions of bamboo mosaic potexvirus satellite RNAs, helper virus, and host plants. *Molecular Plant-Microbe Interactions*, *11*, 1207–1213.

Hsu, H. T., Tseng, Y. H., Chou, Y. L., Su, S. H., Hsu, Y. H., & Chang, B. Y. (2009). Characterization of the RNA-binding properties of the triple-gene-block protein 2 of *Bamboo mosaic virus*. *Virology Journal*, *6*, 50.

Hu, B., Pillai-Nair, N., & Hemenway, C. (2007). Long-distance RNA-RNA interactions between terminal elements and the same subset of internal elements on the *Potato virus X* genome mediate minus- and plus-strand RNA synthesis. *RNA*, *13*, 267–280.

Huang, Y. L., Hsu, Y. H., Han, Y. T., & Meng, M. (2005). mRNA guanylation catalyzed by the S-adenosylmethionine-dependent guanylyltransferase of *Bamboo mosaic virus*. *The Journal of Biological Chemistry*, *280*, 13153–13162.

Huang, Y. W., Hu, C. C., Lin, C. A., Liu, Y. P., Tsai, C. H., Lin, N. S., et al. (2009). Structural and functional analyses of the 3' untranslated region of *Bamboo mosaic virus* satellite RNA. *Virology*, *386*, 139–153.

Huang, Y. W., Hu, C. C., Liou, M. R., Chang, B. Y., Tsai, C. H., Meng, M., et al. (2012). Hsp90 interacts specifically with viral RNA and differentially regulates replication initiation of *Bamboo mosaic virus* and associated satellite RNA. *PLoS Pathogens*, *8*, e1002726.

Huang, C. Y., Huang, Y. L., Meng, M., Hsu, Y. H., & Tsai, C. H. (2001). Sequences at the 3' untranslated region of bamboo mosaic potexvirus RNA interact with the viral RNA-dependent RNA polymerase. *Journal of Virology*, *75*, 2818–2824.

Iglesias, V. A., & Meins, F., Jr. (2000). Movement of plant viruses is delayed in a beta-1,3-glucanase-deficient mutant showing a reduced plasmodesmatal size exclusion limit and enhanced callose deposition. *The Plant Journal*, *21*, 157–166.

Jankowsky, E., & Fairman, M. E. (2007). RNA helicases—One fold for many functions. *Current Opinion in Structural Biology*, *17*, 316–324.

Jaubert, M., Bhattacharjee, S., Mello, A. F., Perry, K. L., & Moffett, P. (2011). ARGONAUTE2 mediates RNA-silencing antiviral defenses against *Potato virus X* in *Arabidopsis*. *Plant Physiology*, *156*, 1556–1564.

Jones, J. D., & Dangl, J. L. (2006). The plant immune system. *Nature*, *444*, 323–329.

Ju, H. J., Brown, J. E., Ye, C. M., & Verchot-Lubicz, J. (2007). Mutations in the central domain of *Potato virus X* TGBp2 eliminate granular vesicles and virus cell-to-cell trafficking. *Journal of Virology*, *81*, 1899–1911.

Ju, H. J., Samuels, T. D., Wang, Y. S., Blancaflor, E., Payton, M., Mitra, R., et al. (2005). The *Potato virus X* TGBp2 movement protein associates with endoplasmic reticulum-derived vesicles during virus infection. *Plant Physiology*, *138*, 1877–1895.

Kadare, G., & Haenni, A. L. (1997). Virus-encoded RNA helicases. *Journal of Virology*, *71*, 2583–2590.

Karpova, O. V., Zayakina, O. V., Arkhipenko, M. V., Sheval, E. V., Kiselyova, O. I., Poljakov, V. Y., et al. (2006). *Potato virus X* RNA-mediated assembly of single-tailed ternary 'coat protein-RNA-movement protein' complexes. *The Journal of General Virology*, *87*, 2731–2740.

Kawakami, S., Watanabe, Y., & Beachy, R. N. (2004). *Tobacco mosaic virus* infection spreads cell to cell as intact replication complexes. *Proceedings of the National Academy of Sciences of the United States of America*, *101*, 6291–6296.

Kendall, A., Bian, W., Junn, J., McCullough, I., Gore, D., & Stubbs, G. (2007). Radial density distribution and symmetry of a *Potexvirus, Narcissus mosaic virus*. *Virology, 357*, 158–164.

Kendall, A., Bian, W., Maris, A., Azzo, C., Groom, J., Williams, D., et al. (2013). A common structure for the potexviruses. *Virology, 436*, 173–178.

Kendall, A., McDonald, M., Bian, W., Bowles, T., Baumgarten, S. C., Shi, J., et al. (2008). Structure of flexible filamentous plant viruses. *Journal of Virology, 82*, 9546–9554.

Kim, K. H., & Hemenway, C. (1996). The 5' nontranslated region of *Potato virus X* RNA affects both genomic and subgenomic RNA synthesis. *Journal of Virology, 70*, 5533–5540.

Kim, K. H., & Hemenway, C. (1997). Mutations that alter a conserved element upstream of the *Potato virus X* triple block and coat protein genes affect subgenomic RNA accumulation. *Virology, 232*, 187–197.

Kim, K. H., & Hemenway, C. L. (1999). Long-distance RNA-RNA interactions and conserved sequence elements affect *Potato virus X* plus-strand RNA accumulation. *RNA, 5*, 636–645.

Kim, M. J., Huh, S. U., Ham, B. K., & Paek, K. H. (2008). A novel methyltransferase methylates *Cucumber mosaic virus* 1a protein and promotes systemic spread. *Journal of Virology, 82*, 4823–4833.

Kim, S. H., Macfarlane, S., Kalinina, N. O., Rakitina, D. V., Ryabov, E. V., Gillespie, T., et al. (2007). Interaction of a plant virus-encoded protein with the major nucleolar protein fibrillarin is required for systemic virus infection. *Proceedings of the National Academy of Sciences of the United States of America, 104*, 11115–11120.

Kim, S. H., Ryabov, E. V., Kalinina, N. O., Rakitina, D. V., Gillespie, T., MacFarlane, S., et al. (2007). Cajal bodies and the nucleolus are required for a plant virus systemic infection. *The EMBO Journal, 26*, 2169–2179.

Kragler, F., Curin, M., Trutnyeva, K., Gansch, A., & Waigmann, E. (2003). MPB2C, a microtubule-associated plant protein binds to and interferes with cell-to-cell transport of *Tobacco mosaic virus* movement protein. *Plant Physiology, 132*, 1870–1883.

Krishnamurthy, K., Heppler, M., Mitra, R., Blancaflor, E., Payton, M., Nelson, R. S., et al. (2003). The *Potato virus X* TGBp3 protein associates with the ER network for virus cell-to-cell movement. *Virology, 309*, 135–151.

Kwon, S. J., & Kim, K. H. (2006). The SL1 stem-loop structure at the 5'-end of *Potato virus X* RNA is required for efficient binding to host proteins and for viral infectivity. *Molecules and Cells, 21*, 63–75.

Laliberte Gagne, M. E., Lecours, K., Gagne, S., & Leclerc, D. (2008). The F13 residue is critical for interaction among the coat protein subunits of *Papaya mosaic virus*. *The FEBS Journal, 275*, 1474–1484.

Lecours, K., Tremblay, M. H., Gagne, M. E., Gagne, S. M., & Leclerc, D. (2006). Purification and biochemical characterization of a monomeric form of papaya mosaic potexvirus coat protein. *Protein Expression and Purification, 47*, 273–280.

Lee, C. C., Ho, Y. N., Hu, R. H., Yen, Y. T., Wang, Z. C., Lee, Y. C., et al. (2011). The interaction between *Bamboo mosaic virus* replication protein and coat protein is critical for virus movement in plant hosts. *Journal of Virology, 85*, 12022–12031.

Lee, S. C., Wu, C. H., & Wang, C. W. (2010). Traffic of a viral movement protein complex to the highly curved tubules of the cortical endoplasmic reticulum. *Traffic, 11*, 912–930.

Li, Y. I., Chen, Y. J., Hsu, Y. H., & Meng, M. (2001). Characterization of the AdoMet-dependent guanylyltransferase activity that is associated with the N terminus of *Bamboo mosaic virus* replicase. *Journal of Virology, 75*, 782–788.

Li, Y. I., Cheng, Y. M., Huang, Y. L., Tsai, C. H., Hsu, Y. H., & Meng, M. (1998). Identification and characterization of the *Escherichia coli*-expressed RNA-dependent RNA polymerase of *Bamboo mosaic virus*. *Journal of Virology, 72*, 10093–10099.

Li, Y. I., Shih, T. W., Hsu, Y. H., Han, Y. T., Huang, Y. L., & Meng, M. (2001). The helicase-like domain of plant potexvirus replicase participates in formation of RNA 5' cap structure by exhibiting RNA 5'-triphosphatase activity. *Journal of Virology, 75*, 12114–12120.

Lim, H. S., Vaira, A. M., Bae, H., Bragg, J. N., Ruzin, S. E., Bauchan, G. R., et al. (2010). Mutation of a chloroplast-targeting signal in *Alternanthera mosaic virus* TGB3 impairs cell-to-cell movement and eliminates long-distance virus movement. *The Journal of General Virology, 91*, 2102–2115.

Lim, H. S., Vaira, A. M., Reinsel, M. D., Bae, H., Bailey, B. A., Domier, L. L., et al. (2010). Pathogenicity of *Alternanthera mosaic virus* is affected by determinants in RNA-dependent RNA polymerase and by reduced efficacy of silencing suppression in a movement-competent TGB1. *The Journal of General Virology, 91*, 277–287.

Lin, J. W., Chiu, H. N., Chen, I. H., Chen, T. C., Hsu, Y. H., & Tsai, C. H. (2005). Structural and functional analysis of the cis-acting elements required for plus-strand RNA synthesis of *Bamboo mosaic virus*. *Journal of Virology, 79*, 9046–9053.

Lin, N. S., & Hsu, Y. H. (1994). A satellite RNA associated with bamboo mosaic potexvirus. *Virology, 202*, 707–714.

Lin, M. K., Hu, C. C., Lin, N. S., Chang, B. Y., & Hsu, Y. H. (2006). Movement of potexviruses requires species-specific interactions among the cognate triple gene block proteins, as revealed by a trans-complementation assay based on the *Bamboo mosaic virus* satellite RNA-mediated expression system. *The Journal of General Virology, 87*, 1357–1367.

Lough, T. J., Netzler, N. E., Emerson, S. J., Sutherland, P., Carr, F., Beck, D. L., et al. (2000). Cell-to-cell movement of potexviruses: Evidence for a ribonucleoprotein complex involving the coat protein and first triple gene block protein. *Molecular Plant-Microbe Interactions, 13*, 962–974.

Lukashina, E., Badun, G., Fedorova, N., Ksenofontov, A., Nemykh, M., Serebryakova, M., et al. (2009). Tritium planigraphy study of structural alterations in the coat protein of *Potato virus X* induced by binding of its triple gene block 1 protein to virions. *The FEBS Journal, 276*, 7006–7015.

Lukashina, E., Ksenofontov, A., Fedorova, N., Badun, G., Mukhamedzhanova, A., Karpova, O., et al. (2012). Analysis of the role of the coat protein N-terminal segment in *Potato virus X* virion stability and functional activity. *Molecular Plant Pathology, 13*, 38–45.

Mackenzie, J. (2005). Wrapping things up about virus RNA replication. *Traffic, 6*, 967–977.

Mathioudakis, M. M., Veiga, R., Ghita, M., Tsikou, D., Medina, V., Canto, T., et al. (2012). *Pepino mosaic virus* capsid protein interacts with a tomato heat shock protein cognate 70. *Virus Research, 163*, 28–39.

Matsushita, Y., Miyakawa, O., Deguchi, M., Nishiguchi, M., & Nyunoya, H. (2002). Cloning of a tobacco cDNA coding for a putative transcriptional coactivator MBF1 that interacts with the *Tomato mosaic virus* movement protein. *Journal of Experimental Botany, 53*, 1531–1532.

Mayor, A., Martinon, F., De Smedt, T., Petrilli, V., & Tschopp, J. (2007). A crucial function of SGT1 and HSP90 in inflammasome activity links mammalian and plant innate immune responses. *Nature Immunology, 8*, 497–503.

Meier, I. (2007). Composition of the plant nuclear envelope: Theme and variations. *Journal of Experimental Botany, 58*, 27–34.

Melcher, U. (2000). The '30K' superfamily of viral movement proteins. *The Journal of General Virology, 81*, 257–266.

Miller, E. D., Kim, K. H., & Hemenway, C. (1999). Restoration of a stem-loop structure required for *Potato virus X* RNA accumulation indicates selection for a mismatch and a GNRA tetraloop. *Virology, 260*, 342–353.

Miller, E. D., Plante, C. A., Kim, K. H., Brown, J. W., & Hemenway, C. (1998). Stem-loop structure in the 5' region of *Potato virus X* genome required for plus-strand RNA accumulation. *Journal of Molecular Biology, 284*, 591–608.

Mitra, R., Krishnamurthy, K., Blancaflor, E., Payton, M., Nelson, R. S., & Verchot-Lubicz, J. (2003). The *Potato virus X* TGBp2 protein association with the endoplasmic reticulum plays a role in but is not sufficient for viral cell-to-cell movement. *Virology, 312*, 35–48.

Modena, N. A., Zelada, A. M., Conte, F., & Mentaberry, A. (2008). Phosphorylation of the TGBp1 movement protein of *Potato virus X* by a *Nicotiana tabacum* CK2-like activity. *Virus Research, 137*, 16–23.

Moffett, P., Farnham, G., Peart, J., & Baulcombe, D. C. (2002). Interaction between domains of a plant NBS-LRR protein in disease resistance-related cell death. *The EMBO Journal, 21*, 4511–4519.

Morozov, S. Y., & Solovyev, A. G. (2003). Triple gene block: Modular design of a multifunctional machine for plant virus movement. *The Journal of General Virology, 84*, 1351–1366.

Namba, K., Pattanayek, R., & Stubbs, G. (1989). Visualization of protein-nucleic acid interactions in a virus. Refined structure of intact tobacco mosaic virus at 2.9 A resolution by X-ray fiber diffraction. *Journal of Molecular Biology, 208*, 307–325.

Nemykh, M. A., Efimov, A. V., Novikov, V. K., Orlov, V. N., Arutyunyan, A. M., Drachev, V. A., et al. (2008). One more probable structural transition in *Potato virus X* virions and a revised model of the virus coat protein structure. *Virology, 373*, 61–71.

Noel, L. D., Cagna, G., Stuttmann, J., Wirthmuller, L., Betsuyaku, S., Witte, C. P., et al. (2007). Interaction between SGT1 and cytosolic/nuclear HSC70 chaperones regulates *Arabidopsis* immune responses. *The Plant Cell, 19*, 4061–4076.

Obermeyer, G., & Tyerman, S. D. (2005). NH4+ currents across the peribacteroid membrane of soybean. Macroscopic and microscopic properties, inhibition by Mg^{2+}, and temperature dependence indicate a SubpicoSiemens channel finely regulated by divalent cations. *Plant Physiology, 139*, 1015–1029.

Ozeki, J., Hashimoto, M., Komatsu, K., Maejima, K., Himeno, M., Senshu, H., et al. (2009). The N-terminal region of the *Plantago asiatica mosaic virus* coat protein is required for cell-to-cell movement but is dispensable for virion assembly. *Molecular Plant-Microbe Interactions, 22*, 677–685.

Park, M. R., Kwon, S. J., Choi, H. S., Hemenway, C. L., & Kim, K. H. (2008). Mutations that alter a repeated ACCA element located at the 5' end of the *Potato virus X* genome affect RNA accumulation. *Virology, 378*, 133–141.

Park, M. R., Park, S. H., Cho, S. Y., & Kim, K. H. (2009). Nicotiana benthamiana protein, NbPCIP1, interacting with *Potato virus X* coat protein plays a role as susceptible factor for viral infection. *Virology, 386*, 257–269.

Parker, L., Kendall, A., & Stubbs, G. (2002). Surface features of *Potato virus X* from fiber diffraction. *Virology, 300*, 291–295.

Pillai-Nair, N., Kim, K. H., & Hemenway, C. (2003). Cis-acting regulatory elements in the *Potato virus X* 3' non-translated region differentially affect minus-strand and plus-strand RNA accumulation. *Journal of Molecular Biology, 326*, 701–720.

Prasanth, K. R., Huang, Y. W., Liou, M. R., Wang, R. Y., Hu, C. C., Tsai, C. H., et al. (2011). Glyceraldehyde 3-phosphate dehydrogenase negatively regulates the replication of *Bamboo mosaic virus* and its associated satellite RNA. *Journal of Virology, 85*, 8829–8840.

Qiao, Y., Li, H. F., Wong, S. M., & Fan, Z. F. (2009). Plastocyanin transit peptide interacts with *Potato virus X* coat protein, while silencing of plastocyanin reduces coat protein accumulation in chloroplasts and symptom severity in host plants. *Molecular Plant-Microbe Interactions, 22*, 1523–1534.

Qu, F., & Morris, T. J. (2005). Suppressors of RNA silencing encoded by plant viruses and their role in viral infections. *FEBS Letters*, *579*, 5958–5964.

Radwan, M. M., Wilson, H. R., & Duncan, G. H. (1981). Diffraction studies of tulip virus X particles. *The Journal of General Virology*, *56*, 297–302.

Raffaele, S., Bayer, E., Lafarge, D., Cluzet, S., German Retana, S., Boubekeur, T., et al. (2009). Remorin, a solanaceae protein resident in membrane rafts and plasmodesmata, impairs *Potato virus X* movement. *The Plant Cell*, *21*, 1541–1555.

Rairdan, G. J., Collier, S. M., Sacco, M. A., Baldwin, T. T., Boettrich, T., & Moffett, P. (2008). The coiled-coil and nucleotide binding domains of the potato Rx disease resistance protein function in pathogen recognition and signaling. *The Plant Cell*, *20*, 739–751.

Rairdan, G. J., & Moffett, P. (2006). Distinct domains in the ARC region of the potato resistance protein Rx mediate LRR binding and inhibition of activation. *The Plant Cell*, *18*, 2082–2093.

Rajendran, L., & Simons, K. (2005). Lipid rafts and membrane dynamics. *Journal of Cell Science*, *118*, 1099–1102.

Rodionova, N. P., Karpova, O. V., Kozlovsky, S. V., Zayakina, O. V., Arkhipenko, M. V., & Atabekov, J. G. (2003). Linear remodeling of helical virus by movement protein binding. *Journal of Molecular Biology*, *333*, 565–572.

Rubtsov, A. M., & Lopina, O. D. (2000). Ankyrins. *FEBS Letters*, *482*, 1–5.

Ryabov, E. V., Robinson, D. J., & Taliansky, M. E. (1999). A plant virus-encoded protein facilitates long-distance movement of heterologous viral RNA. *Proceedings of the National Academy of Sciences of the United States of America*, *96*, 1212–1217.

Salonen, A., Ahola, T., & Kaariainen, L. (2005). Viral RNA replication in association with cellular membranes. *Current Topics in Microbiology and Immunology*, *285*, 139–173.

Samuels, T. D., Ju, H. J., Ye, C. M., Motes, C. M., Blancaflor, E. B., & Verchot-Lubicz, J. (2007). Subcellular targeting and interactions among the *Potato virus X* TGB proteins. *Virology*, *367*, 375–389.

Sanfacon, H., & Zhang, G. (2008). Analysis of interactions between viral replicase proteins and plant intracellular membranes. *Methods in Molecular Biology*, *451*, 361–375.

Santa Cruz, S., Roberts, A. G., Prior, D. A., Chapman, S., & Oparka, K. J. (1998). Cell-to-cell and phloem-mediated transport of *Potato virus X*. The role of virions. *The Plant Cell*, *10*, 495–510.

Schepetilnikov, M. V., Manske, U., Solovyev, A. G., Zamyatnin, A. A., Jr., Schiemann, J., & Morozov, S. Y. (2005). The hydrophobic segment of *Potato virus X* TGBp3 is a major determinant of the protein intracellular trafficking. *The Journal of General Virology*, *86*, 2379–2391.

Schonknecht, G., Brown, J. E., & Verchot-Lubicz, J. (2008). Plasmodesmata transport of GFP alone or fused to *Potato virus X* TGBp1 is diffusion driven. *Protoplasma*, *232*, 143–152.

Schwach, F., Vaistij, F. E., Jones, L., & Baulcombe, D. C. (2005). An RNA-dependent RNA polymerase prevents meristem invasion by *Potato virus X* and is required for the activity but not the production of a systemic silencing signal. *Plant Physiology*, *138*, 1842–1852.

Schwartz, M., Chen, J., Lee, W. M., Janda, M., & Ahlquist, P. (2004). Alternate, virus-induced membrane rearrangements support positive-strand RNA virus genome replication. *Proceedings of the National Academy of Sciences of the United States of America*, *101*, 11263–11268.

Sedgwick, S. G., & Smerdon, S. J. (1999). The ankyrin repeat: A diversity of interactions on a common structural framework. *Trends in Biochemical Sciences*, *24*, 311–316.

Semashko, M. A., Gonzalez, I., Shaw, J., Leonova, O. G., Popenko, V. I., Taliansky, M. E., et al. (2012). The extreme N-terminal domain of a hordeivirus TGB1 movement protein mediates its localization to the nucleolus and interaction with fibrillarin. *Biochimie, 94*, 1180–1188.

Semashko, M. A., Rakitina, D. V., Gonzalez, I., Canto, T., Kalinina, N. O., & Taliansky, M. E. (2012). Movement protein of hordeivirus interacts in vitro and in vivo with coilin, a major structural protein of Cajal bodies. *Doklady. Biochemistry and Biophysics, 442*, 57–60.

Shimizu, T., Yoshii, A., Sakurai, K., Hamada, K., Yamaji, Y., Suzuki, M., et al. (2009). Identification of a novel tobacco DnaJ-like protein that interacts with the movement protein of *Tobacco mosaic virus*. *Archives of Virology, 154*, 959–967.

Solovyev, A. G., Kalinina, N. O., & Morozov, S. Y. (2012). Recent advances in research of plant virus movement mediated by triple gene block. *Frontiers in Plant Science, 3*, 276.

Sriskanda, V. S., Pruss, G., Ge, X., & Vance, V. B. (1996). An eight-nucleotide sequence in the *Potato virus X* 3' untranslated region is required for both host protein binding and viral multiplication. *Journal of Virology, 70*, 5266–5271.

Taliansky, M. E., Brown, J. W., Rajamaki, M. L., Valkonen, J. P., & Kalinina, N. O. (2010). Involvement of the plant nucleolus in virus and viroid infections: Parallels with animal pathosystems. *Advances in Virus Research*, 77, 119–158.

Tamai, A., & Meshi, T. (2001). Cell-to-cell movement of *Potato virus X*: The role of p12 and p8 encoded by the second and third open reading frames of the triple gene block. *Molecular Plant-Microbe Interactions, 14*, 1158–1167.

Tameling, W. I., & Baulcombe, D. C. (2007). Physical association of the NB-LRR resistance protein Rx with a Ran GTPase-activating protein is required for extreme resistance to *Potato virus X*. *The Plant Cell, 19*, 1682–1694.

Tilsner, J., Linnik, O., Wright, K. M., Bell, K., Roberts, A. G., Lacomme, C., et al. (2012). The TGB1 movement protein of *Potato virus X* reorganizes actin and endomembranes into the X-body, a viral replication factory. *Plant Physiology, 158*, 1359–1370.

Tollin, P., Bancroft, J. B., Richardson, J. F., Payne, N. C., & Beveridge, T. J. (1979). Diffraction studies of papaya mosaic virus. *Virology, 98*, 108–115.

Tollin, P., Wilson, H. R., & Bancroft, J. B. (1980). Further observations on the structure of particles of potato virus X. *The Journal of General Virology, 49*, 407–410.

Tollin, P., Wilson, H. R., & Mowat, W. P. (1975). Optical diffraction from particles of narcissus mosaic virus. *The Journal of General Virology, 29*, 331–333.

Tsai, C. H., Cheng, C. P., Peng, C. W., Lin, B. Y., Lin, N. S., & Hsu, Y. H. (1999). Sufficient length of a poly(A) tail for the formation of a potential pseudoknot is required for efficient replication of bamboo mosaic potexvirus RNA. *Journal of Virology, 73*, 2703–2709.

Verchot, J. (2012). Cellular chaperones and folding enzymes are vital contributors to membrane bound replication and movement complexes during plant RNA virus infection. *Frontiers in Plant Science, 3*, 275.

Verchot-Lubicz, J., Torrance, L., Solovyev, A. G., Morozov, S. Y., Jackson, A. O., & Gilmer, D. (2010). Varied movement strategies employed by triple gene block-encoding viruses. *Molecular Plant-Microbe Interactions, 23*, 1231–1247.

Verchot-Lubicz, J., Ye, C. M., & Bamunusinghe, D. (2007). Molecular biology of potexviruses: Recent advances. *The Journal of General Virology, 88*, 1643–1655.

Voinnet, O. (2005). Induction and suppression of RNA silencing: Insights from viral infections. *Nature Reviews. Genetics, 6*, 206–220.

Voinnet, O., Lederer, C., & Baulcombe, D. C. (2000). A viral movement protein prevents spread of the gene silencing signal in *Nicotiana benthamiana*. *Cell, 103*, 157–167.

Walter, P., & Ron, D. (2011). The unfolded protein response: From stress pathway to homeostatic regulation. *Science, 334*, 1081–1086.

Wang, W., Vinocur, B., Shoseyov, O., & Altman, A. (2004). Role of plant heat-shock proteins and molecular chaperones in the abiotic stress response. *Trends in Plant Science, 9*, 244–252.

Wilson, H. R., Al-Mukhtar, J., Tollin, P., & Hutcheson, A. (1978). Observations on the structure of particles of white clover mosaic virus. *The Journal of General Virology*, *39*, 361–364.

Wu, C. H., Lee, S. C., & Wang, C. W. (2011). Viral protein targeting to the cortical endoplasmic reticulum is required for cell-cell spreading in plants. *The Journal of Cell Biology*, *193*, 521–535.

Yamanaka, T., Ohta, T., Takahashi, M., Meshi, T., Schmidt, R., Dean, C., et al. (2000). TOM1, an *Arabidopsis* gene required for efficient multiplication of a tobamovirus, encodes a putative transmembrane protein. *Proceedings of the National Academy of Sciences of the United States of America*, *97*, 10107–10112.

Yan, F., Lu, Y., Lin, L., Zheng, H., & Chen, J. (2012). The ability of PVX p25 to form RL structures in plant cells is necessary for its function in movement, but not for its suppression of RNA silencing. *PLoS One*, *7*, e43242.

Yang, S., Wang, T., Bohon, J., Gagne, M. E., Bolduc, M., Leclerc, D., et al. (2012). Crystal structure of the coat protein of the flexible filamentous *Papaya mosaic virus*. *Journal of Molecular Biology*, *422*, 263–273.

Ye, C., Dickman, M. B., Whitham, S. A., Payton, M., & Verchot, J. (2011). The unfolded protein response is triggered by a plant viral movement protein. *Plant Physiology*, *156*, 741–755.

Ye, C. M., Kelly, V., Payton, M., Dickman, M. B., & Verchot, J. (2012). SGT1 is induced by the *Potato virus X* TGBp3 and enhances virus accumulation in *Nicotiana benthamiana*. *Molecular Plant*, *5*, 1151–1153.

Zabka, M., Lesniak, W., Prus, W., Kuznicki, J., & Filipek, A. (2008). Sgt1 has co-chaperone properties and is up-regulated by heat shock. *Biochemical and Biophysical Research Communications*, *370*, 179–183.

Zamyatnin, A. A., Jr., Solovyev, A. G., Savenkov, E. I., Germundsson, A., Sandgren, M., Valkonen, J. P., et al. (2004). Transient coexpression of individual genes encoded by the triple gene block of *Potato mop-top virus* reveals requirements for TGBp1 trafficking. *Molecular Plant-Microbe Interactions*, *17*, 921–930.

Zayakina, O., Arkhipenko, M., Kozlovsky, S., Nikitin, N., Smirnov, A., Susi, P., et al. (2008). Mutagenic analysis of *Potato virus X* movement protein (TGBp1) and the coat protein (CP): In vitro TGBp1-CP binding and viral RNA translation activation. *Molecular Plant Pathology*, *9*, 37–44.

CHAPTER FOUR

Influenza Virus Transcription and Replication

Jaime Martín-Benito*, Juan Ortín*,†,1
*Centro Nacional de Biotecnología (CSIC), Madrid, Spain
†CIBER de Enfermedades Respiratorias (ISCIII), Madrid, Spain
[1]Corresponding author: e-mail address: jortin@cnb.csic.es

Contents

Abstract

The influenza A viruses cause yearly epidemics and occasional pandemics of respiratory disease, which constitute a serious health and economic burden. Their genome consists of eight single-stranded, negative-polarity RNAs that associate to the RNA polymerase and many nucleoprotein monomers to form ribonucleoprotein complexes (RNPs). Here, we focus on the organization of these RNPs, as well as on the structure and interactions of its constitutive elements and we discuss the mechanisms by which the RNPs transcribe and replicate the viral genome.

Advances in Virus Research, Volume 87
ISSN 0065-3527
http://dx.doi.org/10.1016/B978-0-12-407698-3.00004-1

1. INTRODUCTION

The influenza viruses are classified into three antigenic types, A, B, and C. Among them, the influenza A viruses are the causative agents of yearly epidemics of respiratory disease and the origin of occasional more severe pandemics. The impact of influenza A infections varies depending on the virulence of the particular virus responsible for the epidemic or pandemic. The latest pandemic started in 2009 (Neumann, Noda, & Kawaoka, 2009) and was relatively mild, while the most dramatic pandemic recorded occurred in 1918 and accounted for 20–40 million deaths worldwide (Horimoto & Kawaoka, 2005). In addition, influenza infections cause an important economic impact due to health-care costs and absenteeism from work (Molinari et al., 2007).

The influenza A viruses are endemic in wild aquatic and terrestrial avian species, which constitute their natural reservoir, and are extremely diverse genetically and antigenically. Many viral subtypes can be defined according to the antigenic nature of their main surface glycoproteins, hemagglutinin (HA) and neuraminidase (NA). Thus, 17 different HA and 10 NA classes have been identified (Palese & Shaw, 2007; Tong et al., 2012), and the subtype of a particular virus is defined by the combination of these antigens. From their avian reservoir, in which most subtypes have been detected, influenza viruses can be transferred to humans or other mammals. In addition, genes from avian influenza viruses can be incorporated into human viruses by reassortment. Hence, influenza viruses can start new lineages in humans or other mammals and eventually cause disease (Baigent & McCauley, 2003). When the human population is mostly naïve for any of these new viruses, a pandemic can occur and, in fact, the threat persists that highly pathogenic avian H5N1 influenza viruses that have caused more than 600 sporadic human cases and 300 deaths since 2003 may originate a severe pandemic in the future (http://www.who.int/influenza/human_animal_interface/H5N1_avian_influenza_update150612N.pdf).

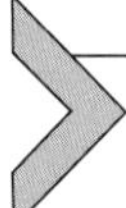

2. MOLECULAR BIOLOGY OF INFLUENZA VIRUS TRANSCRIPTION AND REPLICATION

2.1. Organization of the viral genome

The influenza A viruses have a negative-polarity RNA genome fragmented into eight single-stranded molecules. These are never present as naked RNA in virus particles or infected cells but exist as ribonucleoprotein complexes (RNPs) ranging from 2.5 to 6.3 MDa in size, generated by interaction of each RNA segment with the viral polymerase and a number of nucleoprotein (NP) molecules, depending on the RNA length (Neumann, Brownlee, Fodor, & Kawaoka, 2004; Resa-Infante, Jorba, Coloma, & Ortín, 2011; Ruigrok, Crepin, Hart, & Cusack, 2010). Each RNP encodes a viral protein in a collinear mRNA transcript and several of them also encode an additional gene product generated by a variety of strategies. Segments 7 and 8 not only encode M1 and NS1 proteins, respectively, but also give raise to alternatively spliced mRNAs that encode M2 and nuclear export protein (NEP) (Palese & Shaw, 2007). Segment 2 encodes the polymerase subunit PB1 and also the N40 protein by using an alternative translation initiation site (Wise et al., 2009). In addition, it generates PB1-F2 protein from a second reading frame (Chen et al., 2001). Finally, segment 3 encodes the PA polymerase subunit and also generates protein PA-X by ribosomal frame shifting (Jagger et al., 2012).

2.2. The viral transcription–replication cycle

As mentioned above, the active forms of viral RNAs are the RNPs and each one acts as an independent functional unit during transcription and replication (Fig. 4.1). This property is in line with the high frequency of genetic reassortment among influenza viruses in double infections (Palese & Shaw, 2007). Likewise, the end product of an RNP replication is not a naked RNA but a new RNP (Elton, Digard, Tiley, & Ortín, 2005; Neumann et al., 2004; Resa-Infante et al., 2011). Therefore, the polymerase acts as an enzyme during transcription and replication and is also needed stoichiometrically for the assembly of the progeny RNPs during virus replication.

A property that distinguishes influenza viruses from most other RNA-containing viruses is that they replicate and transcribe in the nucleus of infected cells. Hence, influenza viruses rely on the cellular nucleocytoplasmic transport system for multiplication (Boulo, Akarsu, Ruigrok, & Baudin, 2007). Early after infection, the parental incoming viral RNPs (vRNPs) are transported into the nucleus where they first get engaged in transcription. This is called primary transcription, as parental vRNPs are used as templates, and is an essential step to proceed to RNA replication. Thus, infection in the presence of protein synthesis inhibitors leads to prolonged primary transcription but no RNA replication (Minor & Dimmock, 1975). A landmark of influenza transcription is the

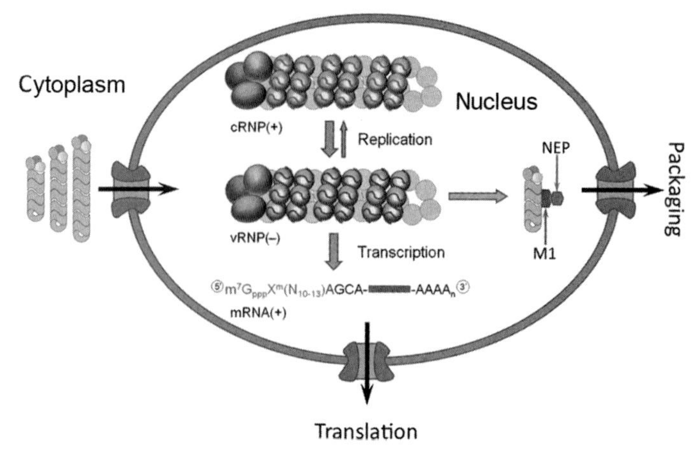

Figure 4.1 The viral transcription–replication cycle. The figure indicates the active transport of parental vRNPs into the nucleus, the process of transcription, the two-step process of RNP replication, and the export of progeny RNPs to the cytoplasm. The polarity of the viral RNAs is indicated in red (negative polarity) or blue (positive polarity). The presence of the cell-derived capped RNA primer at the 5′-end of viral mRNA is indicated in green. (For interpretation of the references to color in this figure legend, the reader is referred to the online version of this chapter.)

requirement for ongoing cellular mRNA synthesis due to a cap-stealing mechanism for initiation (Krug, Broni, & Bouloy, 1979) whereby the virus utilizes capped oligonucleotides derived from cellular nuclear pre-mRNAs as primers to synthesize viral mRNAs. The production of early virus proteins, at least the polymerase and NP, is essential for virus RNA replication (Huang, Palese, & Krystal, 1990). This process occurs in two distinct steps: first, the copy of the parental vRNPs leads to the generation of the complementary RNPs (cRNPs) and second, cRNPs serve as replication intermediates and direct the synthesis of large amounts of progeny vRNPs. These vRNPs can be transcribed (secondary transcription) to allow the expression of late virus proteins and are eventually exported from the nucleus and encapsidated at the plasma membrane into progeny virions (Elton et al., 2005; Neumann et al., 2004; Resa-Infante et al., 2011).

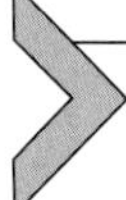

3. STRUCTURAL ELEMENTS OF THE INFLUENZA VIRUS TRANSCRIPTION–REPLICATION MACHINE

3.1. The polymerase

The influenza virus polymerase is a heterotrimer composed by the PB1, PB2, and PA proteins (Fig. 4.2A), which has a total molecular mass of about

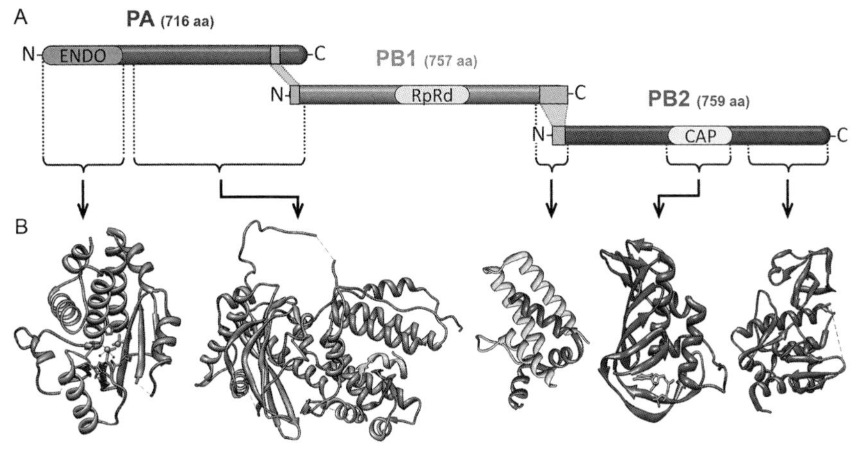

Figure 4.2 Crystallographic structure of influenza A virus polymerase domains. (A) Outline of the polymerase structure showing the described crystallographic interactions; the fragments involved of each subunit are highlighted in yellow. PA, PB1, and PB2 are shown in green, blue, and brown, respectively. (B) Atomic structures determined by X-ray crystallography. From left to right: PA endonuclease domain residues 1–209 (PDB ID 2W69; Dias et al., 2009); interaction between PA C-terminus and PB1 N-terminus, residues 256–716 and 1–16, respectively (PDB ID 3CM8; He et al., 2008); interaction between PB1 C-terminus and PB2 N-terminus, residues 679–757 and 1–37, respectively (PDB ID 2ZTT; Sugiyama et al., 2009); PB2 CAP-binding domain residues 318–483 (PDB ID 2VQZ; Guilligay et al., 2008), and two domains of PB2 C-terminus residues 538–667 and 686–741 (PDB ID 2VY6; Tarendeau et al., 2008). (For interpretation of the references to color in this figure legend, the reader is referred to the online version of this chapter.)

250 kDa, similar to that of the L protein of other negative-stranded RNA viruses (Kranzusch & Whelan, 2012). The results from a variety of *in vivo* and *in vitro* approaches indicate that the heterotrimer is the functional enzyme for viral transcription and replication, although it exerts its action in the context of the RNP, forming a more complex molecular machine that includes the template and the NP and interacts with several cellular factors (see below). Along the years, specific roles have been assigned to each subunit in the polymerase. PB1 was identified as responsible for RNA polymerization by sequence comparisons (Poch, Sauvaget, Delarue, & Tordo, 1990), in agreement with the phenotype of classical temperature-sensitive mutants and site-directed mutations (Biswas & Nayak, 1994; Mahy, 1983). The PB2 subunit was shown to interact with the capped RNAs (Blaas, Patzelt, & Keuchler, 1982; Ulmanen, Broni, & Krug, 1981) and to be required for transcription initiation (Plotch, Bouloy, Ulmanen, & Krug, 1981). The

role of PB2 in this step of viral gene expression was verified genetically (Coloma et al., 2009; Fechter et al., 2003; Mahy, 1983). However, other site-directed mutations in PB2 indicated that this subunit is also involved in RNA replication (Gastaminza, Perales, Falcón, & Ortín, 2003; Jorba, Coloma, & Ortin, 2009). On the other hand, early genetic analyses identified the PA subunit of the polymerase as involved in RNA replication (Mahy, 1983), but the phenotype of site-directed mutants indicated that PA plays an essential role in cap snatching (Fodor et al., 2002) and is also involved in virus-induced proteolysis (Hara et al., 2001; Huarte et al., 2003; Perales et al., 2000; Sanz-Ezquerro, Zürcher, de la Luna, Ortin, & Nieto, 1996).

Early studies indicated that PB1 protein is the core of the complex (Digard, Blok, & Inglis, 1989) and a number of approaches have shown that the architecture of the polymerase is N–PA–PB1–PB2–C with main connections between PA C-terminus and PB1 N-terminus and between PB1 C-terminus and PB2 N-terminus (Fig. 4.2A, yellow) (Ghanem et al., 2007; González, Zürcher, & Ortín, 1996; Perales, de la Luna, Palacios, & Ortín, 1996; Pérez & Donis, 1995; Toyoda, Adyshev, Kobayashi, Iwata, & Ishihama, 1996; Zürcher, de la Luna, Sanz-Ezquerro, Nieto, & Ortín, 1996). These interactions have been verified by cocrystallization of the corresponding protein domains (Fig. 4.2B) and additional interactions have been described that may be transient or involved in the regulation of polymerase activity (Hemerka et al., 2009; Poole, Elton, Medcalf, & Digard, 2004). The pathway for the polymerase complex formation appears closely linked to its import into the nucleus, where the main role of the polymerase takes place (reviewed in Hutchinson & Fodor, 2012). Although each of the polymerase subunits contains functional nuclear localization signals (NLSs) (Mukaigawa & Nayak, 1991; Nath & Nayak, 1990; Nieto, de la Luna, Bárcena, Portela, & Ortín, 1994), several experimental approaches support a model whereby the heterodimer PB1–PA is formed in the cytoplasm and transported to the nucleus by the association of RanBP5 to PB1, whereas PB2 is imported via the classical alpha-importin pathway (Deng, Sharps, Fodor, & Brownlee, 2005; Fodor & Smith, 2004; Huet et al., 2010; reviewed in Hutchinson & Fodor, 2012). In addition, heterotrimer formation appears to require not only classical protein chaperones as Hsp90 (Chase et al., 2008; Naito, Momose, Kawaguchi, & Nagata, 2007) but also the nuclear import machinery itself, which may help in the formation of fully functional polymerase (Hutchinson, Orr, Man Liu, Engelhardt, & Fodor, 2011; Resa-Infante et al., 2008).

At present, no atomic structure of the polymerase heterotrimer is available, but low-resolution structures obtained by electron microscopy (EM) and three-dimensional reconstruction have been reported for the template-free enzyme (Torreira et al., 2007), as well as for the polymerase associated to a mini-RNA template (Resa-Infante, Recuero-Checa, Zamarreño, Llorca, & Ortín, 2010), to a mini-RNP (see below) (Area et al., 2004; Coloma et al., 2009), and to virion RNPs (see below). These studies revealed a quite compact yet flexible structure that could change its conformation upon interaction with the RNA template or the NP, as well as presumably by binding to other viral or cellular factors. Although the complete polymerase heterotrimer could not be crystallized so far, several domains of the polymerase subunits have been solved at atomic resolution (Fig. 4.2B). The intersubunit contacts (PA C-terminus + PB1 N-terminus and PB1 C-terminus + PB2 N-terminus) have been cocrystallized and solved by X-ray diffraction (He et al., 2008; Obayashi et al., 2008; Sugiyama et al., 2009). Likewise, some of the polymerase functional domains have been identified and characterized structurally: The cap-binding domain lays at the center of the PB2 subunit and interacts with a cap analogue much in the same fashion as other cap-binding proteins, although its folding is different (Guilligay et al., 2008). However, the endonuclease responsible for cap snatching resides in the N-terminal domain of PA protein and has a fold analogous to type II restriction endonucleases (Dias et al., 2009; Yuan et al., 2009). This enzymatic activity has been analyzed biochemically (Crepin et al., 2010; Noble, Cox, Deval, & Kim, 2012) and is a potential target for antiviral chemotherapy (Kowalinski et al., 2012). Surprisingly, however, no detailed structural information is yet available for the polymerase domain located in PB1. In addition, the structure of the C-terminal regions of PB2 has been determined, providing information about the interaction of PB2 NLS with alpha-importins that are important for nuclear import (Resa-Infante et al., 2008; Tarendeau et al., 2007) and about a polymerase domain particularly relevant for avian influenza virus adaptation to mammalian hosts (Boivin & Hart, 2011; Tarendeau et al., 2008).

3.2. The nucleoprotein

The influenza virus NP is an essential protein for virus transcription and replication (Huang et al., 1990) that associates to the virus genomic RNAs. The NP binds RNA with high affinity and no sequence specificity (Baudin,

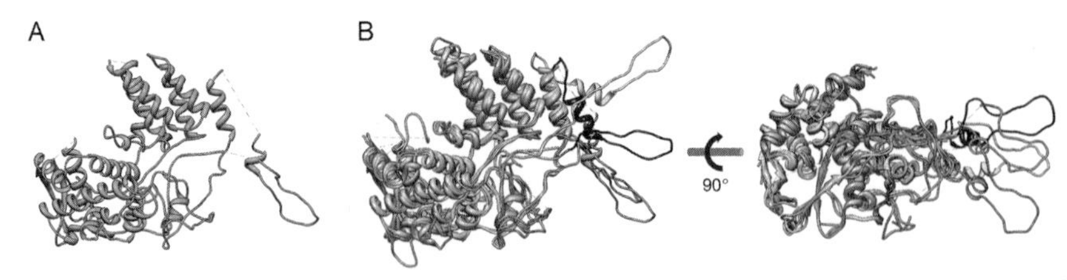

Figure 4.3 Structure of the NP. (A) Structure of the NP protein (PDB ID 2IQH; Ye, Krug, & Tao, 2006). (B) Superposition of the NP monomers extracted from the different crystallographic structures reported. The head-body domains from different NP structures have been aligned using Chimera software. For clarity, only the loop has been colored according to the following code: yellow (PDB ID 2IQH; Ye et al., 2006), pink (PDB ID 3RO5; Gerritz et al., 2011), cyan (PDB ID 2Q06; Ng et al., 2008), and influenza B NP in red (PDB ID 3TJ0; Ng et al., 2012). (For interpretation of the references to color in this figure legend, the reader is referred to the online version of this chapter.)

Bach, Cusack, & Ruigrok, 1994). It has an overall structure similar to the N protein from Mononegavirales (Albertini et al., 2006; Green, Zhang, Wertz, & Luo, 2006; Rudolph et al., 2003), with distinct head and body domains, but its protein topology is different since these domains are not collinear with the primary sequence of the protein (Fig. 4.3A) (Ng et al., 2008; Ye et al., 2006). The binding site for template RNA has been predicted at a protein cleft between the head and body domains, that is rich in surface-exposed basic residues (Ng et al., 2008; Ye et al., 2006), some of which have been shown to be important for RNA binding *in vitro* (Ng et al., 2008). Additionally, the NP shows an intrinsic ability to form oligomeric structures due to the nature of the NP–NP interaction. This interaction is primarily dependent on the intermolecular swapping of a protein loop (positions 402–428) that is flexibly connected to the NP head and is inserted into the neighboring NP body domain (Ng et al., 2008; Ye et al., 2006). The flexibility of the connection between the loop and the head of the NP (Fig. 4.3B) is responsible for the high plasticity that produces a wide range of oligomers, from the trimeric structures obtained in the crystals (Ng et al., 2008; Ye et al., 2006) to the multimeric (Ruigrok & Baudin, 1995) or helicoidal structures (Arranz et al., 2012; Ruigrok & Baudin, 1995) described using EM.

3.3. Accessory viral proteins

The nuclear export protein (NEP/NS2) is encoded by the spliced mRNA generated from segment 8, and its primordial role is to drive the export of progeny RNPs from the nucleus by interaction with M1 protein and the

CRM1-dependent export system (Boulo et al., 2007). However, a number of evidences suggest that it also plays a role in the modulation of the virus transcription versus replication balance in the nucleus (Bullido, Gomez-Puertas, Saiz, & Portela, 2001; Robb, Smith, Vreede, & Fodor, 2009). In addition, it has been shown that NEP is required for the generation of short virus-like RNAs in infected cells (Perez et al., 2010) (see below). Furthermore, adaptation of avian viruses to replicate in human cells involves specific mutations in the *NEP* gene (Manz, Brunotte, Reuther, & Schwemmle, 2012). Altogether, these results suggest that NEP is important for the replicative activity of the virus polymerase.

On the other hand, the NS1 protein was shown to associate to vRNPs by interaction with NP (Marión, Zürcher, de la Luna, & Ortín, 1997; Robb et al., 2011), and temperature-sensitive mutants in the NS1 gene showed defective vRNA but not cRNA accumulations (Falcón et al., 2004; Wolstenholme, Barrett, Nichol, & Mahy, 1980), suggesting that NS1 may play a role in the second step of virus RNA replication.

3.4. The ribonucleoprotein complex

Early studies documented that the RNPs present in virus particles are ribbon-like, closed helicoidal structures showing high flexibility (Compans, Content, & Duesberg, 1972; Heggeness, Smith, Ulmanen, Krug, & Chopin, 1982; Jennings, Finch, Winter, & Robertson, 1983; Pons, Schulze, & Hirst, 1969). Similar structures were observed when analyzing NP–vRNA complexes or purified NP (Ruigrok & Baudin, 1995; Yamanaka, Ishihama, & Nagata, 1990), suggesting that the NP contains structural information important for the overall configuration of a RNP. Although not directly perceptible by EM, immuno-EM showed the presence of the viral polymerase at one end of the RNPs (Murti, Webster, & Jones, 1988), which is important to maintain the closed conformation of the structure (Klumpp, Ruigrok, & Baudin, 1997).

Studies with recombinant RNPs generated by *in vivo* replication in virus-free systems indicated that their general appearance was dependent on the length of the template RNA. The minimal efficient RNP replicon contained about 150 nt, and within the 150–350 nt range they had circular or elliptical structure, whereas they were helicoidal above this length (Ortega et al., 2000). These recombinant RNPs were transcriptionally active *in vitro* and could serve as template for RNA replication when transfected into cultured cells (Jorba et al., 2009), that is, it could be considered a model for the RNPs present in virions or the progeny RNPs generated in infected cells. The structure of one of such RNPs containing 248 nt was solved by EM and image reconstruction and revealed a circular shape with nine NP monomers (Fig. 4.4A), two of them interacting directly with the polymerase complex (Area et al., 2004; Coloma et al., 2009; Martín-Benito et al., 2001). In the higher resolution structure, it is possible to dock the NP atomic structure (Ng et al., 2008; Ye et al., 2006) showing the interaction among NP monomers and between the terminal NPs and the polymerase. The NP arrangement in this mini-RNP shows how the molecular swapping could form large closed structures, involving many NPs, by means of the same swapping loop interaction described for smaller crystallographic structures (Fig. 4.4B) (Coloma et al., 2009). The interactions between the polymerase and the adjacent NP monomers are RNAse sensitive and could not be correlated with the NP–polymerase interactions described biochemically (Biswas, Boutz, & Nayak, 1998; Poole et al., 2004), but specific mutations in the head domain of NP altered its interaction with the polymerase and the mini-RNP activity (Marklund, Ye, Dong, Tao, & Krug, 2012). Although the cryo-EM volume of a mini-RNP is the best structure so far available, much higher resolution is needed to generate a quasi-atomic model that includes information of the polymerase, NP, and template RNA.

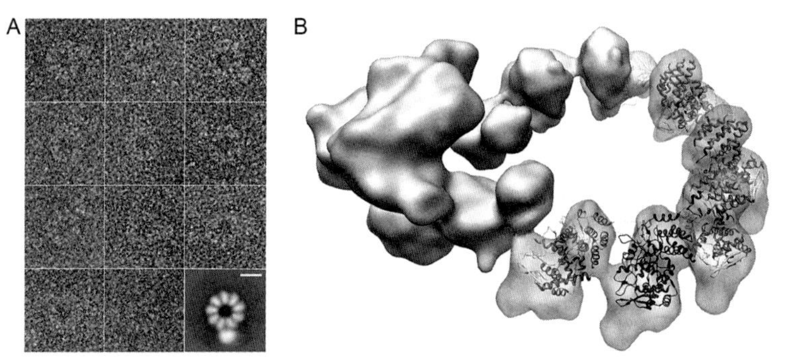

Figure 4.4 Structure of a recombinant mini-RNP. (A) Gallery of images of a recombinant mini-RNP containing a RNA template of 248 nt. In the lower right corner, an average image is shown; the structure presents a circular shape containing nine NP monomers and the polymerase as an extra mass protruding from the ring. Scale bar 100 Å (Martín-Benito et al., 2001). (B) Cryo-EM three-dimensional reconstruction of the mini-RNP (Coloma et al., 2009). The atomic structure of the NP (PDB ID 2IQH; Ye et al., 2006) has been docked. To show the molecular swapping in this structure, one monomer of NP has been colored in red. (For interpretation of the references to color in this figure legend, the reader is referred to the online version of this chapter.)

More recently, the structure of native, helical virion RNPs (Fig. 4.5A) has been determined by a combination of cryo-EM and cryoelectron tomography (Arranz et al., 2012) and revealed a dihedral helical structure with two antiparallel NP strands connected along the helix by dimeric NP–NP interactions (Fig. 4.5B). An analogous RNP organization was also described by analysis of recombinant RNPs generated in the absence of virus infection (Moeller, Kirchdoerfer, Potter, Carragher, & Wilson, 2012). The molecular swapping could be modeled for the intrastrand NP–NP interaction, whereas the interstrand interaction was shown to involve the N-terminal regions of NP monomers of the antiparallel strands (Fig. 4.5C). The two RNP termini were separately analyzed and showed a small NP loop connecting both helix strands at one end and the polymerase complex at the other (Fig. 4.5B, ends). The structure reported provide clues to understand the mechanisms responsible for transcription and replication of full-length helicoidal RNPs and other important aspects of influenza virus biology, like the generation of defective-interfering particles or the ordered packaging of RNPs into virions.

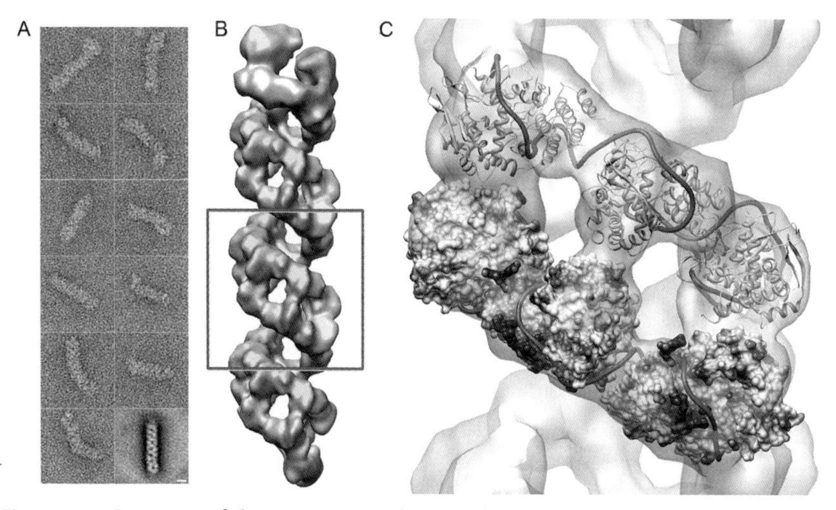

Figure 4.5 Structure of the native RNP obtained from virions. (A) Gallery of images of native RNPs. In the lower right corner, an average image obtained from 233 straight and homogeneous in size RNPs is shown. Scale bar 100 Å. (B) Model of segment 8 RNP. The polymerase is shown in gray, the antiparallel NP strands are shown in red and blue, and the closing loop in yellow. (C) Docking of the NP monomer in the helical region (square in B). In the lower strand NP, monomers are represented as potential surface and in the upper strand as ribbons; the modeled RNA template is depicted as a red thread. (For interpretation of the references to color in this figure legend, the reader is referred to the online version of this chapter.)

4. MECHANISMS OF INFLUENZA VIRUS TRANSCRIPTION AND REPLICATION

4.1. RNA synthesis initiation

The mechanisms for the initiation of viral transcription and replication are drastically distinct. Thus, mRNA synthesis requires the use of cap-containing oligonucleotide primers that the viral polymerase generates by cap snatching from nascent RNA polymerase transcripts (Bouloy, Plotch, & Krug, 1978; Krug et al., 1979; Plotch et al., 1981; Ulmanen et al., 1981), whereas cRNAs contain a 5′-terminal triphosphate indicative of *de novo* initiation (Hay, Skehel, & McCauley, 1982). Transcription initiation involves the concerted action of the PB2-associated cap-binding activity (Guilligay et al., 2008) and the PA-associated endonuclease activity (Dias et al., 2009; Fodor et al., 2002; Yuan et al., 2009) to produce a capped oligonucleotide primer that is preferentially elongated by the polymerase when its 3′-terminal sequence is CA (Rao, Yuan, & Krug, 2003). In contrast to transcription initiation with a capped primer derived from a host-cell mRNA, *de novo* initiation *in vitro* requires high concentrations of not only ATP but also GTP and CTP that would be incorporated at positions +2 and +3 (Vreede, Gifford, & Brownlee, 2008), consistent with an internal initiation site (Vreede et al., 2008; Zhang, Wang, Wang, & Toyoda, 2010).

It has been proposed that vRNPs can initiate stochastically either *de novo* or by cap priming, and the proportion of final products would be dependent on the stabilization of the cRNA by interaction with additional polymerase and NP synthesized during the infection (Vreede & Brownlee, 2007; Vreede, Jung, & Brownlee, 2004). However, it is possible that these alternative initiation modes are modulated by the interaction of the vRNPs with the cellular RNA polymerase II (Engelhardt, Smith, & Fodor, 2005) (for transcription) or with *trans*-acting polymerase complexes (Jorba, Area, & Ortin, 2008) and/or soluble NP (Newcomb et al., 2009) (for replication).

The initiation step on cRNP templates to generate progeny vRNPs also requires high concentrations of NTPs (Vreede et al., 2008) and takes place at position +4 in the template to first generate a pppApG product that is then realigned to the +1 position for elongation (Deng, Vreede, & Brownlee, 2006; Zhang et al., 2010), suggesting considerable conformational changes in the polymerase complex when bound to either vRNA or cRNA promoters.

4.2. Assembly of ribonucleoproteins

As stated above, the interaction of NP with RNA is not sequence dependent and hence NP cannot provide specificity in the assembly of newly synthesized RNA into a RNP. As the viral polymerase interacts with high affinity and specificity with the 5′-terminus of vRNA (González & Ortín, 1999; Tiley, Hagen, Mathews, & Krystal, 1994), it is presumed that the first step in RNP assembly is the protection of the nascent RNA chain by binding to a soluble polymerase complex *in trans*, as supported by genetic data for cRNA (Vreede et al., 2004) and vRNA synthesis (Jorba et al., 2009). Further, assembly steps would depend on specific polymerase–NP interactions (Biswas et al., 1998; Marklund et al., 2012; Poole et al., 2004) and NP–NP oligomerization and RNA binding (Coloma et al., 2009; Ng et al., 2008; Ye et al., 2006). Recent experiments using RNA binding-defective NP mutants and mutants affected in the swapping loop or groove have documented that oligomerization is crucial for NP recruitment and the polarity of the assembly is "tail-loop first," that is, the incoming NP monomer inserts its loop into the groove of the RNP growing end (Turrell, Lyall, Tiley, Fodor, & Vreede, 2013). At present, it is not clear how the nascent RNP adopts the closed conformation in which both ends of the RNA are linked together and bound to the polymerase (Flick & Hobom, 1999; Fodor, Pritlove, & Brownlee, 1994; Hsu, Parvin, Gupta, Krystal, & Palese, 1987; Lee et al., 2002; Tiley et al., 1994), but an attractive hypothesis would be that the polymerase complex that protects the nascent RNA during assembly piggybacks onto the polymerase that actually copies the template, allowing for an easy interaction between the first and last RNA sequences generated during replication.

4.3. RNA synthesis termination

As for RNA synthesis initiation, fully distinct mechanisms operate for the termination steps during virus transcription or replication. The synthesis of viral mRNAs terminates by repetitive copy of an oligo-U signal located close to the 5′-terminus of the vRNA template (Li & Palese, 1994; Poon, Pritlove, Fodor, & Brownlee, 1999; Robertson, Schubert, & Lazzarini, 1981) and hence viral mRNAs are polyadenylated and lack the sequence complementary to the 5′-end of the vRNA. Genetic evidence indicates that binding of the vRNP-associated polymerase to the 5′-terminus of the template is important to accomplish polyadenylation (Poon, Pritlove,

Sharps, & Brownlee, 1998; Pritlove, Poon, Fodor, Sharps, & Brownlee, 1998), and lack of *in vivo* complementation with a genetically marked *trans* polymerase suggests that transcription itself is carried out *in cis* by the vRNP-associated polymerase (Jorba et al., 2009). In contrast to mRNAs, cRNAs are full complementary copies of the respective vRNA templates, indicating that the polyadenylation signal is overridden in the replication process. This implies that the vRNP-associated polymerase is displaced from the 5′-end but at present the mechanism responsible is not clear. A *trans*-mechanism for replication would be an attractive possibility (see below) but no evidence is available at present for the synthesis of cRNA.

4.4. A *cis/trans* model for transcription versus replication

The mechanisms for initiation and termination of RNA synthesis during transcription or replication are not only different but are also necessarily linked to determine that only mRNAs (but not cRNAs or vRNAs) are capped and polyadenylated and only cRNAs and vRNAs (but not mRNAs) are full copies of each other and are assembled into RNPs. How this is achieved has been a matter of experimental analyses for a long time and several models have been proposed. Obviously, both soluble polymerase complexes and NP monomers are necessary *in trans* to carry out the assembly of new cRNPs or vRNPs. But on top of such a "structural" role, free NP monomers have also been proposed to alter the capacity of the vRNP for transcription versus replication by interaction with the vRNP-associated polymerase or with the vRNP template (Biswas et al., 1998; Fodor et al., 1994; Hsu et al., 1987; Klumpp et al., 1997; Mena et al., 1999; Poole et al., 2004). However, it has been recently shown that efficient *in vivo* replication of short viral RNA templates can be achieved in the absence of NP (Resa-Infante et al., 2010; F. Vreede, personal communication), indicating that this protein is not essential for the initiation or termination steps of RNA replication, although it is essential as elongation factor for long RNA templates (Honda, Ueda, Nagata, & Ishihama, 1988; Kawaguchi, Momose, & Nagata, 2011). On the other hand, the ability of the NEP protein to increase the replication of recombinant RNPs, at the expense of their transcription activity (Perez et al., 2012; Robb et al., 2009), and its requirement to generate short virion RNAs *in vivo* (Perez et al., 2010, 2012) suggest that NEP may act as a replicase cofactor during infection. In addition, the capacity of the virus polymerase to oligomerize (Jorba, Area, et al., 2008) and the results of *in vivo* complementation using genetically marked polymerase complexes (Jorba et al., 2009) suggested that free polymerase could be crucial to drive RNP replication. Thus, these experiments showed that vRNP assembly takes place *in trans* by a free polymerase distinct from the replicating polymerase and, furthermore, indicated that the polymerase actually performing vRNA synthesis was not the RNP-associated polymerase but a nonresident polymerase complex acting *in trans*. These results suggest a *cis/trans* model for transcription/replication whereby transcription would be carried out *in cis* by the RNP-associated polymerase. The transcribing polymerase would be activated by cap binding and would carry out polyadenylation due to the steric hindrance derived from its own binding to the 5′-terminus of the template. In contrast, vRNP synthesis would be carried out *in trans* by a soluble polymerase that would be activated by interaction with the RNP-associated polymerase. Furthermore, the *trans*-replication model would allow synthesis of a full copy of the template by displacement of the RNP-associated polymerase. Recently, complex associations of vRNPs have been observed by EM (Moeller et al., 2012) that would be compatible with the earlier proposed *trans* model of replication (Jorba et al., 2009).

5. ROLE OF HOST CELL FACTORS

As all other, influenza viruses use cellular factors, structures, and pathways to accomplish their replication cycles, as has been shown by genome-wide screenings using a variety of experimental approaches (Brass et al., 2009; Karlas et al., 2010; Konig et al., 2009; Naito, Kiyasu, et al., 2007; Shapira et al., 2009). Some of these cellular factors might be involved in virus RNA replication or transcription and have been identified by biochemical techniques, like functional complementation *in vitro* (Nagata, Takeuchi, & Ishihama, 1989), two-hybrid screening in yeast (Honda, Okamoto, & Ishihama, 2007; Huarte, Sanz-Ezquerro, Roncal, Ortin, & Nieto, 2001; Tafforeau et al., 2011), or proteomic analyses of polymerase- or RNP-containing complexes (Fislova, Thomas, Graef, & Fodor, 2010; Jorba, Area, et al., 2008; Jorba, Juarez, et al., 2008; Mayer et al., 2007; Resa-Infante et al., 2008). A number of the factors identified are normally involved in protein folding and have been shown to play a role in the normal accumulation and/or functionality of the polymerase or the NP. Thus, Hsp90 and CCT are important for normal polymerase assembly or function (Chase et al., 2008; Fislova et al., 2010; Naito, Momose, et al., 2007). However, interaction of Hsp70 with the polymerase reduces virus RNA replication and transcription (Li, Zhang, Tong, Liu, & Ye, 2010). On the other hand, various factors of the nuclear import machinery seem to help establishing the proper conformation of the polymerase subunits and the formation of active complex (Hutchinson et al., 2011; Resa-Infante et al., 2008). Other cellular factors like UAP56 and Tat-SF1, not normally involved in protein folding, appear to participate in maintaining the normal conformation and avoiding aggregation of the NP during influenza RNA replication (Kawaguchi et al., 2011; Momose, Handa, & Nagata, 1996; Naito, Kiyasu, et al., 2007). On the contrary, interaction of NP with other cellular proteins, like cyclophilin E or NF90, inhibits the accumulation of viral RNAs in the infection and leads to reductions in final viral titer (Wang et al., 2011, 2009). Likewise, the interaction of the viral polymerase with the Ebp1 human protein leads to an inhibition of the RNA synthesis activity of the enzyme, but not in the cap-binding or cap-snatching activities, and a reduction in virus titer (Honda, 2008; Honda et al., 2007). Whether these inhibitory interacting factors play a regulatory role in the infection or are an aspect of the antiviral cell response is unknown at present.

As indicated above, viral transcription is linked to the process of cellular mRNA synthesis, and not surprisingly, an association of the viral polymerase with cellular RNA polymerase II has been reported (Engelhardt et al., 2005), as well as with other cellular proteins involved in mRNA transcription, like the hCLE protein (Perez-Gonzalez, Rodriguez, Huarte, Salanueva, & Nieto, 2006; Rodriguez, Perez-Gonzalez, & Nieto, 2011) and the chromatin remodeler CHD6 (Alfonso et al., 2011). Whereas RNA polymerase II and hCLE proteins are required for viral normal multiplication, CHD6 acts as a negative regulator.

In addition, the efficiency of the polyadenylation of the viral transcripts at the polyadenylation signal located close to the 5′-terminus of the RNA template is affected by the presence of human SFPQ/PSF protein, a nuclear factor normally involved in cellular transcription and splicing (Landeras-Bueno, Jorba, Pérez-Cidoncha, & Ortín, 2011).

6. OUTLOOK

Our knowledge on the influenza A virus replication and transcription processes has improved substantially in the recent years, mainly due to the contribution of new structural information on the vRNPs, either recombinant or native virion RNPs, and the elements included in this virus molecular machine. However, no atomic structure of the trimeric RNA polymerase complex is yet available and, although the structure of some subunit domains has been determined, information on the polymerase domain is still lacking. The docking of the published atomic structures for the polymerase and the NP, and those to come, within the density maps for the various forms of the polymerase and the RNP obtained by EM will permit the generation of quasi-atomic structures for these large macromolecular complexes. The compilation of this structural information with biochemical and genetic data will allow the proposal of improved models for transcription, replication, and ordered encapsidation of influenza RNPs and help designing proper experiments to test them.

ACKNOWLEDGMENTS

Research described from the author's laboratories was supported by the Spanish Ministry of Science and Innovation (Ministerio de Ciencia e Innovación) grants BFU2010-17540/BMC (J. O.) and BFU2011-25090/BMC (J. M.-B.), by the FLUPHARM strep project (FP7-259751) (J. O.), and by Fundación Marcelino Botín (J. O.).

REFERENCES

Albertini, A. A., Wernimont, A. K., Muziol, T., Ravelli, R. B., Clapier, C. R., Schoehn, G., et al. (2006). Crystal structure of the rabies virus nucleoprotein-RNA complex. *Science*, *313*(5785), 360–363.

Alfonso, R., Lutz, T., Rodriguez, A., Chavez, J. P., Rodriguez, P., Gutierrez, S., et al. (2011). CHD6 chromatin remodeler is a negative modulator of influenza virus replication that relocates to inactive chromatin upon infection. *Cellular Microbiology*, *13*(12), 1894–1906.

Area, E., Martín-Benito, J., Gastaminza, P., Torreira, E., Valpuesta, J. M., Carrascosa, J. L., et al. (2004). Three-dimensional structure of the influenza virus RNA polymerase: Localization of subunit domains. *Proceedings of the National Academy of Sciences of the United States of America*, *101*, 308–313.

Arranz, R., Coloma, R., Chichón, F. J., Conesa, J. J., Carrascosa, J. L., Valpuesta, J. M., et al. (2012). The structure of native influenza virion ribonucleoproteins. *Science*, *338*, 1634–1637.

Baigent, S. J., & McCauley, J. W. (2003). Influenza type A in humans, mammals and birds: Determinants of virus virulence, host-range and interspecies transmission. *BioEssays*, *25*(7), 657–671.

Baudin, F., Bach, C., Cusack, S., & Ruigrok, R. W. (1994). Structure of influenza virus RNP. I. Influenza virus nucleoprotein melts secondary structure in panhandle RNA and exposes the bases to the solvent. *The EMBO Journal*, *13*(13), 3158–3165.

Biswas, S. K., Boutz, P. L., & Nayak, D. P. (1998). Influenza virus nucleoprotein interacts with influenza virus polymerase proteins. *Journal of Virology*, *72*(7), 5493–5501.

Biswas, S. K., & Nayak, D. P. (1994). Mutational analysis of the conserved motifs of influenza A virus polymerase basic protein 1. *Journal of Virology*, *68*, 1819–1826.

Blaas, D., Patzelt, E., & Keuchler, E. (1982). Identification of the cap binding protein of influenza virus. *Nucleic Acids Research*, *10*, 4803–4812.

Boivin, S., & Hart, D. J. (2011). Interaction of the influenza A virus polymerase PB2 C-terminal region with importin alpha isoforms provides insights into host adaptation and polymerase assembly. *The Journal of Biological Chemistry*, *286*(12), 10439–10448.

Boulo, S., Akarsu, H., Ruigrok, R. W., & Baudin, F. (2007). Nuclear traffic of influenza virus proteins and ribonucleoprotein complexes. *Virus Research*, *124*(1–2), 12–21.

Bouloy, M., Plotch, S. J., & Krug, R. M. (1978). Globin mRNAs are primers for the transcription of influenza viral RNA in vitro. *Proceedings of the National Academy of Sciences of the United States of America*, *75*, 4886–4890.

Brass, A. L., Huang, I. C., Benita, Y., John, S. P., Krishnan, M. N., Feeley, E. M., et al. (2009). The IFITM proteins mediate cellular resistance to influenza A H1N1 virus, West Nile virus, and dengue virus. *Cell*, *139*(7), 1243–1254.

Bullido, R., Gomez-Puertas, P., Saiz, M. J., & Portela, A. (2001). Influenza A virus NEP (NS2 protein) downregulates RNA synthesis of model template RNAs. *Journal of Virology*, *75*(10), 4912–4917.

Chase, G., Deng, T., Fodor, E., Leung, B. W., Mayer, D., Schwemmle, M., et al. (2008). Hsp90 inhibitors reduce influenza virus replication in cell culture. *Virology*, *377*(2), 431–439.

Chen, W., Calvo, P. A., Malide, D., Gibbs, J., Schubert, U., Bacik, I., et al. (2001). A novel influenza A virus mitochondrial protein that induces cell death. *Nature Medicine*, 7(12), 1306–1312.

Coloma, R., Valpuesta, J. M., Arranz, R., Carrascosa, J. L., Ortin, J., & Martin-Benito, J. (2009). The structure of a biologically active influenza virus ribonucleoprotein complex. *PLoS Pathogens*, *5*(6), e1000491.

Compans, R. W., Content, J., & Duesberg, P. H. (1972). Structure of the ribonucleoprotein of influenza virus. *Journal of Virology*, *4*, 795–800.

Crepin, T., Dias, A., Palencia, A., Swale, C., Cusack, S., & Ruigrok, R. W. (2010). Mutational and metal binding analysis of the endonuclease domain of the influenza virus polymerase PA subunit. *Journal of Virology*, *84*(18), 9096–9104.

Deng, T., Sharps, J., Fodor, E., & Brownlee, G. G. (2005). In vitro assembly of PB2 with a PB1-PA dimer supports a new model of assembly of influenza A virus polymerase subunits into a functional trimeric complex. *Journal of Virology*, *79*(13), 8669–8674.

Deng, T., Vreede, F. T., & Brownlee, G. G. (2006). Different de novo initiation strategies are used by influenza virus RNA polymerase on its cRNA and viral RNA promoters during viral RNA replication. *Journal of Virology*, *80*(5), 2337–2348.

Dias, A., Bouvier, D., Crepin, T., McCarthy, A. A., Hart, D. J., Baudin, F., et al. (2009). The cap-snatching endonuclease of influenza virus polymerase resides in the PA subunit. *Nature*, *458*(7240), 914–918.

Digard, P., Blok, V. C., & Inglis, S. C. (1989). Complex formation between influenza virus polymerase proteins expressed in Xenopus oocytes. *Virology*, *171*(1), 162–169.

Elton, D., Digard, P., Tiley, L., & Ortín, J. (2005). Structure and function of the influenza virus RNP. In Y. Kawaoka (Ed.), *Current topics in influenza virology* (pp. 1–92). Norfolk: Horizon Scientific Press.

Engelhardt, O. G., Smith, M., & Fodor, E. (2005). Association of the influenza A virus RNA-dependent RNA polymerase with cellular RNA polymerase II. *Journal of Virology*, *79*(9), 5812–5818.

Falcón, A. M., Marión, R. M., Zürcher, T., Gómez, P., Portela, A., Nieto, A., et al. (2004). Defective RNA replication and late gene expression in temperature-sensitive (A/Victoria/3/75) influenza viruses expressing deleted forms of NS1 protein. *Journal of Virology*, *78*, 3880–3888.

Fechter, P., Mingay, L., Sharps, J., Chambers, A., Fodor, E., & Brownlee, G. G. (2003). Two aromatic residues in the PB2 subunit of influenza A RNA polymerase are crucial for cap binding. *The Journal of Biological Chemistry*, *278*(22), 20381–20388.

Fislova, T., Thomas, B., Graef, K. M., & Fodor, E. (2010). Association of the influenza virus RNA polymerase subunit PB2 with the host chaperonin CCT. *Journal of Virology*, *84*(17), 8691–8699.

Flick, R., & Hobom, G. (1999). Interaction of influenza virus polymerase with viral RNA in the 'corkscrew' conformation. *The Journal of General Virology*, *80*(Pt. 10), 2565–2572.
Fodor, E., Crow, M., Mingay, L. J., Deng, T., Sharps, J., Fechter, P., et al. (2002). A single amino acid mutation in the PA subunit of the influenza virus RNA polymerase inhibits endonucleolytic cleavage of capped RNAs. *Journal of Virology*, *76*, 8989–9001.
Fodor, E., Pritlove, D. C., & Brownlee, G. G. (1994). The influenza virus panhandle is involved in the initiation of transcription. *Journal of Virology*, *68*(6), 4092–4096.
Fodor, E., & Smith, M. (2004). The PA subunit is required for efficient nuclear accumulation of the PB1 subunit of the influenza A virus RNA polymerase complex. *Journal of Virology*, *78*(17), 9144–9153.
Gastaminza, P., Perales, B., Falcón, A. M., & Ortín, J. (2003). Influenza virus mutants in the N-terminal region of PB2 protein are affected in virus RNA replication but not transcription. *Journal of Virology*, *76*, 5098–5108.
Gerritz, S. W., Cianci, C., Kim, S., Pearce, B. C., Deminie, C., Discotto, L., et al. (2011). Inhibition of influenza virus replication via small molecules that induce the formation of higher-order nucleoprotein oligomers. *Proceedings of the National Academy of Sciences of the United States of America*, *108*(37), 15366–15371.
Ghanem, A., Mayer, D., Chase, G., Tegge, W., Frank, R., Kochs, G., et al. (2007). Peptide-mediated interference with influenza A virus polymerase. *Journal of Virology*, *81*(14), 7801–7804.
González, S., & Ortín, J. (1999). Characterization of the influenza virus PB1 protein binding to vRNA: Two separate regions of the protein contribute to the interaction domain. *Journal of Virology*, *73*, 631–637.
González, S., Zürcher, T., & Ortín, J. (1996). Identification of two separate domains in the influenza virus PB1 protein responsible for interaction with the PB2 and PA subunits: A model for the viral RNA polymerase structure. *Nucleic Acids Research*, *24*, 4456–4463.
Green, T. J., Zhang, X., Wertz, G. W., & Luo, M. (2006). Structure of the vesicular stomatitis virus nucleoprotein-RNA complex. *Science*, *313*(5785), 357–360.
Guilligay, D., Tarendeau, F., Resa-Infante, P., Coloma, R., Crepin, T., Sehr, P., et al. (2008). The structural basis for cap binding by influenza virus polymerase subunit PB2. *Nature Structural and Molecular Biology*, *15*(5), 500–506.
Hara, K., Shiota, M., Kido, H., Ohtsu, Y., Kashiwagi, T., Iwahashi, J., et al. (2001). Influenza virus RNA polymerase PA subunit is a novel serine protease with Ser624 at the active site. *Genes to Cells*, *6*, 87–97.
Hay, A. J., Skehel, J. J., & McCauley, J. (1982). Characterization of influenza virus RNA complete transcripts. *Virology*, *116*, 517–522.
He, X., Zhou, J., Bartlam, M., Zhang, R., Ma, J., Lou, Z., et al. (2008). Crystal structure of the polymerase PA(C)-PB1(N) complex from an avian influenza H5N1 virus. *Nature*, *454*(7208), 1123–1126.
Heggeness, M. H., Smith, P. R., Ulmanen, I., Krug, R. M., & Chopin, P. W. (1982). Studies on the helical nucleocapsid of influenza virus. *Virology*, *118*, 466–470.
Hemerka, J. N., Wang, D., Weng, Y., Lu, W., Kaushik, R. S., Jin, J., et al. (2009). Detection and characterization of influenza A virus PA-PB2 interaction through a bimolecular fluorescence complementation assay. *Journal of Virology*, *83*(8), 3944–3955.
Honda, A. (2008). Role of host protein Ebp1 in influenza virus growth: Intracellular localization of Ebp1 in virus-infected and uninfected cells. *Journal of Biotechnology*, *133*(2), 208–212.
Honda, A., Okamoto, T., & Ishihama, A. (2007). Host factor Ebp1: Selective inhibitor of influenza virus transcriptase. *Genes to Cells*, *12*(2), 133–142.
Honda, A., Ueda, K., Nagata, K., & Ishihama, A. (1988). RNA polymerase of influenza virus: Role of NP in RNA chain elongation. *The Journal of Biochemistry (Tokyo)*, *104*(6), 1021–1026.

Horimoto, T., & Kawaoka, Y. (2005). Influenza: Lessons from past pandemics, warnings from current incidents. *Nature Reviews. Microbiology, 3*(8), 591–600.

Hsu, M. T., Parvin, J. D., Gupta, S., Krystal, M., & Palese, P. (1987). Genomic RNAs of influenza viruses are held in a circular conformation in virions and in infected cells by a terminal panhandle. *Proceedings of the National Academy of Sciences of the United States of America, 84*(22), 8140–8144.

Huang, T. S., Palese, P., & Krystal, M. (1990). Determination of influenza virus proteins required for genome replication. *Journal of Virology, 64*(11), 5669–5673.

Huarte, M., Falcon, A., Nakaya, Y., Ortin, J., Garcia-Sastre, A., & Nieto, A. (2003). Threonine 157 of influenza virus PA polymerase subunit modulates RNA replication in infectious viruses. *Journal of Virology, 77*(10), 6007–6013.

Huarte, M., Sanz-Ezquerro, J. J., Roncal, F., Ortin, J., & Nieto, A. (2001). PA subunit from influenza virus polymerase complex interacts with a cellular protein with homology to a family of transcriptional activators. *Journal of Virology, 75*, 8597–8604.

Huet, S., Avilov, S., Ferbitz, L., Daigle, N., Cusack, S., & Ellenberg, J. (2010). Nuclear import and assembly of the influenza A virus RNA polymerase studied in live cells by fluorescence cross correlation spectroscopy. *Journal of Virology, 84*(3), 1254–1264.

Hutchinson, E. C., & Fodor, E. (2012). Nuclear import of the influenza A virus transcriptional machinery. *Vaccine, 30*(51), 7353–7358.

Hutchinson, E. C., Orr, O. E., Man Liu, S., Engelhardt, O. G., & Fodor, E. (2011). Characterization of the interaction between the influenza A virus polymerase subunit PB1 and the host nuclear import factor Ran-binding protein 5. *The Journal of General Virology, 92*(Pt. 8), 1859–1869.

Jagger, B. W., Wise, H. M., Kash, J. C., Walters, K. A., Wills, N. M., Xiao, Y. L., et al. (2012). An overlapping protein-coding region in influenza A virus segment 3 modulates the host response. *Science, 337*(6091), 199–204.

Jennings, P. A., Finch, J. T., Winter, G., & Robertson, J. S. (1983). Does the higher order structure of the influenza virus ribonucleoprotein guide sequence rearrangements in influenza viral RNA? *Cell, 34*(2), 619–627.

Jorba, N., Area, E., & Ortin, J. (2008). Oligomerization of the influenza virus polymerase complex in vivo. *The Journal of General Virology, 89*(Pt. 2), 520–524.

Jorba, N., Coloma, R., & Ortin, J. (2009). Genetic trans-complementation establishes a new model for influenza virus RNA transcription and replication. *PLoS Pathogens, 5*(5), e1000462.

Jorba, N., Juarez, S., Torreira, E., Gastaminza, P., Zamarreno, N., Albar, J. P., et al. (2008). Analysis of the interaction of influenza virus polymerase complex with human cell factors. *Proteomics, 8*(10), 2077–2088.

Karlas, A., Machuy, N., Shin, Y., Pleissner, K. P., Artarini, A., Heuer, D., et al. (2010). Genome-wide RNAi screen identifies human host factors crucial for influenza virus replication. *Nature, 463*(7282), 818–822.

Kawaguchi, A., Momose, F., & Nagata, K. (2011). Replication-coupled and host factor-mediated encapsidation of the influenza virus genome by viral nucleoprotein. *Journal of Virology, 85*(13), 6197–6204.

Klumpp, K., Ruigrok, R. W., & Baudin, F. (1997). Roles of the influenza virus polymerase and nucleoprotein in forming a functional RNP structure. *The EMBO Journal, 16*(6), 1248–1257.

Konig, R., Stertz, S., Zhou, Y., Inoue, A., Hoffmann, H. H., Bhattacharyya, S., et al. (2009). Human host factors required for influenza virus replication. *Nature, 463*(7282), 813–817.

Kowalinski, E., Zubieta, C., Wolkerstorfer, A., Szolar, O. H., Ruigrok, R. W., & Cusack, S. (2012). Structural analysis of specific metal chelating inhibitor binding to the endonuclease domain of influenza pH1N1 (2009) polymerase. *PLoS Pathogens, 8*(8), e1002831.

Kranzusch, P., & Whelan, S. (2012). Architecture and regulation of negative-strand viral enzymatic machinery. *RNA Biology*, *9*, 941–948.

Krug, R. M., Broni, B. A., & Bouloy, M. (1979). Are the 5′-ends of influenza viral mRNAs synthesized in vivo donated by host mRNAs? *Cell*, *18*, 329–334.

Landeras-Bueno, S., Jorba, N., Pérez-Cidoncha, M., & Ortín, J. (2011). The splicing factor proline-glutamine rich (SFPQ/PSF) is involved in influenza virus transcription. *PLoS Pathogens*, 7, e1002397.

Lee, M. T., Bishop, K., Medcalf, L., Elton, D., Digard, P., & Tiley, L. (2002). Definition of the minimal viral components required for the initiation of unprimed RNA synthesis by influenza virus RNA polymerase. *Nucleic Acids Research*, *30*(2), 429–438.

Li, X., & Palese, P. (1994). Characterization of the polyadenylation signal of influenza virus RNA. *Journal of Virology*, *68*(2), 1245–1249.

Li, G., Zhang, J., Tong, X., Liu, W., & Ye, X. (2010). Heat shock protein 70 inhibits the activity of Influenza A virus ribonucleoprotein and blocks the replication of virus in vitro and in vivo. *PLoS One*, *6*(2), e16546.

Mahy, B. W. J. (1983). Mutants of influenza virus. In P. Palese, & D. W. Kingsbury (Eds.), *Genetics of influenza viruses* (pp. 192–253). Wien: Springer Verlag.

Manz, B., Brunotte, L., Reuther, P., & Schwemmle, M. (2012). Adaptive mutations in NEP compensate for defective H5N1 RNA replication in cultured human cells. *Nature Communications*, *3*, 802.

Marión, R. M., Zürcher, T., de la Luna, S., & Ortín, J. (1997). Influenza virus NS1 protein interacts with viral transcription-replication complexes in vivo. *The Journal of General Virology*, *78*, 2447–2451.

Marklund, J. K., Ye, Q., Dong, J., Tao, Y. J., & Krug, R. M. (2012). Sequence in the influenza a virus nucleoprotein required for viral polymerase binding and RNA synthesis. *Journal of Virology*, *86*(13), 7292–7297.

Martín-Benito, J., Area, E., Ortega, J., Llorca, O., Valpuesta, J. M., Carrascosa, J. L., et al. (2001). Three dimensional reconstruction of a recombinant influenza virus ribonucleoprotein particle. *EMBO Reports*, *2*, 313–317.

Mayer, D., Molawi, K., Martinez-Sobrido, L., Ghanem, A., Thomas, S., Baginsky, S., et al. (2007). Identification of cellular interaction partners of the influenza virus ribonucleoprotein complex and polymerase complex using proteomic-based approaches. *Journal of Proteome Research*, *6*(2), 672–682.

Mena, I., Jambrina, E., Albo, C., Perales, B., Ortin, J., Arrese, M., et al. (1999). Mutational analysis of influenza A virus nucleoprotein: Identification of mutations that affect RNA replication. *Journal of Virology*, *73*, 1186–1194.

Minor, P. D., & Dimmock, N. J. (1975). Inhibition of synthesis of influenza virus proteins: Evidence of two host-cell-dependent events during multiplication. *Virology*, *67*(1), 114–123.

Moeller, A., Kirchdoerfer, R. N., Potter, C. S., Carragher, B., & Wilson, I. A. (2012). Organization of the influenza virus replication machinery. *Science*, *338*, 1631–1634.

Molinari, N. A., Ortega-Sanchez, I. R., Messonnier, M. L., Thompson, W. W., Wortley, P. M., Weintraub, E., et al. (2007). The annual impact of seasonal influenza in the US: Measuring disease burden and costs. *Vaccine*, *25*(27), 5086–5096.

Momose, F., Handa, H., & Nagata, K. (1996). Identification of host factors that regulate the influenza virus RNA polymerase activity. *Biochimie*, *78*(11–12), 1103–1108.

Mukaigawa, J., & Nayak, D. P. (1991). Two signals mediate nuclear localization of influenza virus (A/WSN/33) polymerase basic protein 2. *Journal of Virology*, *65*, 245–253.

Murti, K. G., Webster, R. G., & Jones, I. M. (1988). Localization of RNA polymerases of influenza viral ribonucleoproteins by immunogold labeling. *Virology*, *164*, 562–566.

Nagata, K., Takeuchi, K., & Ishihama, A. (1989). In vitro synthesis of influenza viral RNA: Biochemical complementation assay of factors required for influenza virus replication. *Journal of Biochemistry*, *106*(2), 205–208.

Naito, T., Kiyasu, Y., Sugiyama, K., Kimura, A., Nakano, R., Matsukage, A., et al. (2007). An influenza virus replicon system in yeast identified Tat-SF1 as a stimulatory host factor for viral RNA synthesis. *Proceedings of the National Academy of Sciences of the United States of America, 104*(46), 18235–18240.

Naito, T., Momose, F., Kawaguchi, A., & Nagata, K. (2007). Involvement of Hsp90 in assembly and nuclear import of influenza virus RNA polymerase subunits. *Journal of Virology, 81*(3), 1339–1349.

Nath, S. T., & Nayak, D. P. (1990). Function of two discrete regions is required for nuclear localization of polymerase basic protein 1 of A/WSN/33 influenza virus (H1N1). *Molecular and Cellular Biology, 10*(8), 4139–4145.

Neumann, G., Brownlee, G. G., Fodor, E., & Kawaoka, Y. (2004). Orthomyxovirus replication, transcription, and polyadenylation. *Current Topics in Microbiology and Immunology, 283*, 121–143.

Neumann, G., Noda, T., & Kawaoka, Y. (2009). Emergence and pandemic potential of swine-origin H1N1 influenza virus. *Nature, 459*(7249), 931–939.

Newcomb, L. L., Kuo, R. L., Ye, Q., Jiang, Y., Tao, Y. J., & Krug, R. M. (2009). Interaction of the influenza a virus nucleocapsid protein with the viral RNA polymerase potentiates unprimed viral RNA replication. *Journal of Virology, 83*(1), 29–36.

Ng, A. K., Lam, M. K., Zhang, H., Liu, J., Au, S. W., Chan, P. K., et al. (2012). Structural basis for RNA binding and homo-oligomer formation by influenza B virus nucleoprotein. *Journal of Virology, 86*(12), 6758–6767.

Ng, A. K., Zhang, H., Tan, K., Li, Z., Liu, J. H., Chan, P. K., et al. (2008). Structure of the influenza virus A H5N1 nucleoprotein: Implications for RNA binding, oligomerization, and vaccine design. *The FASEB Journal, 22*(10), 3638–3647.

Nieto, A., de la Luna, S., Bárcena, J., Portela, A., & Ortín, J. (1994). Complex structure of the nuclear translocation signal of the influenza virus polymerase PA subunit. *The Journal of General Virology, 75*, 29–36.

Noble, E., Cox, A., Deval, J., & Kim, B. (2012). Endonuclease substrate selectivity characterized with full-length PA of influenza A virus polymerase. *Virology, 433*(1), 27–34.

Obayashi, E., Yoshida, H., Kawai, F., Shibayama, N., Kawaguchi, A., Nagata, K., et al. (2008). The structural basis for an essential subunit interaction in influenza virus RNA polymerase. *Nature, 454*(7208), 1127–1131.

Ortega, J., Martín-Benito, J., Zürcher, T., Valpuesta, J. M., Carrascosa, J. L., & Ortín, J. (2000). Ultrastructural and functional analyses of recombinant influenza virus ribonucleoproteins suggest dimerization of nucleoprotein during virus amplification. *Journal of Virology, 74*, 156–163.

Palese, P., & Shaw, M. (2007). Orthomyxoviridae: The viruses and their replication. In D. M. Knipe, & P. M. Howley (Eds.), (5th ed.). *Fields virology*, Vol. 1, (pp. 1647–1689). Philadelphia: Lippincott Williams & Wilkins.

Perales, B., de la Luna, S., Palacios, I., & Ortín, J. (1996). Mutational analysis identifies functional domains in the Influenza A PB2 polymerase subunit. *Journal of Virology, 70*, 1678–1686.

Perales, B., Sanz-Ezquerro, J. J., Gastaminza, P., Ortega, J., Fernández-Santarén, J., Ortín, J., et al. (2000). The replication activity of influenza virus polymerase is linked to the capacity of the PA subunit to induce proteolysis. *Journal of Virology, 74*, 1307–1312.

Pérez, D. R., & Donis, R. O. (1995). A 48-amino-acid region of influenza A virus PB1 protein is sufficient for complex formation with PA. *Journal of Virology, 69*, 6932–6939.

Perez, J. T., Varble, A., Sachidanandam, R., Zlatev, I., Manoharan, M., Garcia-Sastre, A., et al. (2010). Influenza A virus-generated small RNAs regulate the switch from transcription to replication. *Proceedings of the National Academy of Sciences of the United States of America, 107*(25), 11525–11530.

Perez, J. T., Zlatev, I., Aggarwal, S., Subramanian, S., Sachidanandam, R., Kim, B., et al. (2012). A small-RNA enhancer of viral polymerase activity. *Journal of Virology*, *86*(24), 13475–13485.

Perez-Gonzalez, A., Rodriguez, A., Huarte, M., Salanueva, I. J., & Nieto, A. (2006). hCLE/CGI-99, a human protein that interacts with the influenza virus polymerase, is a mRNA transcription modulator. *Journal of Molecular Biology*, *362*(5), 887–900.

Plotch, S. J., Bouloy, M., Ulmanen, I., & Krug, R. M. (1981). A unique cap(m7GpppXm)-dependent influenza virion endonuclease cleaves capped RNAs to generate the primers that initiate viral RNA transcription. *Cell*, *23*, 847–858.

Poch, O., Sauvaget, I., Delarue, M., & Tordo, N. (1990). Identification of four conserved motifs among the RNA-dependent polymerase encoding elements. *The EMBO Journal*, *8*, 3867–3874.

Pons, M. W., Schulze, I. T., & Hirst, G. K. (1969). Isolation and characterization of the ribonucleoprotein of influenza virus. *Virology*, *39*, 250–259.

Poole, E., Elton, D., Medcalf, L., & Digard, P. (2004). Functional domains of the influenza A virus PB2 protein: Identification of NP- and PB1-binding sites. *Virology*, *321*(1), 120–133.

Poon, L. L. M., Pritlove, D. C., Fodor, E., & Brownlee, G. G. (1999). Direct evidence that the poly(A) tail of influenza A virus mRNA is synthesized by reiterative copying of a U track in the virion RNA template. *Journal of Virology*, *73*(4), 3473–3476.

Poon, L. L., Pritlove, D. C., Sharps, J., & Brownlee, G. G. (1998). The RNA polymerase of influenza virus, bound to the 5′ end of virion RNA, acts in cis to polyadenylate mRNA. *Journal of Virology*, *72*, 8214–8219.

Pritlove, D. C., Poon, L. L., Fodor, E., Sharps, J., & Brownlee, G. G. (1998). Polyadenylation of influenza virus mRNA transcribed in vitro from model virion RNA templates: Requirement for 5' conserved sequences. *Journal of Virology*, *72*, 1280–1286.

Rao, P., Yuan, W., & Krug, R. M. (2003). Crucial role of CA cleavage sites in the cap-snatching mechanism for initiating viral mRNA synthesis. *The EMBO Journal*, *22*(5), 1188–1198.

Resa-Infante, P., Jorba, N., Coloma, R., & Ortín, J. (2011). The influenza virus RNA synthesis machine: Advances in its structure and function. *RNA Biology*, *8*(2), 1–9.

Resa-Infante, P., Jorba, N., Zamarreno, N., Fernandez, Y., Juarez, S., & Ortin, J. (2008). The host-dependent interaction of alpha-importins with influenza PB2 polymerase subunit is required for virus RNA replication. *PLoS One*, *3*(12), e3904.

Resa-Infante, P., Recuero-Checa, M. A., Zamarreño, N., Llorca, O., & Ortín, J. (2010). Structural and functional characterisation of an influenza virus RNA polymerase-genomic RNA complex. *Journal of Virology*, *84*, 10477–10487.

Robb, N. C., Chase, G., Bier, K., Vreede, F. T., Shaw, P. C., Naffakh, N., et al. (2011). The influenza A virus NS1 protein interacts with the nucleoprotein of viral ribonucleoprotein complexes. *Journal of Virology*, *85*(10), 5228–5231.

Robb, N. C., Smith, M., Vreede, F. T., & Fodor, E. (2009). NS2/NEP protein regulates transcription and replication of the influenza virus RNA genome. *The Journal of General Virology*, *90*(Pt. 6), 1398–1407.

Robertson, J. S., Schubert, M., & Lazzarini, R. A. (1981). Polyadenylation sites for influenza mRNA. *Journal of Virology*, *38*, 157–163.

Rodriguez, A., Perez-Gonzalez, A., & Nieto, A. (2011). Cellular human CLE/C14orf166 protein interacts with influenza virus polymerase and is required for viral replication. *Journal of Virology*, *85*(22), 12062–12066.

Rudolph, M. G., Kraus, I., Dickmanns, A., Eickmann, M., Garten, W., & Ficner, R. (2003). Crystal structure of the borna disease virus nucleoprotein. *Structure*, *11*(10), 1219–1226.

Ruigrok, R. W., & Baudin, F. (1995). Structure of influenza virus ribonucleoprotein particles. II. Purified RNA-free influenza virus ribonucleoprotein forms structures that

are indistinguishable from the intact influenza virus ribonucleoprotein particles. *The Journal of General Virology*, *76*, 1009–1014.
Ruigrok, R. W., Crepin, T., Hart, D. J., & Cusack, S. (2010). Towards an atomic resolution understanding of the influenza virus replication machinery. *Current Opinion in Structural Biology*, *20*(1), 104–113.
Sanz-Ezquerro, J. J., Zürcher, T., de la Luna, S., Ortin, J., & Nieto, A. (1996). The amino-terminal one-third of the influenza virus PA protein is responsible for the induction of proteolysis. *Journal of Virology*, *70*, 1905–1911.
Shapira, S. D., Gat-Viks, I., Shum, B. O., Dricot, A., de Grace, M. M., Wu, L., et al. (2009). A physical and regulatory map of host-influenza interactions reveals pathways in H1N1 infection. *Cell*, *139*(7), 1255–1267.
Sugiyama, K., Obayashi, E., Kawaguchi, A., Suzuki, Y., Tame, J. R., Nagata, K., et al. (2009). Structural insight into the essential PB1-PB2 subunit contact of the influenza virus RNA polymerase. *The EMBO Journal*, *28*(12), 1803–1811.
Tafforeau, L., Chantier, T., Pradezynski, F., Pellet, J., Mangeot, P. E., Vidalain, P. O., et al. (2011). Generation and comprehensive analysis of an influenza virus polymerase cellular interaction network. *Journal of Virology*, *85*(24), 13010–13018.
Tarendeau, F., Boudet, J., Guilligay, D., Mas, P. J., Bougault, C. M., Boulo, S., et al. (2007). Structure and nuclear import function of the C-terminal domain of influenza virus polymerase PB2 subunit. *Nature Structural and Molecular Biology*, *14*(3), 229–233.
Tarendeau, F., Crepin, T., Guilligay, D., Ruigrok, R. W., Cusack, S., & Hart, D. J. (2008). Host determinant residue lysine 627 lies on the surface of a discrete, folded domain of influenza virus polymerase PB2 subunit. *PLoS Pathogens*, *4*(8), e1000136.
Tiley, L. S., Hagen, M., Mathews, J. T., & Krystal, M. (1994). Sequence-specific binding of the influenza virus RNA polymerase to sequences located at the 5'-end of the viral RNAs. *Journal of Virology*, *68*, 5108–5116.
Tong, S., Li, Y., Rivailler, P., Conrardy, C., Castillo, D. A., Chen, L. M., et al. (2012). A distinct lineage of influenza A virus from bats. *Proceedings of the National Academy of Sciences of the United States of America*, *109*(11), 4269–4274.
Torreira, E., Schoehn, G., Fernandez, Y., Jorba, N., Ruigrok, R. W., Cusack, S., et al. (2007). Three-dimensional model for the isolated recombinant influenza virus polymerase heterotrimer. *Nucleic Acids Research*, *35*(11), 3774–3783.
Toyoda, T., Adyshev, D. M., Kobayashi, M., Iwata, A., & Ishihama, A. (1996). Molecular assembly of the influenza virus RNA polymerase: Determination of the subunit-subunit contact sites. *The Journal of General Virology*, 77, 2149–2157.
Turrell, L., Lyall, J. W., Tiley, L. S., Fodor, E., & Vreede, F. T. (2013). The role and assembly mechanism of nucleoprotein in influenza A virus ribonucleoprotein complexes. *Nature Communications*, *4*, 1591.
Ulmanen, I., Broni, B. A., & Krug, R. M. (1981). The role of two of the influenza virus core P proteins in recognizing cap 1 structures (m7GpppNm) on RNAs and in initiating viral RNA transcription. *Proceedings of the National Academy of Sciences of the United States of America*, *78*, 7355–7359.
Vreede, F. T., & Brownlee, G. G. (2007). Influenza virion-derived viral ribonucleoproteins synthesize both mRNA and cRNA in vitro. *Journal of Virology*, *81*(5), 2196–2204.
Vreede, F. T., Gifford, H., & Brownlee, G. G. (2008). Role of initiating nucleoside triphosphate concentrations in the regulation of influenza virus replication and transcription. *Journal of Virology*, *82*(14), 6902–6910.
Vreede, F. T., Jung, T. E., & Brownlee, G. G. (2004). Model suggesting that replication of influenza virus is regulated by stabilization of replicative intermediates. *Journal of Virology*, *78*(17), 9568–9572.

Wang, Z., Liu, X., Zhao, Z., Xu, C., Zhang, K., Chen, C., et al. (2011). Cyclophilin E functions as a negative regulator to influenza virus replication by impairing the formation of the viral ribonucleoprotein complex. *PLoS One*, *6*(8), e22625.

Wang, P., Song, W., Mok, B. W., Zhao, P., Qin, K., Lai, A., et al. (2009). Nuclear factor 90 negatively regulates influenza virus replication by interacting with viral nucleoprotein. *Journal of Virology*, *83*(16), 7850–7861.

Wise, H. M., Foeglein, A., Sun, J., Dalton, R. M., Patel, S., Howard, W., et al. (2009). A complicated message: Identification of a novel PB1-related protein translated from influenza A virus segment 2 mRNA. *Journal of Virology*, *83*(16), 8021–8031.

Wolstenholme, A. J., Barrett, T., Nichol, S. T., & Mahy, B. W. (1980). Influenza virus-specific RNA and protein syntheses in cells infected with temperature-sensitive mutants defective in the genome segment encoding nonstructural proteins. *Journal of Virology*, *35*(1), 1–7.

Yamanaka, K., Ishihama, A., & Nagata, K. (1990). Reconstitution of influenza virus RNA-nucleoprotein complexes structurally resembling native viral ribonucleoprotein cores. *The Journal of Biological Chemistry*, *265*(19), 11151–11155.

Ye, Q., Krug, R. M., & Tao, Y. J. (2006). The mechanism by which influenza A virus nucleoprotein forms oligomers and binds RNA. *Nature*, *444*(7122), 1078–1082.

Yuan, P., Bartlam, M., Lou, Z., Chen, S., Zhou, J., He, X., et al. (2009). Crystal structure of an avian influenza polymerase PA(N) reveals an endonuclease active site. *Nature*, *458*(7240), 909–913.

Zhang, S., Wang, J., Wang, Q., & Toyoda, T. (2010). Internal initiation of influenza virus replication of viral RNA and complementary RNA in vitro. *The Journal of Biological Chemistry*, *285*(52), 41194–41201.

Zürcher, T., de la Luna, S., Sanz-Ezquerro, J. J., Nieto, A., & Ortín, J. (1996). Mutational analysis of the influenza virus A/Victoria/3/75 PA protein: Studies of interaction with PB1 protein and identification of a dominant negative mutant. *The Journal of General Virology*, 77, 1745–1749.

CHAPTER FIVE

The Molecular Biology of Ilarviruses

Vicente Pallas[*,1], Frederic Aparicio[*], Mari C. Herranz[*], Jesus A. Sanchez-Navarro[*], Simon W. Scott[†,1]

[*]Instituto de Biología Molecular y Celular de Plantas (IBMCP), Universidad Politécnica de Valencia-Consejo Superior de Investigaciones Científicas, Valencia, Spain
[†]Department of Biological Sciences, Clemson University, Clemson, South Carolina, USA
[1]Corresponding authors: e-mail address: vpallas@ibmcp.upv.es; sscott@clemson.edu

Contents

Abstract

Ilarviruses were among the first 16 groups of plant viruses approved by ICTV. Like *Alfalfa mosaic virus* (AMV), bromoviruses, and cucumoviruses they are isometric viruses and possess a single-stranded, tripartite RNA genome. However, unlike these other three groups, ilarviruses were recognized as being recalcitrant subjects for research (their ready lability is reflected in the sigla used to create the group name) and were renowned as unpromising subjects for the production of antisera. However, it was recognized that they shared properties with AMV when the phenomenon of genome activation, in which the coat protein (CP) of the virus is required to be present to initiate infection, was demonstrated to cross group boundaries. The CP of AMV could activate

Advances in Virus Research, Volume 87
ISSN 0065-3527
http://dx.doi.org/10.1016/B978-0-12-407698-3.00005-3

the genome of an ilarvirus and vice versa. Development of the molecular information for ilarviruses lagged behind the knowledge available for the more extensively studied AMV, bromoviruses, and cucumoviruses. In the past 20 years, genomic data for most known ilarviruses have been developed facilitating their detection and allowing the factors involved in the molecular biology of the genus to be investigated. Much information has been obtained using *Prunus necrotic ringspot virus* and the more extensively studied AMV. A relationship between some ilarviruses and the cucumoviruses has been defined with the recognition that members of both genera encode a 2b protein involved in RNA silencing and long distance viral movement. Here, we present a review of the current knowledge of both the taxonomy and the molecular biology of this genus of agronomically and horticulturally important viruses.

1. INTRODUCTION

Members of the genus *Ilarvirus* have been known for 100 years. The first report of a disease that was later associated with an ilarvirus (*Apple mosaic virus*, ApMV) involved the description of a graft-transmissible variegation in apple (Stewart, 1910). *Tobacco streak virus* (TSV), the type member of the genus, was first described in 1936 (Johnson, 1936) but, for the most part, the ilarviruses reported in the interim since 1910 have infected woody species (King, Adams, Carstens, & Lefkowitz, 2012). The name Ilarvirus was created as a sigla from the properties of the viruses (***is***ometric, ***la***bile pa**r**ticles which are associated with **r**ingspot symptoms) and was first used as a specific reference to *Prune dwarf virus* (PDV) and *Necrotic ringspot virus*—now referred to as *Prunus necrotic ringspot virus* (PNRSV; Fulton, 1968) but later adopted for wider use by the ICTV (Fenner, 1976). Ilarviruses have a reputation of being difficult to purify and discouraging subjects for the production of antibodies. However, initial subdivisions within the group were based on serological relationships. The development of genomic sequences for the majority of ilarviruses in the past 20 years has allowed the generation of a taxonomy based on molecular traits and a reorganization of the individual viruses placed within the subgroups. Prior to the development of the sequence data, there had been evidence of a close relationship between the ilarviruses and AMV. The presence of either a few molecules of coat protein (CP) and/or the subgenomic RNA 4 is necessary for infection with ilarviruses to occur. However, the CP of AMV can substitute for the CP of ilarviruses and vice versa in this phenomenon known as "genome activation." This activation results from conserved arginine motifs in the CPs and AUGC motifs in the 3′-terminal regions of the genomic RNAs. The link between ilarviruses and AMV has allowed the molecular models developed for AMV to be examined in relation to some of the ilarviruses. Despite these close links, AMV is still considered to be a member of a separate genus based on the fact that it is aphid transmissible and ilarviruses are transmitted via pollen and thrips (King et al., 2012). Sequence data have also shown links with the cucumoviruses. The 2b open reading frame (ORF) found in the cucumoviruses is present in subgroup 1 and subgroup 2 ilarviruses (see Section 4.2) but not in other subgroups (King et al., 2012). Although the development of monoclonal antibodies and antibodies produced against recombinant proteins (Abou-Jawdah et al., 2004; Aparício et al., 2009; Boari et al., 1997; Halk, Hsu, Aebig, & Franke, 1984; Imed et al., 1997; Menzel, Hamacher, Weissbrodt, & Winter, 2012) has addressed some of the problems previously associated with the serological detection of ilarviruses, it is only in the past two decades, and with the advent of PCR and other molecular methodologies, that ilarviruses have been detected more frequently and have become more amenable subjects for research. In this review, we provide current information on the molecular biology of this expanding genus of viruses that had previously been considered intractable subjects for research.

2. VIRION AND GENOME STRUCTURE

Virions of most members of the genus are quasi-spherical, 26–36 nm in diameter and exhibit icosahedral symmetry ($T=3$). Bacilliform particles occur in plants infected by PDV, *Tulare apple mosaic virus* (TAMV), PNRSV, ApMV, and *Spinach latent virus* (SpLV) (Fig. 5.1A). Cryoelectron microscopy of preparations of TSV revealed isometric particles permeated by numerous pores (Fig. 5.1B). This latter observation perhaps explains the lability of these viruses. Ilarviruses possess a tripartite genome of positive-sense, single-stranded RNAs (ssRNAs) without a poly A tail (Fig. 5.2) (King et al., 2012) and are included in the alpha-like superfamily of viruses (Goldbach, Le Gall, & Wellinck, 1991). The genome of the type species, TSV, has a total length of 8622 nucleotides(nt) with other members having genomes of approximately the same size. The RNA 1 (~3400 nt) is monocistronic coding for a viral replicase. The RNA 2 (~2900 nt) is bicistronic in some members of the genus. The 5′ proximal ORF codes for the viral polymerase. The second, smaller ORF, found in some members of the genus, is located toward the 3′-terminus of the molecule and expressed through a subgenomic RNA that codes for a 2b protein

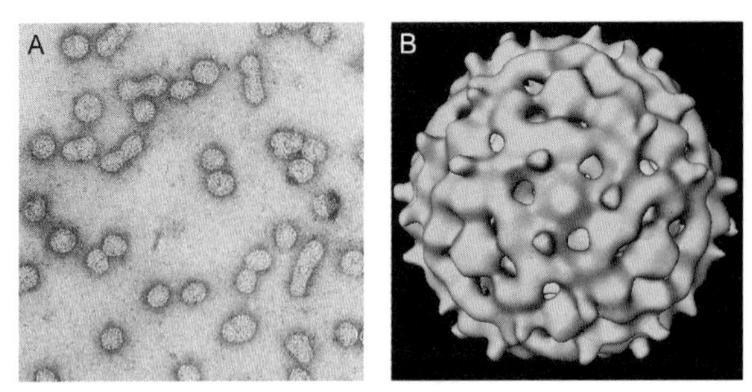

Figure 5.1 (A) Virions of Prunus necrotic ringspot virus prepared in uranyl acetate. (B) Three-dimensional reconstruction of Tobacco streak virus by electron cryomicroscopy. The TSV was frozen in vitrified ice on an EM grid and the sample was imaged using a JEOL JEM-1200 electron cryomicroscope operating at 30,000 × and 100 keV. Images were recorded on Kodak SO-163 film and scanned to a resolution of 5.33 Å per pixel. Image reconstruction was as described in *Journal of Virology, 72, 1534–1541.* Image courtesy of A. Paredes, Director of Electron Microscopy Group, NCTR/ORA Nanotechnology Core Facility, Office of Scientific Coordination, NCTR, FDA, 3900 NCTR Road, Jefferson, Arkansas 72079. (For color version of this figure, the reader is referred to the online version of this chapter.)

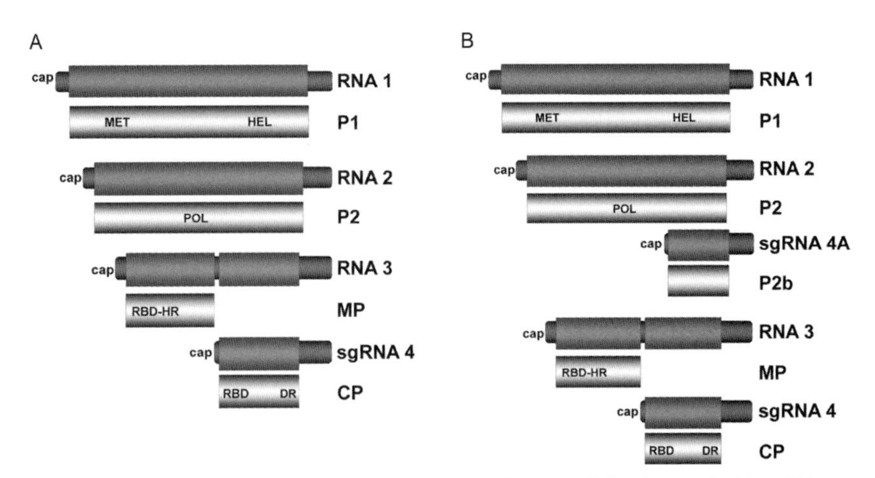

Figure 5.2 Diagram showing the genome organization of ilarviruses lacking (A) and presenting (B), the ORF 2b at the 3′end of the RNA 2. Genomic RNAs 1, 2, 3, and subgenomic (sg) RNA 4 and 4A are presented as green and gray boxes for ORF RNAs and nontranslated regions, respectively. Yellow boxes represent viral proteins P1, P2, P2b, MP, and CP. Characteristic domains methyltransferase (MET), helicase (HEL), RNA-dependent RNA polymerase (POL) RNA-binding domain (RBD), hydrophobic region (HR), and dimerization region (DR) are indicated. (For interpretation of the references to color in this figure legend, the reader is referred to the online version of this chapter.)

(Xin, Ji, Scott, Symons, & Ding, 1998). The function of this protein has yet to be demonstrated in ilarviruses but, based on similarities with the cucumoviruses, it is inferred to be involved in viral movement and gene silencing (Lucy, Guo, Li, & Ding, 2000). This latter function has been demonstrated for AV-2 by work from Japan (Shimura, H., Masuta, C., Yoshida, Sueda, & Suzuki, 2013). The 2b ORF is found only in subgroup 1 and subgroup 2 ilarviruses. The RNA 3 (~2200 nt) codes for the movement protein (MP) (proximal ORF), and the CP (distal ORF). The distal gene is expressed via a subgenomic RNA 4 (King et al., 2012). This subgenomic RNA is transcribed from an internal promoter region within the genomic RNA 3, the secondary structure of which closely resembles that of the internal promoter previously identified in *Brome mosaic virus* as being required for the synthesis of the subgenomic RNA 4 (Aparicio & Pallás, 2002; Jaspars, 1988). In *Fragaria chiloensis latent virus* (FCiLV), a third ORF has been identified on the RNA 3 (Tzanetakis & Martin, 2005). The 5′UTR of the RNA 3 of some ilarvirus species contains several sequence elements similar to the internal control region 2 of eukaryotic tRNA gene promoters (Aparicio & Pallás, 2002; Shiel, Alrefai, Domier, Korban, & Berger, 1995). For the equivalent ICR motifs present in AMV RNA 3, a relevant function in plus-strand promoter activity has been proposed (Van der Vossen, Neeleman, & Bol, 1993). In some ilarvirus species, small subgenomic RNAs have been detected (Francki, Milne, & Hatta, 1985). No function has been assigned to these but for PNRSV an approximately 200 bp fragment was determined to be a nonfunctional copy of the 3′UTR (Di Terlizzi, Skrzeczkowski, Mink, Scott, & Zimmerman, 2001).

3. TAXONOMY AND GENETIC DIVERSITY

The ilarvirus group was officially recognized by ICTV in its second report (Fenner, 1976). Because this group, and viruses in the Bromovirus and Cucumovirus groups, and AMV all possessed isometric particles and tripartite genomes, it was later proposed that they should belong to a single viral family with the name *Tricornaviridae* (Van Vloten Doting, Francki, Fulton, Kaper, & Lane, 1981). This proposal was never approved by ICTV but, in 1992, the formal family name *Bromoviridae* was adopted for this grouping with the previously named groups becoming genera. Subsequently, two other genera (*Anulavirus* and *Oleavirus*) have been added to the family.

Initially, ilarviruses were subdivided into groups A and B (Table 5.1) based on serological relationships among members within the group (Shepherd et al., 1976). As more ilarviruses were identified, but did not display serological relationships with the members of the other two groups, additional subgroups were created (Table 5.2). In addition, the two major groups were revised (Uyeda & Mink, 1983). TSV was removed from group A and placed in a distinct subgroup 1 and it was recognized that *Black raspberry latent virus* was an isolate of TSV. The remaining members of group A formed the nucleus of subgroup 2. The viruses in group B were reduced

Table 5.1 The original viruses considered to be members of the ilarvirus group

Group A	Group B
Tobacco streak virus	Prunus necrotic ringspot virus
Tulare apple mosaic virus	Cherry rugose mosaic virus
Elm mottle virus	Rose mosaic virus
Citrus leaf rugose virus	Apple mosaic virus
Citrus variegation virus	Danish plum line pattern virus
Black raspberry latent virus	Hop virus A
	Hop virus C

Prune dwarf virus and American plum line pattern virus were regarded as potential members of the group. After Shepherd et al. (1976).

Table 5.2 Species in the genus *Ilarvirus* as listed in the sixth report of the ICTV in 1995 (Murphy et al., 1995).
Type species: Tobacco streak virus

Subgroup 1
Tobacco streak virus
Subgroup 2
Asparagus virus 2
Blueberry shock virus
Citrus leaf rugose virus
Citrus variegation virus
Elm mottle virus
Tulare apple mosaic virus
Subgroup 3
Apple mosaic virus
Prunus necrotic ringspot virus
Subgroup 4
Prune dwarf virus
Subgroup 5
American plum line pattern virus
Subgroup 6
Spinach latent virus
Subgroup 7
Lilac ring mottle virus
Subgroup 8
Hydrangea mosaic virus
Subgroup 9
Humulus japonicus virus
Subgroup 10
Parietaria mottle virus

in number. *Cherry rugose mosaic virus*, *Hop virus C*, and *Danish plum line pattern virus* became to be regarded as isolates of PNRSV. *Rose mosaic virus* and *Hop virus A* were regarded as isolates of ApMV. This grouping was renamed as subgroup 3 and included PDV. From then on, a number of ungrouped and provisional members were described. However, discussion on the serology of ilarviruses was complicated as viruses with different names may in fact have been closely related strains. For example: Cherry rugose mosaic disease is now recognized as being caused by an isolate of PNRSV and the name Cherry rugose mosaic virus is no longer valid. Similarly, Rose mosaic virus is no longer a valid name and the symptoms associated with rose mosaic disease have been shown to be induced by any one of the following viruses: ApMV, PNRSV, and TSV, and *Arabis mosaic virus* (a nepovirus) and *Strawberry latent ringspot virus* (an unassigned species in the family *Secoviridae*).

The data in Table 5.2 were published in 1995 (Murphy et al., 1995) and represent the taxonomic arrangements of ilarviruses prior to the development of extensive genomic sequence data for members of this genus. At that time, GenBank accessions, X00435, L03726, U03857, and L28145, represented the entire collection of full-length sequences for the RNA 3s of TSV, two isolates of ApMV, and PDV, respectively. Although the isolate associated with U03857 was designated as ApMV it is in fact regarded as PNRSV cf. GenBank accession L38823. The currently accepted species of ilarviruses (http://www.ictvonline.org/virusTaxonomy.asp?version=2011) are shown in Table 5.3. Many of the changes between Tables 5.2 and 5.3 are the result of the development of molecular sequence data for the different viruses. However, as viral species are considered a polythetic class (Van Regenmortel et al., 2013), and no single property should be used to define a species, the molecular data have been used in conjunction with other characters to verify, revise, or redefine previously accepted relationships.

Strawberry necrotic shock disease had been regarded as being caused by a unique isolate of TSV, but sequence data suggest that it is sufficiently different to be regarded as a distinct ilarvirus, *Strawberry necrotic shock virus* (SNSV) in subgroup 1 (Tzanetakis, Martin, & Scott, 2010). The initial description of *Hydrangea mosaic virus* (HdMV) indicated that it was a new ilarvirus (Thomas, Barton, & Tuszynski, 1983), However, later work demonstrated a serological relationship with *Elm mottle virus* (Jones, 1985). Sequencing of the original isolates of both viruses showed them to be the same virus. Because of the dates on which the EMoV and HdMV were first reported (Jones & Mayo, 1973; Thomas et al., 1983, respectively), HdMV was relegated to the status of a synonym for EMoV (Scott, Zimmerman, & Ge, 2003).

Table 5.3 Species in the genus ilarvirus as listed in the ninth report of the ICTV in 2012
Type species: Tobacco streak virus

Subgroup 1	Subgroup 3
Blackberry chlorotic ringspot virus*	Apple mosaic virus
Parietaria mottle virus	Blueberry shock virus
Strawberry necrotic shock virus*	Lilac leaf chlorosis virus*
Tobacco streak virus	Prunus necrotic ringspot virus
Potential members	
Bacopa chlorosis virus	
Tomato necrotic spot virus	
Subgroup 2	**Subgroup 4**
Asparagus virus 2	Fragaria chiloensis latent virus
Citrus leaf rugose virus	Prune dwarf virus
Citrus variegation virus	Potential members
Elm mottle virus	Viola white distortion virus
Lilac ring mottle virus	
Spinach latent virus	
Tulare apple mosaic virus	

Including species newly accepted by the ICTV in 2012 (*) and other viruses that may be included in the genus after ratification. *No known relationships with existing subgroups*: American plum line pattern virus; Humulus japonicus latent virus. Other names that have been ascribed to viruses with ilarvirus-like properties/sequence but for which there is limited information/and or sequence data: Grapevine angular mosaic virus, Grapevine line pattern virus, and Potato yellowing virus. Partial sequences for all three genomic molecules of Raphanus latent virus are available in GenBank accessions JN107637–JN107639.

Similarly, sequencing the genome of SpLV showed it to be a distinct subgroup 2 virus, and a serological relationship between SpLV and subgroup 2 viruses was detected (Ge, Scott, & Zimmerman, 1997). Partial genomic sequence data showed that *Lilac ring mottle virus* (LiRMV) was also best considered to be a subgroup 2 virus (Scott & Zimmerman, 2008). Sequence data have been developed for *Blueberry shock virus* (BlShV) but remain to be published and show the virus to be related to PNRSV and ApMV (subgroup 3) (R.R. Martin, personal communication). Relationships at the molecular level have been demonstrated for FCiLV and PDV (Tzanetakis & Martin, 2005) and they are now accepted as being the members of subgroup 4 (King et al., 2012). *American plum line pattern virus* (APLPV) and *Humulus japonicus latent virus* (HJLV) possess the properties of ilarviruses but they do not cluster with any of the existing subgroups (Scott & Zimmerman, 2001, 2006). There are a several viruses that share properties with the ilarviruses but do not yet have complete genomic sequence data and thus have neither been grouped definitively with existing members nor awarded species status by ICTV: *Bacopa chlorosis virus* (Maroon-Lango, Aebig, & Hammond,

2006; Menzel et al., 2012), *Tomato necrotic spot virus* (Batuman et al., 2009), and *Viola white distortion virus* (GenBank accession GU168941). In addition a recent report of a virus, *Raphanus latent virus* (GenBank accessions JN107637–JN107639), is available. A characterization of *Grapevine angular mosaic virus* has been completed but there is only a small amount of sequence information available (Girgis et al., 2009), *Grapevine line pattern virus* (Lehoczky et al., 1987), and *Potato yellowing virus* (Silvestre, Untiveros, & Cuellar, 2011) also share properties with the ilarviruses but there is even less information known about them.

Unlike some other viral genera and families (*Alphaflexiviridae*, *Potyviridae*; King et al., 2012), no guidelines as to what constitutes a "New" ilarvirus species at the molecular level have been proposed. It might be expected that the RNA 1 and RNA 2 of ilarviruses coding for the helicase (HEL) and polymerase, respectively, would be conserved to some extent (see supplemental figures: http://www.elsevierdirect.com/companions/9780124076983; see Tamura et al., 2011). Indeed they are, with various amino acid motifs in the putative translation products of these two genomic molecules being conserved throughout the genus. However, it is possible to distinguish between, but also group, ilarviruses using these motifs (Boulila, 2009; Maliogka, Dovas, & Katis, 2007). The polymerase signature (Candresse, Morch, & Dunez, 1990) of most subgroup 2 ilarviruses contains 6 amino acids (aa) conserved among TAMV, *Citrus leaf rugose virus* (CiLRV), CVV, EMoV AV-2, and SpLV, but different from the corresponding amino acids in other ilarviruses (Scott et al., 2003). LiRMV shares only 2 of these 6 conserved aa but if the 2b proteins of subgroup 2 viruses are examined, a number of conserved motifs are present that distinguish the 2b proteins of subgroup 2 viruses from the 2b proteins of subgroup 1 viruses and can clearly place LiRMV as being a member of subgroup 2 (Scott & Zimmerman, 2008).

A number of phylogenetic studies have been completed on ilarviruses either referring to single species within the genus (Aparicio, Myrta, Di Terlizzi, & Pallas, 1999; Aparicio & Pallás, 2002; Fiore et al., 2008; Fonseca, Neto, Martins, & Nolasco, 2005; Glasa, Betinová, Kúdela, & Subr, 2002; Hammond, 2003; Lakshmi et al., 2011; Oliver et al., 2009; Vaskova, Petrzik, & Karesova, 2000; Vaskova, Petrzik, & Spak, 2000) or using the genus and its members as part of the family *Bromoviridae* (Codoñer, Cuevas, Sanchez-Navarro, Pallas, & Elena, 2005; Codoñer & Elena, 2006, 2008; Codoñer, Fares, & Elena, 2006). Analyses have used data for complete genomes (Codoñer & Elena, 2006), CPs (Hammond, 2003) and MPs (Fiore et al., 2008) with attempts being made to group ilarviruses, particularly in relation to plant host, country of origin, and serotype (Pallas et al., 2012). In hindsight, the transport of fruit tree cultivars worldwide perhaps needs to be added to the equation. Although movement of fruit tree cultivars is regulated, viz., the importation of prunus germplasm into the United States and many other countries, there have been recorded movements of material that had not been virus indexed, of material that was imported prior to the enforcement of regulations, and of material that was symptomless but infected. Thus, the worldwide distribution of germplasm and of desirable horticultural and agronomic cultivars may have contributed to the dissemination of specific virus isolates. A prime example from outside the ilarviruses involves *Plum pox virus*. Although transmitted by aphids and spread locally by this means, the spread of this virus worldwide has been due entirely to the movement of budwood/trees. Peach latent mosaic viroid appears to have been endemic in the United States but unrecognized and then imported into Europe in certified virus indexed material.

The majority of the ilarviruses are represented by single isolates. However, more than one isolate exists for viruses that infect agronomic and horticultural crops (ApMV, BCRV, CVV, PNRSV, PDV, SNSV, and TSV). These isolates are most commonly represented by sequences for the RNA 3 (MP and CP) and offer the opportunity to examine the diversity present in a viral species.

3.1. Subgroup 1

TSV affects an increasingly wide range of crop species (Bhat et al., 2002) but the CP of isolates from different hosts and locations appears highly conserved, displaying >95% identity among the aa of the CP. The MP, as might be expected, is less conserved, but even so, the aa sequence of the MP shares >84% identity among isolates of TSV. Similar degrees of identity exist among the aa sequences for the CP and MP of isolates of other subgroup one viruses. Amino acid identity within the MP of seven PMoV-T isolates ranged from 96% to 100%, but varied from 94% to 96% when PMoV-T isolates were compared to the original isolate of PMoV. Most differences were located in the C-terminal region of the protein, where the identity shared between the last 40 aa varied between 82% and 100% (Galipienso, Herranz, Lopez, Pallas, & Aramburu, 2008). Other subgroup 1 ilarviruses (BCRV, PMoV, SNSV) share <80% identity with the CP of TSV.

3.2. Subgroup 2

Most members of subgroup 2 share a number of motifs within the aa sequence of the CP (MSGNAIE at the amino terminal region, TWxSFPGEQWH from

67 aa, TKVYSWVM from 112 aa and PLPGxKPPxNFLV at the carboxy terminal). However, CiLRV and TAMV have a 4-aa insertion between the initial methionine and the N residue in the first motif. LiRMV shares some of the residues in these motifs but is currently the most divergent of the subgroup 2 ilarviruses. CVV represents the greatest number of complete sequences for the RNA 3 of a subgroup 2 ilarvirus. The CP of the CVV isolates share greater than 95% identity but share less than 85% identity with other members of subgroup 2. The MP of CVV shares 76% identity with the MP of CiLRV (as might be expected) but 49% or less identity with other members of subgroup 2.

3.3. Subgroup 3, subgroup 4, and unclassified members

The diversity of the subgroup 3 viruses, PNRSV and ApMV, has been reviewed recently by Pallas et al. (2012). This same review covered genetic diversity of PDV and the unclassified APLPV. However, the earlier comment regarding the worldwide distribution of germplasm may apply to this latter virus. Herranz et al. (2008) identified low genetic variability within isolates of APLPV isolated from *Prunus salicina*. Although this may be an accurate depiction of viral variation when grown in the same host cultivar over an extended period of time, it may also reflect the extensive distribution of the plums (*P. salicina*: cvs. Shiro, Burbank, Santa Rosa) bred and released by Burbank at the beginning of the twentieth century. Similar worldwide exploitation of certain fruit tree cultivars potentially containing ilarviruses has occurred in all pome fruit, stone fruit, and citrus crops. Seed of asparagus cultivars (potentially containing AV-2) and spinach cultivars (potentially containing SPLV) has also been widely distributed.

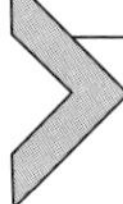

4. GENOME EXPRESSION

4.1. Replicase complex proteins

Ilarvirus RNAs 1 and 2 encode for the P1 and P2 replicase proteins, respectively (Fig. 5.2). Analogous to the situation that exists in other genera in the family *Bromoviridae*, ilarvirus P1 and P2 proteins are considered to be the subunits of the viral replicase complex (Bol, 2005). P1 is implicated in viral RNA recruitment and anchorage to the vacuolar membrane where replication complexes are assembled. Moreover, P1 provides methyltransferase and HEL activities. P2 carries out RNA-dependent RNA polymerization. Subcellular localization studies using PNRSV P1 and P2 tagged with green fluorescent protein (GFP) have shown that GFP:P1 accumulates in punctuate structures, probably associated with the vacuolar membrane, whereas P2: GFP remains as a soluble cytoplasmic protein (M.C. Herranz & V. Pallas, unpublished data). Genomic RNAs and replicase proteins have to be directed to host membranes structures where replication complexes are assembled. Studies on the replication of BMV have revealed that the BMV 1a protein induces the formation of spherical invaginations of the endoplasmic reticulum (ER), recruits genomic RNAs from translation to replication, and directs RNAs, together with the 2a protein, to the spherules at the ER (Ahlquist, Noueiry, Lee, Kushner, & Dye, 2003). Other viruses in the family *Bromoviridae* may adopt a similar strategy. Available evidence, however, has indicated that the replication complexes of alfamo-, ilar-, and cucumoviruses assemble at the vacuolar membrane (Van der Heijden, Carette, Reinhoud, Haegi, & Bol, 2001). Recently, Ibrahim, Hutchens, Berg, and Loesch-Fries (2012) have shown that AMV replicase proteins, P1 and P2, localize to the tonoplast in the presence of virus RNA.

The sequences of the RNA 1 of all the ilarviruses (including tentative members and AMV) sequenced so far range between 3332 (PNRSV) and 3644 (AMV) nt. For RNA 2, nucleotide sequences vary between 2404 (APLPV) and 2990 (CiLRV) nt in length. The RNA 1 of all the members of the genus *Ilarvirus* is monocistronic and encodes the P1 replicase whose size varies between 1037 (*Lilac leaf chlorosis virus*) and 1098 (PMoV) aa. The N-terminal region of the P1 protein contains a methyltransferase domain (MET). This domain presents several of the motifs described in the alphavirus-like superfamily (Rozanov, Koonin, & Gorbalenya, 1992). Motifs I, Ia1, Ia2, II, and III are conserved in all ilarviruses, whereas motif IIa1 (CxxxxxxC) has not been detected in APLPV and motif IV has not been found in both APLPV and HJLV (Supplemental Fig. 1: http://www.elsevierdirect.com/companions/9780124076983). All the P1 N-terminal motifs present conserved residues which, in some cases, have proved critical for P1 functionality. Thus, the analysis performed with the AMV P1 protein has revealed that the mutation of H100A (I) or C189S (IIa1) displays no accumulation of negative- or positive-strand RNAs (Vlot, Menard, & Bol, 2002). However, the BMV P1 protein carrying the equivalent H100A mutation replicates in *Saccharomyces cerevisiae* at 10% of the wild-type (wt) level (Ahola, den Boon, & Ahlquist, 2000). The single mutation of conserved residues D154A (II), R157G (II), or Y266A (IV) in AMV P1 support negative-strand synthesis at 30–40% of the wt level, whereas accumulation of positive-strand RNA is relatively low (2% or less) (Vlot et al., 2002). The mutation of the equivalent R157 residue in the BMV P1 protein (R136 in BMV) reduces positive-strand RNA accumulation in yeast to less than 0.5% of the wt level, and this P1 mutant seems to affect the BMV RNAs recruitment function (Ahola et al., 2000).

The P1 C-terminal part contains the NTPase/HEL-like domain (Gorbalenya & Koonin, 1993) with strictly conserved motifs I, Ia, II, III, IV, V, and VI (Supplemental Fig. 2: http://www.elsevierdirect.com/companions/9780124076983). As observed for the P1 N-terminus, the majority of the motifs located at the C-terminus contain conserved residues that have been analyzed in the P1 protein of BMV and AMV. The mutation of a histidine (H) residue to alanine (A) placed in motif V in P1 of BMV blocks RNA 3 replication (Wang et al., 2005). At this position, PNRSV, ApMV, and LLCV contain an A, whereas the rest of ilarviruses have an H. The possible effects of these different residues on replication mechanisms have not yet been analyzed. In the crystal structures of related HELs (PcrA and Rep) (e.g., Dillingham, Soultanas, Wiley, Webb, & Wigley, 2001; Velankar, Soultanas, Dillingham, Subramanya, & Wigley, 1999), the aa corresponding to conserved residue K (I) interacts with the ATP phosphates, while the aa corresponding to the conserved residue D (II) coordinates Mg^{2+}, and that of Q (III) contacts the ATP phosphate (Caruthers & McKay, 2002; Velankar et al., 1999). Mutations at equivalent amino acids block NTPase activities in multiple HELs (Gu et al., 2000; Li et al., 2001). Similarly, the ATPase activity of the corresponding P1 BMV mutants is only 10–25% of that of the wt 1a protein and blocks RNA replication (Wang et al., 2005), indicating that similar activity can be assigned to the ilarvirus conserved residues. In this sense, the mutation of the K844 (I), D934 (III), and R1085 (VI) residues of the AMV P1 protein blocked the accumulation of negative- and positive-strand AMV RNAs (Vlot, Laros, & Bol, 2003). Moreover, mutation of the conserved TYR (motif IV) and R (motif VI) residues of the BMV P1 protein does not induce detectable spherule formation and possesses near-wt ATPase activity (Wang et al., 2005). These mutants also block the RNA replication of BMV, and the corresponding motifs have been postulated to be involved in spherule formation. Finally, all the above conserved residues from motifs I, II, III, IV, V, and VI have proven to be defective in recruiting and stabilizing BMV RNA 3 (Wang et al., 2005). One explanation for this phenotype can be attributed to the alteration of the RNA-binding activity of the 1a protein. This idea is quite possible for the conserved residue R in motif III since the corresponding amino acid in the PcrA (R260) and Rep helicases (R251) crystal structures contacts ssDNA (Velankar et al., 1999).

The RNA 2 of all the ilarviruses encodes a P2 replicase subunit in their 5′ region whose size ranges between 740 aa (APLPV) and 876 aa (ApMV) (Fig. 5.2). P2 carries out RNA-dependent RNA polymerization and it shows the greatest homology among viruses in its C-terminal part. All eight conserved motifs (I–VIII) described for the Supergroup III of the RNA-dependent RNA polymerases of positive-strand RNA viruses (Koonin, 1991) are present in all the ilarviruses, except for motif VI or motifs VII and VIII, which are not conserved in TAMV or SpLV, respectively (Supplemental Fig. 3: http://www.elsevierdirect.com/companions/9780124076983). The eight motifs (I–VIII) correspond to the six motifs (A–F) that have been identified in the vRdRps of dsRNA and positive-sense RNA viruses (Bruenn, 2003; Gorbalenya et al., 2002; Poch, Sauvaget, Delarue, & Tordo, 1989). The four motifs (A–D, Supplemental Fig. 3: http://www.elsevierdirect.com/companions/9780124076983) are common to all hand-shaped polymerases (Poch et al., 1989) and are located in the palm domain, which consists of four antiparallel β-strands and two α-helices. The four regions contain four strictly conserved residues: an aspartate in motifs IV(A) and VI(C), a glycine in motif V(B), and a lysine in motif VII(D) (Poch et al., 1989). The catalytic aspartate of motif VI(C) is the first aspartate of the GDD sequence that is characteristic of both vRdRps and reverse transcriptases with some modification at the first position. As expected, the mutations of the catalytic aspartate in the GDD motif of the AMV RNA 2-encoded P2 protein blocks minus-strand RNA synthesis and abolishes the infectivity of the virus (Van Rossum, Garcia, & Bol, 1996; Vlot et al., 2003). However, this critical aspartate is not present in the two isolates of PMoV characterized so far. Thus, analysis of other PMoV isolates is necessary to confirm if the presence of an alanine, instead of the critical aspartate, is a property peculiar to PMoV.

Studies on the interchangeability of the noncoding sequences of AMV RNA 3 with those of PNRSV have led to the conclusion that the AMV replicase is able to recognize the 3′UTR of PNRSV and can promote minus-strand RNA synthesis albeit after changing the terminal AAGC motif to that present in AMV (AUGC) (Aparicio, Sanchez-Navarro, Olsthoorn, Pallas, & Bol, 2001). Interestingly, the AMV replicase does not recognize the PNRSV promoter for genomic (plus-strand) RNA synthesis, which indicates that ilarviruses have specific sequences to recognize the cognate replicase (Aparicio et al., 2001).

4.2. 2b protein

RNA 2 is monocistronic in APLPV, ApMV, PDV, PNRSV, HJLV, and Lilac leaf chlorosis virus, but is bicistronic in all the other members of the genus for which sequence information is available (King et al., 2012; Scott et al., 2003). The second ORF of these bicistronic RNAs 2, named ORF 2b, overlaps the 3′end of the P2 gene (Fig. 5.2B). The corresponding protein, 2b, is synthesized through a subgenomic RNA derived from RNA 2 (sgRNA4A). The sgRNA4A has been found to accumulate at high levels in plants infected with SpLV and PMoV (Door Peeters, 2009; Xin et al., 1998). Yet despite the 2b protein never having been detected in infected plants, a protein species of the expected size for SpLV P2b has been translated *in vitro* from virions containing RNA 4A (Xin et al., 1998). It has been suggested that ilar 2b may be functionally similar to the *Cucumber mosaic virus* (CMV) 2b protein, which is involved in viral movement and gene silencing (Ding, Li, & Symons, 1995; Guo & Ding, 2002; Xin et al., 1998), although *in vivo* translation and 2b function in infected plants has yet to be demonstrated in ilarviruses. Indeed, all the attempts made to demonstrate the potential activity as a silencing suppressor of the 2b protein of PMoV have been unsuccessful, at least at the local level (A. Door Peeters, Pallas, & Aparicio, unpublished data). The FCiLV, which is most closely related to PDV, also contains a second ORF, although it is not located on the same portion of the genome as other ilarviruses (Tzanetakis & Martin, 2005).

4.3. Movement protein

The MPs encoded by the members of the genus *Ilarvirus* belong to the 30K-type superfamily (Melcher, 2000). Members of this superfamily, represented by 30K MP of the *Tobacco mosaic virus* (TMV), share common characteristics which include binding nucleic acids, localizing and increasing the size exclusion limit of plasmodesmata (PD) (Fig. 5.3), and interacting with the ER membrane. The PNRSV MP is the best studied in this genus and its involvement in inter- and intracellular virus movement has been reported (Herranz & Pallas, 2004; Herranz, Sanchez-Navarro, Sauri, Mingarro, & Pallas, 2005; Martinez-Gil et al., 2009; Sánchez-Navarro, Carmen Herranz, & Pallas, 2006). The *in vitro* RNA-binding properties of PNRSV MP were studied and

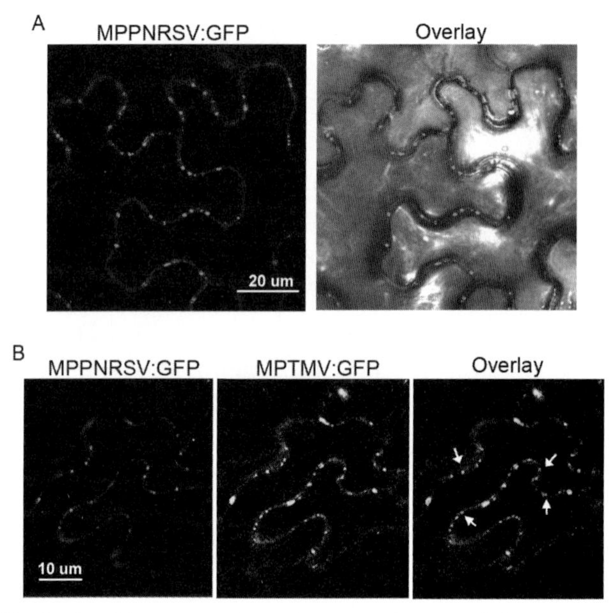

Figure 5.3 (A) Confocal laser-scanning microscopy images of the green channel (left) and overlay with bright-field channel (right) of the subcellular localization in epidermal cells of the PNRSV MP fused to the green fluorescent protein (MPPNRSV:GFP). Images show the typical punctuate pattern at the cell wall corresponding to accumulation in plasmodesmata. (B) Image of an epidermic cell of the green and red channels coexpressing the MPs of PNRSV and TMV fused to the GFP (MPPNRSV:GFP) and the DsRed (MPTMV:dsRed), respectively. Arrows in the overlay image of both red and green channels denote plasmodesmata filled with both proteins. (For interpretation of the references to color in this figure legend, the reader is referred to the online version of this chapter.)

a RNA-binding domain (RBD) was identified (Herranz & Pallas, 2004). These studies reveal that the PNRSV MP protein preferentially binds ssRNA in a nonspecific and cooperative manner with an estimated dissociation constant (K_d) of 1.4 μM, which is in the same order as other sequence-nonspecific RNA-binding proteins (Daros & Carrington, 1997; Fujita, Mise, Kajiura, Dohi, & Furusawa, 1998; Marcos, Vilar, Perez-Paya, & Pallas, 1999; Vilar, Esteve, Pallas, Marcos, & Perez-Paya, 2001). Additionally, a mutational analysis of the protein mapped the RBD between 56 and 88 aa at the N-terminus of PNRSV MP (Herranz & Pallas, 2004). A similar domain has been described within the MP of the LLCV (James, Varga, Leippi, Godkin, & Masters, 2010). This domain contains basic residues that are conserved among LLCV, ApMV, and PNRSV (KxxxxK/RxxxKxxK/

RxxxxK/RxxxxxK/RxK/R). Interestingly, the MP RBDs of PNRSV and AMV are located at the N-terminus of the MP, whereas the MP RBDs of CMV and BMV are positioned at the C-terminus of the MP (Li & Palukaitis, 1996). Although sequence alignment has not revealed any obvious similarities, several common features can be drawn from the analysis: (i) a high proportion of basic amino acids (Arg or Lys) is present at the C-terminus of the four domains and within or preceding this motif, there is an unusually high number of aliphatic amino acids (Ile, Leu); (ii) secondary structure predictions have revealed the presence of two α-helices (red shadow in Fig. 5.4); the presence of α-helices in a basic environment has been described for other proteins that bind RNA (Gomez & Pallas, 2001; Marcos et al., 1999; Tan & Frankel, 1995; Vilar et al., 2001); (iii) similar to the RBD "B" of TMV (Citovsky, Wong, Shaw, Venkataram, & Zambryski, 1992), there is a high probability that the four RBDs in ilarviruses, AMV, CMV, and BMV occur on the surface of the protein.

Transient expression studies with PNRSV MP fused to the GFP have revealed the capacity of these fused proteins for intercellular trafficking and the presence of highly fluorescent punctuate structures between neighboring cells resembling PD (Fig. 5.3). Accordingly, Herranz et al. (2005) have demonstrated that an AMV hybrid virus containing different mutant PNRSV MP genes, either lacking the RBD or with point mutations in this region (in tandem of three residues; see Fig. 5.4, RBD), is unable to move from cell to cell. The synthetic peptides representing the mutants and wt RBDs have been used to study RNA-binding affinities by EMSA assays. RNA-binding affinities were seen to be approximately 20-fold lower in the mutants. In addition, analyses using circular dichroic spectroscopy have revealed that the secondary structure of peptides is not significantly affected by mutations. All together, these results indicate that RNA binding is a critical activity for virus transport in this type of viruses.

Many virus-encoded MPs interact with membranes. For 30K superfamily members, a topological model for TMV MP with two hydrophobic regions (HRs) spanning the membrane has been proposed (Brill, Nunn, Kahn, Yeager, & Beachy, 2000). PNRSV MP bears a single HR located at 89–109 aa, which interacts with the membrane interface. Within this domain, the presence of a proline and a glutamine constrains both the secondary structure of this region and its interaction with the lipid interface. The mutation of this strictly conserved proline residue (Fig. 5.4, HR) excludes viral cell-to-cell movement *in vivo* (Martinez-Gil et al., 2009).

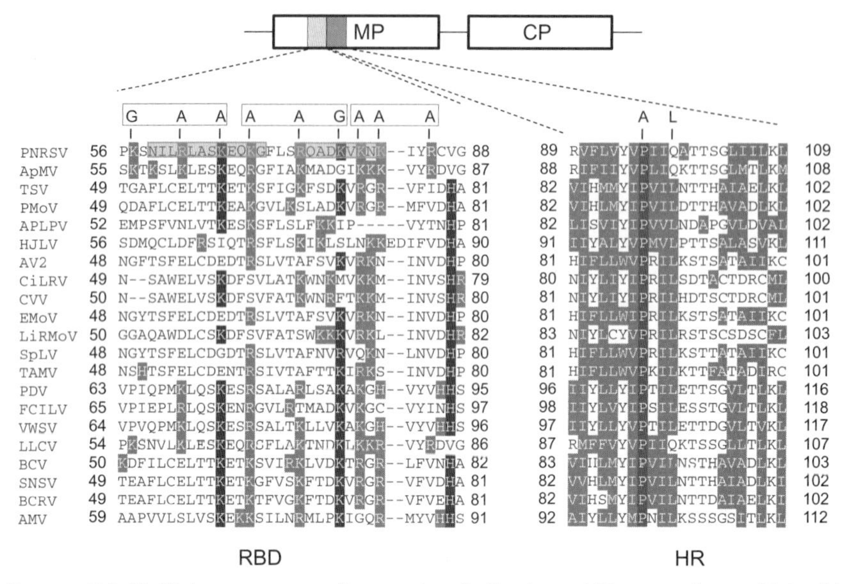

Figure 5.4 Multiple-sequence alignment of ilarvirus MPs members (Bujarski, Figlerowicz, Gallitelli, Roossinck, & Scott, 2012) using CLUSTAL X (1.8) (ftp://ftp-igbmc.u-strasbg.fr/pub/ClustalX/) and GeneDoc Editor. Only the RNA-binding domain (RBD) and the hydrophobic region (HR) are shown. Basic amino acids in RBD appear shaded in gray and those with a conserved percentage over 80% in dark green. Predicted α-helices for PNRSV MP RBD are indicated in orange rectangles. At the top of the RBDs alignment, mutational analysis carried out by Herranz et al. (2005) is shown. The nine basic amino acids were changed into groups of three by noncharged alanine and glycine residues. Right, sequence alignment of the HRs of ilarviruses characterized so far showing the central proline residue (shaded in red) conserved in all them. In violet, the high number of hydrophobic residues in all the HRs analyzed are stand out and the single mutations carried out by Martinez-Gil et al. (2009) marked on the top. (For interpretation of the references to color in this figure legend, the reader is referred to the online version of this chapter.)

4.4. Coat protein

The CP of ilarviruses is expressed from a subgenomic RNA 4, which is transcribed from minus-strand polarity RNA 3. Computer-assisted comparisons have led to the proposal that a stem–loop structure in viral minus-strand RNA 3 located 20–37 nt upstream of the RNA 4 start site was required for RNA 4 subgenomic transcription in all Bromoviridae. This stem–loop contains four nucleotides which have proven essential for synthesis of BMV RNA and are believed to be responsible for the direct interaction with the viral replicase complex (Adkins & Kao, 1998; Siegel, Adkins, & Kao,

1997). For ApMV and PNRSV, both viruses present two stable stem–loops containing the most distal four essential nucleotides (Jaspars, 1988). A comparative analysis of this region done among several PNRSV isolates has shown that some of them lose the proximal hairpin and that they contain only the distal stem–loop. This observation has led to the proposal that PNRSV must have acquired a second functional hairpin through a duplication mechanism (Aparicio & Pallás, 2002).

A considerable proportion of the arginine and/or lysine residues that occur in the CP of ilarviruses are found in the N-terminal region of the molecule (Sánchez-Navarro & Pallás, 1997) (Fig. 5.5A). The amino-terminal peptides containing this basic motif are sufficient to bind to the 3′-nontranslated region (3′-NTR) of its own RNA (Aparicio, Vilar, Perez-Paya, & Pallas, 2003; Baer, Houser, Loesch-Fries, & Gehrke, 1994). Mutational and comparative analysis of the N-terminal CP sequences of AMV, TSV, and CVV, have led to the proposal of an RNA-binding consensus sequence (Q/K/R-P/N-T-X-**R**-S-R/Q-Q/N/S-W/F-A) (Ansel-McKinney, Scott, Swanson, Ge, & Gehrke, 1996). In this sequence, a single arginine (R) is the only residue that is essential and responsible for the specific binding of either those peptides corresponding to this motif or full-length CPs to a terminal fragment of 3′-NTR RNA (Ansel-McKinney et al., 1996; Swanson et al., 1998; Yusibov & Loesch-Fries, 1998) (Fig. 5.5A; the critical proposed R for all the CP ilarviruses is underlined and in dark gray). In recent years, the sequence of the RNA 3 of several ilarvirus has been determined and, interestingly, the corresponding CP sequence of six of them share a second putative RNA-binding consensus sequence (V-T/S-R/N-R-Q-S/R-R-N-A-A/R-**R**-A-A-X-Y/F-R) (Fig. 5.5A). In contrast, the basic motif in the CPs of APLPV, ApMV, HJLV, LLCV, and PNRSV shows higher variability and does not fit either of the proposed consensus sequences, but conserves the putative critical single R residue. For the CP of PNRSV, the basic motif is located between amino acid residues 30 and 50 and contains from three to five R residues (Aparicio, Myrta, et al., 1999; Fiore et al., 2008). This motif has the capacity to bind to the 3′-NTR of its RNA 3 (Aparicio et al., 2003; Pallas, Sanchez-Navarro, & Diez, 1999). However, it does not have an equivalent single-critical R responsible for RNA interaction since the exchange of R at positions 30 or 41 by alanine eliminates the RNA interaction capability of a synthetic peptide corresponding to positions 25–50. These results suggest that, at least in PNRSV, more than one R amino acid residue of the basic motif contributes to the RNA affinity of the CP (Aparicio et al., 2003) (Fig. 5.5A;

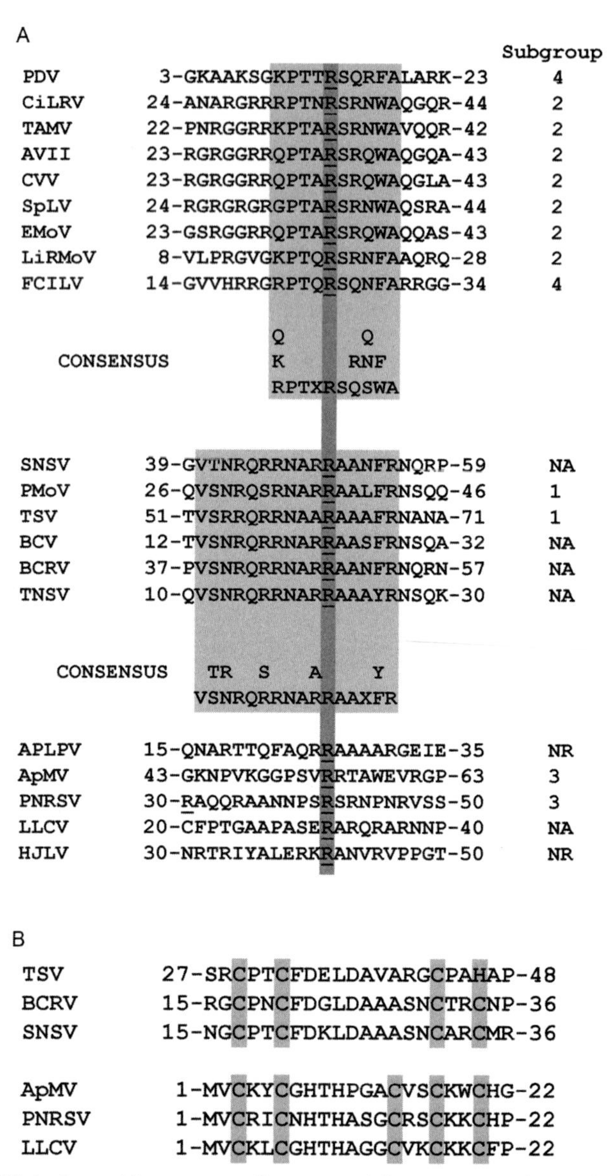

Figure 5.5 (A) Amino acid sequence alignment of the arginine-rich motif of ilarviruses CPs. The proposed critical arginine residue is underlined and in dark gray, whereas those residues forming the putative consensus sequences are in light gray. Colum on the right reflects the grouping taxonomic structure of ilarviruses. NA, not assigned to any subgroup yet; NR, not related to any subgroup. (B) Amino acid sequence alignment of the putative zinc-finger motif present in six ilarviruses CP. The cysteine and histidine residues involved in the formation of this motif are in light gray. Numbers refer to the residue positions in each CP.

both the R residues are underlined in the PNRSV CP sequence). This basic motif is highly conserved in all the PNRSV isolates characterized so far, except for one isolate, which lacks one of the critical R residues (Aparicio, Myrta, et al., 1999; Aparicio, Sanchez-Pina, et al., 1999; Fiore et al., 2008). It is noteworthy that all but one of the ilarviruses containing the RNA-binding sequence consensus proposed by Ansel-McKinney et al. (1996) belongs to subgroups 2 or 4, whereas two of the ilarviruses harboring critical R residues in a different context (PNRSV and ApMV) belong to subgroup 3 and to so far unassigned viruses (Fig. 5.5A).

Moreover, the N-terminal part of the CP of six ilarviruses, also presents a zinc-finger motif, either C_2C_2 or CCCH type (Fig. 5.5B) (James et al., 2010; Sánchez-Navarro & Pallás, 1994; Sánchez-Navarro, Reusken, Bol, & Pallas, 1997). Interestingly, subgroup 1 ilarviruses present a C_2C_2 motif whereas the CCCH motif is present in subgroup 3 ilarviruses. Such motifs are candidates for linking zinc ions through a tetrahedral complex and have been found in several proteins with nucleic acid binding activity (e.g., Berg, 1986; Gramstat, Courtpozanis, & Rohde, 1990). It has been proposed that zinc-finger motifs can also participate in the RNA-binding properties of these ilarvirus CPs by conferring increased binding affinity and/or stability to the protein (Ansel-McKinney et al., 1996; Aparicio et al., 2003; Sánchez-Navarro & Pallás, 1994). In fact for the CP of PNRSV, the K_d observed for the synthetic peptide described above has been estimated to be 170 nM, whereas it is 11.3 nM for the entire protein (Aparicio et al., 2003). The K_d of the CP of AMV, which lacks the zinc-finger motif, is 500 nM (Baer et al., 1994). Finally, a region located between residues 9 and 27 from the C-terminal end of the PNRSV CP has been recently shown to participate in CP dimerization (Aparicio, Sánchez-Navarro, & Pallás, 2006) (see Fig. 5.2).

5. CP–RNA INTERACTION AND GENOME ACTIVATION

In the 1970s, it was observed that inoculum consisting of a mixture of genomic RNAs of AMV and ilarviruses CiLRV, CVV, PDV, PNRSV, or TSV was unable to establish infection, unless the CP was added (Bol, van Vloten-Doting, & Jaspars, 1971; Gonsalves & Fulton, 1977; Gonsalves & Garnsey, 1975; Halk & Fulton, 1978; Van Vloten-Doting, 1975). Moreover, it has also been reported that the TSV CP can induce this process in AMV, and vice versa (Van Vloten-Doting, 1975). This early function of the CP is termed "genome activation" and is specific for AMV and ilarviruses. Recently, it has been found that the replication of the *Raspberry bushy dwarf virus*, a virus that shares some similarities with ilarviruses, but is currently classified as the sole member of the genus *Idaeovirus*, is stimulated by the presence of the CP in a mechanism resembling genome activation (MacFarlane & McDavin, 2009).

Early studies performed with AMV aimed to understand the genomic activation mechanism. They have shown that this phenomenon involves the interaction of the CP with the 3′-NTR of viral RNAs (Houwing & Jaspars, 1980; Smith, Roosien, van Vloten-Doting, & Jaspars, 1981). Initially, it was reported that synthetic peptides corresponding to the sequence of the first 25 or 38 amino acid residues of AMV CP were sufficient for binding to viral RNA and for replacing the entire protein upon the initiation of infection (Baer et al., 1994; Zuidema, Bierhuizen, & Jaspars, 1983). As stated above, a highly conserved R residue was found to be critical for this process in the CPs of AMV, CVV, and TSV (Ansel-McKinney et al., 1996) and at least two R residues in PNRSV (Aparicio et al., 2003). Moreover, the 3′-NTRs of the viral RNAs of AMV and ilarviruses are believed to fold in a highly conserved secondary structure. This may consist in a linear array of several stem–loops flanked by nonpaired AUGC sequences, or variants of this, which represent specific binding sites for the CP (Ansel-McKinney & Gehrke, 1998; Reusken & Bol, 1996; Reusken, Neeleman, & Bol, 1994; Zuidema et al., 1983). Band shift assays demonstrated that the 3′-NTR of AMV and PNRSV RNA 3 contains four and three independent sites to bind the CP, respectively (Aparicio et al., 2003; Houser-Scott, Baer, Liem, Cai, & Gehrke, 1994).

Currently, there are two different models that explain the CP's early role in the viral life cycle, both of which are based on *in vitro* RNA-binding assays and on *in vivo* functional analyses of the mutants of the 3′-NTR of AMV. The conformational switch model suggests that the 3′-NTR of AMV and ilarviruses can adopt, in addition to the secondary linear stem–loop array, an alternative tertiary conformation, which resembles a tRNA-like structure (TLS), through the formation of a pseudoknot (Olsthoorn, Mertens, Brederode, & Bol, 1999). The binding of the CP to 3′-NTR would trigger the conformational change from the TLS, which favors the viral RNA polymerase binding and minus-strand synthesis of viral RNAs, to the linear stem–loop structure, thus permitting plus-strand synthesis and translation. Therefore, the switch between both structures mediated by CP binding would represent a mechanism in the life cycle of AMV and *Ilarvirus*, which would be implicated in the regulation of minus- and plus-strand viral RNA accumulation and viral protein translation, respectively (Chen & Olsthoorn, 2010; Olsthoorn et al., 1999). RNA-binding assays conducted with PNRSV CP have reported that the 3′-NTR of PNRSV may adopt a similar TLS (Aparicio et al., 2003), and recently, phylogenetic comparisons have led to the proposal that all known ilarviruses might adopt the TLS (Chen, Gultyaev, Pleij, & Olsthoorn, 2009). In contrast, the 3′ organization model proposed by Gehrke and coworkers has demonstrated that the 3′-NTR–CP interaction is required for replication. This model suggests that the 3′-NTR of AMV and ilarviruses does not adopt a TLS, but instead the interaction between the CP and RNA would compact the RNA structure rather than extending it. This compacted RNA–CP complex would be the functional equivalent of the TLS and would be the conformational feature required for the recognition and initiation of viral RNA replication (Baer et al., 1994; Guogas, Filman, Hogle, & Gehrke, 2004; Petrillo, Rocheleau, Kelley-Clarke, & Gehrke, 2005).

Recently, it has been observed that AMV CP localizes at the nucleus and nucleolus of infected cells. Nonetheless, the role of this subcellular localization in the viral cycle remains unsolved (Herranz, Pallas, & Aparicio, 2012). There is still no evidence as to whether ilarvirus CPs also enter the nucleus and accumulate in the nucleolus. Finally, in addition to genome activation and virion formation, the CP of ilarviruses might be involved in cell-to-cell movement and in the systemic spread of the virus, as has previously been shown for AMV (see below).

6. VIRUS MOVEMENT WITHIN AND AMONG PLANTS

To systemically infect a plant, viruses must invade adjacent cells via cell wall connections known as PD, a process called cell-to-cell transport (Fernandez-Calviño, Faulkner, & Maule, 2011), and must reach distal plant parts through the vascular tissue, a process named systemic transport (Carrington, Kasschau, Mahajan, & Schaad, 1996; Pallas, Genoves, Sánchez-Pina, & Navarro, 2011; Ueki & Citovsky, 2007; Waigmann, Ueki, Trutnyeva, & Citovsky, 2004). For this purpose, plant viruses express one or more MPs to support virus transport (Lucas, 2006). Viral MPs facilitate virus cell-to-cell movement by different mechanisms, thus permitting the transport of ribonucleoprotein complexes between MPs and viral RNA (e.g., TMV; Waigmann et al., 2004), the CP (e.g., CMV or

AMV; Sánchez-Navarro et al., 2006), or virion particles (e.g., the *Grapevine fanleaf virus*; Ritzenthaler & Hofmann, 2007).

Early observations in the electronic microscope revealed that some ilarviruses aggregate in paracrystalline forms of particle-containing tubules (Gerola, Bassi, & Belli, 1969). A similar observation has also been made for the type member of the ilarviruses TSV (Edwardson & Purciful, 1974). A conclusive demonstration that MPs are the unique viral element required to form these tubular structures stems from the use of a 35S promoter-driven expression system in protoplasts (Wellink et al., 1993). Unlike AMV (Zheng, Wang, & Zhang, 1997), this capacity to form tubules has not been observed for PNRSV MP when the protein was fused to the GFP (Aparicio, Pallas, & Sanchez-Navarro, 2010). However, the possibility that the presence of the GFP at the C-terminus of the protein compromises this function of the MP capacity cannot be eliminated. A lack of infectious clones has compromised the functional analyses of ilarvirus MPs. However, the use of the AMV model system and the observation that the MP gene can be functionally replaced with the corresponding gene of other viruses assigned to the 30K family has helped acquire partial knowledge of the function of the ilarvirus MP. The first results were observed by replacing the CP and MP genes in AMV RNA 3 by the corresponding genes of PNRSV and the chimeric viruses analyzed in transgenic *Nicotiana tabacum* plants, constitutively expressing the P1 and P2 polymerase proteins of AMV (P12 plants) (Taschner, Van der Kuyl, Neeleman, & Bol, 1991). The results showed that the hybrid AMV RNA 3 carrying the MP and CP genes of PNRSV exhibited basic competence for encapsidation and replication in P12 protoplasts, and also for a low level of cell-to-cell movement in P12 plant, a non-host for PNRSV (Sánchez-Navarro, Reusken, Bol and Pallas, 1997). These results suggest that PNRSV MP employs a similar cell-to-cell mechanism to that previously described for AMV. Further information on the cell-to-cell transport of PNRSV has been obtained by using an AMV RNA 3 vector which expresses the GFP (Sánchez-Navarro, Miglino, Ragozzino, & Bol, 2001). In this vector, the GFP gene is located at the 5′end of RNA 3, while the MP gene is expressed from a duplicated subgenomic promoter. The results obtained with this modified RNA 3 have revealed that the MP gene of AMV is functionally interchangeable with different MPs assigned to the 30K superfamily, including the MP of PNRSV (Sánchez-Navarro & Bol, 2001; Sánchez-Navarro et al., 2006, 2010), which allows the cell-to-cell and systemic transport of the corresponding chimera constructs. Heterologous MPs require the fusion at the C-terminus of the C-terminal 44 aa of

AMV MP that are responsible for interacting with the cognate CP (Sánchez-Navarro et al., 2006). The results have indicated that the MP of PNRSV permits the local and systemic transport of the hybrid AMV RNA 3 in P12 plants, a nonhost for PNRSV.

Further results using the AMV CP mutants defective in virion formation have revealed that the MP of PNRSV is also sufficiently competent to transport viral complexes other than virus particles to neighboring cells (Sánchez-Navarro et al., 2006), indicating that the MP of PNRSV, and probably the rest of the ilarvirus MPs, permits the transport of ribonucleoprotein complexes. This observation has also been reported for other MPs of the 30K family, whose viruses are assigned to the genera *Tobamovirus*, *Cucumovirus*, *Bromovirus*, or *Comovirus* (Sánchez-Navarro et al., 2006). The mutational analyses of PNRSV MP have demonstrated that the C-terminal 38 aa are dispensable for the functionality of the protein, although they are required for a compatible interaction with the cognate CP (Aparicio et al., 2010). In the same analysis, the authors have reported that the residues located between 242 and 245 aa are required to target the protein to the PD, and that they were critical for the functionality of the MP. However, such residues are not conserved among the remaining ilarvirus MPs. The results obtained with the MP of PNRSV, together with the previous analysis of the MPs assigned to the 30K family, have indicated an MP scheme in which the major part of the N-terminus of the protein is required to allow virus transport, while the C-terminal region contains the specific determinants involved in interaction with the CP. In this sense, functional MPs carrying C-terminal deletions have been described for CMV (Nagano, Okuno, Mise, & Furusawa, 1997), BMV and AMV (Sánchez-Navarro & Bol, 2001), TMV (Berna et al., 1991), and PNRSV (Aparicio et al., 2010). For CMV, the deletion of the dispensable C-terminal 33 aa of the MP induces an increase in RNA affinity, indicating that both effects, lack of the C-terminus and increasing RNA affinity, may be required to convert the protein into a nonspecific MP with the capacity to transport any viral RNA (Kim et al., 2004). With PNRSV, it has been proven that the dispensable C-terminal 38 aa of the MP interacts directly with the cognate CP, thus supporting the previous *in silico* analysis which determined that the PNRSV MP residues interacting with the CP are located at positions 253, 256, 257, and 261 (Codoñer et al., 2006). As for the remaining MPs of the 30K family that have been analyzed, it can be concluded that the C-terminal region of the MPs is associated directly or indirectly with the CP interaction and that it can probably be applied to the remaining ilarvirus MPs. In this sense, the alignment of ilarvirus MPs has revealed that the C-terminal portion of the protein is the most variable region, probably due to the specific interactions with the cognate CP, as previously proposed for the Bromoviridae family (Codoñer et al., 2005).

After the infection of neighboring cells, the next step during virus transport relies on the capacity of the virus to infect the distal plant parts throughout the vascular tissue. In the case of ilarviruses, the information about this process is very scarce. The use of the AMV system has proved that PNRSV MP permits the systemic transport of the hybrid AMV RNA 3 in P12 plants (Fajardo, Peiro, Pallas, & Sanchez-Navarro, 2013). The data obtained with AMV, but also with the closely related BMV and CMV, have suggested that virions are the critical virus component that is transported through the vascular tissue to the upper plant parts (Blackman, Boevink, Cruz, Palukaitis, & Oparka, 1998; Cooper, Schmitz, Rao, Beachy, & Dodds, 1996; Sánchez-Navarro & Bol, 2001; Van der Vossen, Neeleman, & Bol, 1994). However, the observation that the AMV MP mutant, lacking the CP-interacting C-terminal 44 aa, permits the systemic transport of viral RNA 3 (Sánchez-Navarro & Bol, 2001) suggests the idea that complexes, other than the virus particle, can be systemically transported. In this sense, it will be very interesting to determine whether the cell-to-cell competent MP of PNRSV that lacks the CP-interacting 38 aa (Aparicio et al., 2010) is able to support the systemic transport of hybrid AMV RNA 3.

The final virus cycle step corresponds to the infection of a new healthy plant by appropriate vectors. Seed and pollen transmission are common mechanisms of spread for approximately 20% of plant viruses, which assures their spread from generation to generation (Hull, 2002; Mink, 1993). Seed transmission can occur directly through embryo infection from the mother plant, as in the *Pea seed-borne mosaic virus* or indirectly via pollen or ovule during fertilization; for example, in the *Pea early browning virus* (Maule & Wang, 1996). Thus, the presence of viruses in pollen grains has been reported for PNRSV (Aparicio, Sanchez-Pina, et al., 1999; Greber, Klose, Milne, & Teakle, 1991), TSV (Walter, Kaiser, Klein, & Wyatt, 1992), the BlShV (Bristow & Martin, 1999), ApMV (Gottlieb & Berbee, 1973), PDV (Silva, Tereso, Nolasco, & Oliveira, 2003), and PMoV (Aramburu, Galipienso, Aparicio, Soler, & Lopez, 2010).

Transmission by seed and pollen has been reported for PNRSV (Amari, Burgos, Pallas, & Sanchez-Pina, 2007, 2009; Aparicio, Sanchez-Pina, et al., 1999), PDV (Cation, 1949; Mink & Aichele, 1984), ApMV (Gottlieb & Berbee, 1973), FCiLV (Spiegel, Martin, Leggett, Terborg, & Postman,

1993), LiRMV (Van der Meer, Huttinga, & Maat, 1976), SpLV (Bos, Huttinga, & Maat, 1980; Stefanac & Wrischer, 1983), *Asparagus virus 2* (Uyeda & Mink, 1981), HdMV (Thomas et al., 1983), and TSV (Walter, Wyau, & Kaiser, 1995). Although there is evidence for the presence of APLPV in the seed coats of seeds collected from APLPV-infected cherry and plum fruits (A. Myrta, unpublished data), transmission of the virus through seed has not yet been demonstrated. Walter et al. (1995) compared the RNA population of seed-transmitted TSV isolate Mel 40 with that of the infrequently seed-transmitted isolate Mel F and observed minor RNA species in the former. However, no unequivocal correlation with the transmission properties could be drawn. Interestingly, the mediation of thrips or other flower-visiting arthropods has been reported to be required for plant-to-plant transmission (Aramburu et al., 2010; Bristow & Martin, 1999; Greber, Teakle, & Mink, 1992; Johansen, Edwards, & Hampton, 1994; Klose et al., 1996; Mink, 1993; Walter et al., 1992). It has also been reported that PNRSV invades early pollen grains in nectarine and apricot by infecting the megaspore and generative cell of bicellular pollen grains (Amari et al., 2007; Aparicio, Sanchez-Pina, et al., 1999) trees. The mechanism by which PNRSV is transmitted through the fertilization process has been recently reviewed (Pallas et al., 2012).

7. HOST RESPONSES

A general characteristic of the host response to ilarvirus infection consists of an initial shock reaction, followed by plants initiating a recovery phase which, in some cases, may lead to a relatively normal appearance. For instance, the initial (shock) symptoms of PNRSV consist of chlorotic spots and rings that turn necrotic on recently developed leaves. Necrotic tissues later fall out, leaving leaves with a tattered or shot-hole appearance. After the shock phase, trees recover, even though the virus continues to replicate and reaches a high concentration annually (Uyemoto & Scott, 1992). Bristow and Martin (2002) have reported another example, in which the plants infected with the BlShV recovered after the shock reaction phase of the disease. This phenomenon very much resembles the natural host recovery from viral diseases, which has been described as a form of host resistance response to several plant virus infections (Matthews, 1991). Xin and Ding (2003) have reported the identification and characterization of a spontaneous mutant of TSV, which became defective in triggering recovery in tobacco plants. This mutant (TSVnr) harbors a single mutation located in the intergenic region of RNA 3 upstream of the mapped transcriptional start site of the CP mRNA. Intriguingly, the recovery induced by TSV was not accompanied by virus clearance and TSV consistently accumulated to significantly higher levels than TSVnr did, even though TSVnr-infected plants displayed severe symptoms throughout the course of infection. As stated by the authors, these findings indicate that the recovery of the host can be initiated by minimal genetic changes in a viral genome and may even occur in the absence of virus clearance. More recently, Jovel, Walker, and Sanfaçon (2007) demonstrated an association between the recovery phenomenon and the RNA silencing machinery. Remarkably, however, as observed for the TSV interaction, no reduction in virus titer was required for this process to occur.

Among the different levels of resistance in whole plants, hormones play a key role in regulating and/or mediating host responses. A large body of evidence indicates that salicylic acid (SA) signaling triggers resistance against biotrophic and hemibiotrophic pathogens (see Carr, Lewsey, & Palukaitis, 2010; Pallas & Garcia, 2011; Robert-Seilaniantz, Grant, & Jones, 2011, for a review). Noteworthily, it has been reported that gentisic acid (GA) (2,5-dihydroxybenzoic acid), a metabolic derivative of SA, accumulates in PNRSV-infected cucumber plants (Bellés et al., 2006). Time-course analyses have shown a correlation between the accumulation levels of GA and the development of the disease. In addition, both GA and PNRSV bring about an enhancement of peroxidase (POX) activity, whereas SA and PNRSV, but not GA, trigger the expression of defensive 28 kDa chitinase. These results have led the authors to propose that GA can act as an additional signal to SA for the activation of plant defenses in cucumber plants (Bellés et al., 2006). In another set of experiments, the same authors observed that PNRSV is able to activate the phenylpropanoid pathway (Belles et al., 2008) as a component of the response of cucumber to viruses by producing systemic infections, which thus extends previous results obtained in plants infected with fungi or bacteria.

As stated above, PNRSV is able to invade the immature apricot seed, including the embryo (Amari et al., 2009). Interestingly, PNRSV infection produces oxidative stress in apricot seeds, as indicated by increased lipid peroxidation. This increase in lipid peroxidation is paralleled with an imbalance in seed antioxidant enzymes. A significant decrease in both ascorbate–GSH cycle enzymes and POX activity has been observed in infected seeds, suggesting a reduced ability to eliminate H_2O_2 (Amari et al., 2007). Oxidative stress and the imbalance in the antioxidant systems from

PNRSV-infected apricot seeds resemble the hypersensitive response observed in some virus–host interactions. This defense mechanism would inactivate PNRSV during seed formation and/or the storage period, or even during seed germination. These results can explain the decrease in seed germination and the low transmission of PNRSV by seeds in apricot trees.

High-throughput technologies are providing a more global view of the host responses triggered by plant viruses (Postnikova & Nemchinov, 2012; Whitham et al., 2003). Unfortunately, ilarviruses have only very occasionally been the subjects of such studies. Together with two other stone-fruit viruses, Dardick (2007) included PNRSV in a comparative expression profiling of the systemically infected leaves of *Nicotiana benthamiana* plants. Very few genes have been observed to be differentially expressed, while most are repressed. The above author has suggested that this limited response is due to the mild symptoms caused by PNRSV in this experimental host.

In recent years, budding yeast has been used as a model to unravel the interactions of plant viruses with their hosts and has helped facilitate the identification of replication- and/or translation-associated factors that affect host–virus interactions, virus pathology, virus evolution, and host range (Nagy, 2008). *S. cerevisiae* cells transiently inhibit the initiation of protein synthesis under environmental stress to avoid the misfolding of proteins, which could compromise cell viability. Inhibition of translation is achieved by the phosphorylation of the alpha subunit of eukaryotic translation initiation factor-2 (eIF2a). The sole eIF2a kinase present in budding yeast is encoded by the general control nonrepressible-2 (GCN2) gene. Gcn2p seems to be a master regulator of gene expression in yeast and in response to various kinds of stresses. The fundamental nature of the general control nonrepressible (GCN) pathway is reflected by the fact that the Gcn2p function is conserved throughout evolution. Based on a yeast-based approach, Aparicio et al. (2011) have found that the overexpression of PNRSV MP causes a severe growth defect in yeast cells. cDNA microarrays analyses have been carried out to characterize the growth interference phenotype at the molecular level. They have reported the induction of genes related to amino acid deprivation, suggesting that the expression of PNRSV MP activates the GCN pathway in yeast cells. Accordingly, PNRSV MP triggers the activation of the Gcn2p kinase, judged by the increased eIF2a phosphorylation. Overall, these findings uncover a previously uncharacterized function for an MP viral protein and point out that Gcn2p kinases are candidates of susceptibility factors for plant viral infections.

Recently, a remarkable synergistic effect has been observed at the transcriptomic level in peach fruits doubly infected by PNRSV and *Peach latent mosaic viroid* (PLMVd) (Herranz et al., 2013). Microarray slides containing 4261 unigenes obtained from peach mesocarp tissues have been used to study differentially expressed genes upon single PLMVd or PNRSV infection or doubly infected fruits. The number of statistically significant gene expression changes is relatively low in single infections. In contrast, doubly infected fruits present a large number of differentially regulated genes. Among them, down-regulated genes are prevalent. Although no synergistic effect has been observed at the biological level, these results demonstrate that mixed infections, which frequently occur under field conditions, result in a more complex transcriptional response than that observed in single infections.

8. SYNERGY AMONG ILARVIRUSES

Synergy between viruses has been reported for many viral species (Hull, 2002). In a few cases, these are ilarviruses (Nemeth, 1986). The most reported synergism involving ilarviruses concerns dual infection of peach trees with PDV and PNRSV and results in a distinct dwarfing disease referred to as peach stunt (Schmitt, Williams, & Nyland, 1977). The synergistic effects of the two viruses result in poor tree growth (stunting) and reduced fruit yield (Scott, Zimmerman, Yilmaz, Bachman, & Zehr, 2001) with the extent of the changes being dependent on the isolates of the two viruses that are involved. Examination at the molecular level of the much studied "classical" interaction between and *Potato virus X* and members of the *Potyviridae* (*Potato virus Y*, *Tobacco etch virus*, and *Tobacco vein mottling virus*) indicated a number of changes. The concentration of infectious PVX increased up to 10-fold in coinfected cells compared to the concentration observed in cells infected by PVX alone. The concentration of minus-strand RNA increased even more altering the ratio of plus to minus-strand RNA in an infection (Vance, Berger, Carrington, Hunt, & Shi, 1995). Many of these changes involve the interaction with the HC pro protein of the potyvirus. However, a major difference between the classical PVX/PVY synergism, and that noted for ilarviruses, is that the former involves viruses from two separate families with distinctly different genomic structures and replicative strategies. The PDV/PNRSV synergism involves two viruses from the same family albeit members of two different subgroups. Notwithstanding this major difference, preliminary studies of the PDV/PNRSV interaction at the molecular level indicate effects on the concentration of the RNA 3 of the two viruses. When PNRSV was present in a dual infection with PDV, the concentration of the (+) strand of the RNA 3 of PDV was reduced by up to 17-fold depending on the scion/rootstock combination being examined. The presence of PDV had no effect on the concentration of the (+) strand of the RNA 3 of PNRSV (Scott et al., 2001).

9. SOME CONCLUSIONS AND MANY QUESTIONS

Ilarviruses are distributed worldwide and affect a large number of agronomically relevant crop species including fruit trees, vegetables, and ornamentals. In the 1970s, it was recognized that ilarviruses, together with the closely related AMV, were unique among plant viruses due to their requirement for infection of the presence of a few molecules of the CP in the inoculum. This peculiarity, together with the characteristic lability of the virus particles in tissue extracts delayed the development of knowledge on the molecular biology of ilarviruses in comparison with other less economically relevant viruses. During the decade of the 1990s, a large number of ilarviruses were molecularly characterized and this allowed initial approaches to understanding several steps in the viral life cycle of this group of viruses to be made. However, in most cases, these achievements were in the shadow of the research completed using AMV. The enormous progress made with the AMV model meant that some properties shared with ilarviruses, such as genome expression, viral replication, and genome activation, were addressed bearing in mind in the need for confirmatory research. Nowadays, however, several ilarviruses are the main subject of molecular biological approaches and significant progress has been made in knowing how these viruses are transmitted by seed and pollen and how they move through the plant. However, these areas are still in need of considerable research. For instance, how will the expected increase in global temperatures influence pollen transmission taking into account that it is mediated by thrips or honeybees? How does the virus move through the pollen tube or through the ovary? The demonstration that an ilarvirus infecting tomato (PMoV) is transmitted by pollen can help to unravel these mechanisms. Another characteristic not shared between some ilarviruses and AMV is the presence in the former of an extra ORF that codes for a 2b protein. This protein is not phylogenetically related to the CMV 2b and a study of the evolutionary advantage for those ilarviruses harboring this protein, if any, should be a part of the future research agenda.

Ilarviruses can also be the pathogens of choice to study the molecular mechanisms of the natural host recovery phenomenon since ilarviruses are involved in several examples of apparent recovery occurring in nature. The identification of recovery phenotypes in transgenic pathogen-derived resistance provided the first experimental evidence for a link between RNA silencing and an antiviral defense mechanism. We have recently observed that PNRSV generates a very limited number of small interfering RNAs in infected plants. It is a challenge to demonstrate whether or not this observation, together with the apparent absence of a silencing suppressor encoded by some ilarviruses, has any relationship with the frequent host recovery shown by this group of viruses.

Unfortunately, Arabidopsis is a nonhost for any of the so far characterized ilarviruses, and thus, a genetic approach for studying pathogenicity and/or resistance determinants in this group of viruses is not viable. Although, the recent report of an ilarvirus (Raphanus latent virus) infecting a crucifer may alter the status of Arabidopsis as a nonhost. However, even without this change, the availability of the complete genomes of economically relevant crop species (peach, apricot, tomato, apple, grapevine, etc.) that are hosts for ilarviruses put this group of viruses in a very good position to be used in expanding the knowledge of how viruses modify host metabolism to accomplish their life cycle. High-throughput technologies (e.g., Postnikova & Nemchinov, 2012) might also allow circumvention of the lack of a reverse genetic system for this group of viruses and facilitate the unraveling of the interactome of ilarvirus proteins with host factors. Such an understanding is an essential requisite for designing specific control strategies against these viruses and undoubtedly as more ilarviruses are reported, we will attain a better understanding of the intricacies of the molecular processes used by ilarviruses that fall outside the PNRSV/AMV model.

ACKNOWLEDGMENTS

M. C. H. and F. A. were recipients of a contract from the Juan de la Cierva and the Ramon y Cajal (RYC-2010-06169) programs of the Ministerio de Ciencia e Innovación of Spain. Research in the V. P. lab is funded by grants from the DGICYT agency (grant BIO2011-25018) and from the Generalitat Valenciana (grant Prometeo GV2011/003). Research in the S. W. S. lab is funded by Clemson University Experiment Station, the Peach Council of South Carolina, and USDA-Aphis National Clean Plant Network.

REFERENCES

Abou-Jawdah, Y., Sobh, H., Cordahi, N., Kawtharani, H., Nemer, G., Maxwell, D. P., et al. (2004). Immunodiagnosis of Prune dwarf virus using antiserum produced to its recombinant coat protein. *Journal of Virological Methods*, *121*, 31–38.

Adkins, S., & Kao, C. C. (1998). Subgenomic RNA promoters dictate the mode of recognition by bromoviral RNA-dependent RNA polymerases. *Virology*, *252*, 1–8.

Ahlquist, P., Noueiry, A. O., Lee, W. M., Kushner, D. B., & Dye, B. T. (2003). Host factors in positive-strand RNA virus genome replication. *Journal of Virology*, 77, 8181–8186.

Ahola, T., den Boon, J. A., & Ahlquist, P. (2000). Helicase and capping enzyme active site mutations in brome mosaic virus protein 1a cause defects in template recruitment, negative-strand RNA synthesis, and viral RNA capping. *Journal of Virology, 74*, 8803–8811.

Amari, K., Burgos, L., Pallas, V., & Sanchez-Pina, M. A. (2007). Prunus necrotic ringspot virus early invasion and its effects on apricot pollen grain performance. *Phytopathology, 97*, 892–899.

Amari, K., Burgos, L., Pallas, V., & Sanchez-Pina, M. A. (2009). Vertical transmission of Prunus necrotic ringspot virus: Hitch-hiking from gametes to seedling. *The Journal of General Virology, 90*, 1767–1774.

Ansel-McKinney, P., & Gehrke, L. (1998). RNA determinants of a specific RNA-coat protein peptide interaction in alfalfa mosaic virus: Conservation of homologus features in Ilarvirus RNAs. *Journal of Molecular Biology, 278*, 767–785.

Ansel-McKinney, P., Scott, S. W., Swanson, M., Ge, X., & Gehrke, L. (1996). A plant viral coat protein RNA binding consensus sequence contains a crucial arginine. *The EMBO Journal, 15*, 5077–5084.

Aparicio, F., Aparicio-Sanchis, R., Gadea, J., Sánchez-Navarro, J. A., Pallás, V., & Murguía, J. R. (2011). A plant virus movement protein regulates the Gcn2p kinase in budding yeast. *PLoS One, 6*(11), e27409.

Aparício, F., Aramburu, J., Soler, S., Galipienso, L., Nuez, F., Pallas, V., et al. (2009). Immunodiagnosis of Parietaria mottle virus in tomato crops using a polyclonal antiserum against its coat protein expressed in a bacterial system. *Journal of Phytopathology, 157*, 511–513.

Aparicio, F., Myrta, A., Di Terlizzi, B., & Pallas, V. (1999). Molecular variability among isolates of Prunus necrotic ringspot virus from different *Prunus* spp. *Phytopathology, 89*, 991–999.

Aparicio, F., & Pallás, V. (2002). Molecular variability analysis of the RNA 3 of fifteen isolates of Prunus necrotic ringspot virus sheds light on the minimal requirements for the synthesis of the subgenomic RNA. *Virus Genes, 25*, 75–84.

Aparicio, F., Pallas, V., & Sanchez-Navarro, J. (2010). Implication of the C terminus of the Prunus necrotic ringspot virus movement protein in cell-to-cell transport and in its interaction with the coat protein. *The Journal of General Virology, 91*, 1865–1870.

Aparicio, F., Sanchez-Navarro, J. A., Olsthoorn, R. C., Pallas, V., & Bol, J. F. (2001). Recognition of cis-acting sequences in RNA 3 of Prunus necrotic ringspot virus by the replicase of Alfalfa mosaic virus. *The Journal of General Virology, 82*, 947–951.

Aparicio, F., Sánchez-Navarro, J. A., & Pallás, V. (2006). In vitro and in vivo mapping of the *Prunus necrotic ringspot virus* coat protein C-terminal dimerization domain by bimolecular fluorescence complementation. *The Journal of General Virology, 87*, 1745–1750.

Aparicio, F., Sanchez-Pina, M. A., Sanchez-Navarro, J. A., & Pallas, V. (1999). Location of prunus necrotic ringspot ilarvirus within pollen grains of infected nectarine trees: Evidence from RT-PCR, dot-blot and in situ hybridisation. *European Journal of Plant Pathology, 105*, 623–627.

Aparicio, F., Vilar, M., Perez-Paya, E., & Pallas, V. (2003). The coat protein of prunus necrotic ringspot virus specifically binds to and regulates the conformation of its genomic RNA. *Virology, 313*, 213–223.

Aramburu, J., Galipienso, L., Aparicio, F., Soler, S., & Lopez, C. (2010). Mode of transmission of Parietaria mottle virus. *Journal of Plant Pathology, 92*, 679–684.

Baer, M., Houser, F., Loesch-Fries, L. S., & Gehrke, L. (1994). Specific RNA binding by amino-terminal peptides of alfalfa mosaic virus coat protein. *The EMBO Journal, 13*, 727–735.

Batuman, O., Miyao, G., Kuo, Y.-W., Chen, L.-F., Davis, R. M., & Gilbertson, R. L. (2009). An outbreak of a necrosis disease of tomato in California in 2008 was caused by a new ilarvirus species related to *Parietaria mottle virus*. *Plant Disease, 93*, 546.

Bellés, J. M., Garro, R., Pallás, V., Fayos, J., Rodrigo, I., & Conejero, V. (2006). Accumulation of gentisic acid as associated with systemic infections but not with the hypersensitive response in plant–pathogen interactions. *Planta, 223*, 500–511.

Belles, J. M., Lopez-Gresa, M. P., Fayos, J., Pallás, V., Rodrigo, I., & Conejero, V. (2008). Induction of cinnamate 4-hydroxylase and phenylpropanoids in virus-infected cucumber and melon plants. *Plant Science, 174*, 524–533.

Berg, J. M. (1986). Potential metal binding domains in nucleic acid binding proteins. *Science, 232*, 485–486.

Berna, A., Gafny, R., Wolf, S., Lucas, W. J., Holt, C. A., & Beachy, R. N. (1991). The TMV movement protein: Role of the C-terminal 73 amino acids in subcellular localization and function. *Virology, 182*, 682–689.

Bhat, A. I., Jain, R. K., Chaudhary, V., Krishna Reddy, M., Ramiah, M., Chattannavar, S. N., et al. (2002). Sequence conservation in the coat protein gene of Tobacco streak virus isolates causing necrosis in cotton, mungbean, sunflower and sun-hemp in India. *Indian Journal of Biotechnology, 1*, 350–356.

Blackman, L. M., Boevink, P., Cruz, S. S., Palukaitis, P., & Oparka, K. J. (1998). The movement protein of cucumber mosaic virus traffics into sieve elements in minor veins of Nicotiana clevelandii. *The Plant Cell, 10*, 525–537.

Boari, A., Boscia, D., Yurtmen, M., Potere, O., Torturo, C., & Savino, V. (1997). Production of monoclonal antibodies to prune dwarf ilarvirus and their use in the serological characterization of almond virus isolates. *EPPO Bulletin, 27*, 555–556.

Bol, J. F. (2005). Replication of alfamo- and ilarviruses: Role of the coat protein. *Annual Review of Phytopathology, 43*, 39–62.

Bol, J. F., van Vloten-Doting, L., & Jaspars, E. M. J. (1971). A functional equivalence of top component a RNA and coat protein in the initiation of infection by alfalfa mosaic virus. *Virology, 46*, 73–85.

Bos, L., Huttinga, H., & Maat, D. Z. (1980). Spinach latent virus, a new ilarvirus seed-borne in Spinacia oleracea. *The Netherlands Journal of Plant Pathology, 86*, 79–98.

Boulila, M. (2009). Recombination structure and genetic relatedness among members of the family *Bromoviridae* based on their RNAs 1 and 2 sequence analyses. *Virus Genes, 38*, 435–444.

Brill, L. M., Nunn, R. S., Kahn, T. W., Yeager, M., & Beachy, R. N. (2000). Recombinant tobacco mosaic virus movement protein is an RNA-binding, alpha-helical membrane protein. *Proceedings of the National Academy of Science United States of America, 97*, 7112–7117.

Bristow, P. R., & Martin, R. R. (1999). Transmission and the role of honeybees in field spread of blueberry shock ilarvirus, a pollen-borne virus of highbush blueberry. *Phytopathology, 89*, 124–130.

Bristow, P. R., & Martin, R. R. (2002). Recovery of plants infected with blueberry shock ilarvirus (BlShV). *Acta Horticulturae, 574*, 85–89.

Bruenn, J. A. (2003). A structural and prima ry sequence comparison of the viral RNA-dependent RNA polymerases. *Nucleic Acids Research, 31*, 1821–1829.

Bujarski, J., Figlerowicz, M., Gallitelli, D., Roossinck, M. J., & Scott, S. W. (2012). Family *Bromoviridae*. In: *Virus taxonomy. Ninth report of the international committee on taxonomy of viruses* (pp. 965–976), San Diego, CA: Elsevier Academic Press.

Candresse, T., Morch, M. D., & Dunez, J. (1990). Multiple alignment and hierarchical clustering of conserved amino acid sequences in the replication-associated proteins of plant RNA viruses. *Research in Virology, 141*, 315–329.

Carr, J. P., Lewsey, M. G., & Palukaitis, P. (2010). Signaling in induced resistance. *Advances in Virus Research, 76*, 57–121.

Carrington, J. C., Kasschau, K. D., Mahajan, S. K., & Schaad, M. C. (1996). Cell-to-cell and long-distance transport of viruses in plants. *The Plant Cell, 8*, 1669–1681.

Caruthers, J. M., & McKay, D. B. (2002). Helicase structure and mechanism. *Current Opinion in Structural Biology, 12*, 123–133.

Cation, D. (1949). Transmission of viruses through cherry seeds. *Phytopathology, 39*, 4.

Chen, S. C., Gultyaev, A. P., Pleij, C. W. A., & Olsthoorn, R. C. L. (2009). A secondary structure model for the 3'-untranslated region of ilarvirus RNAs. In Z. Feng & M. Long (Eds.), *Viral genomes: Diversity, properties and parameters*. Hauppage, NY: Nova Science Publishers.

Chen, S. C., & Olsthoorn, R. C. L. (2010). In vitro and in vivo studies of the RNA conformational switch in Alfalfa mosaic virus. *Journal of Virology, 84*, 1423–1429.

Citovsky, V., Wong, M. L., Shaw, A. L., Venkataram, P. B. V., & Zambryski, P. (1992). Visualisation and characterisation of Tobacco mosaic virus movement protein binding to single-stranded nucleic acids. *The Plant Cell, 4*, 397–411.

Codoñer, F. M., Cuevas, J. M., Sanchez-Navarro, J. A., Pallas, V., & Elena, S. F. (2005). Molecular evolution of the plant virus family Bromoviridae based on RNA3-encoded proteins. *Journal of Molecular Evolution, 61*, 697–705.

Codoñer, F. M., & Elena, S. F. (2006). Evolutionary relationships among members of the Bromoviridae deduced from whole proteome analysis. *Archives of Virology, 151*, 299–307.

Codoñer, F. M., & Elena, S. F. (2008). The promiscuous evolutionary history of the *Bromoviridae*. *The Journal of General Virology, 89*, 1739–1747.

Codoñer, F. M., Fares, M. A., & Elena, S. F. (2006). Adaptive covariation between the coat and movement proteins of Prunus necrotic ringspot virus. *Journal of Virology, 80*, 5833–5840.

Cooper, B., Schmitz, I., Rao, A. L. N., Beachy, R. N., & Dodds, J. A. (1996). Cell-to-cell transport of movement-defective cucumber mosaic and tobacco mosaic viruses in transgenic plants expressing heterologous movement protein genes. *Virology, 216*, 208–213.

Dardick, C. (2007). Comparative expression profiling of *Nicotiana benthamiana* leaves systemically infected with three fruit tree viruses. *Molecular Plant-Microbe Interactions, 20*, 1004–1017.

Daros, J. A., & Carrington, J. C. (1997). RNA binding activity of NIa proteinase of tobacco etch potyvirus. *Virology, 237*, 327–336.

Dillingham, M. S., Soultanas, P., Wiley, P., Webb, M. R., & Wigley, D. B. (2001). Defining the roles of individual residues in the single-stranded DNA binding site of PcrA helicase. *Proceedings of the National Academy of Science United States of America, 98*, 8381–8387.

Ding, S. W., Li, W. X., & Symons, R. H. (1995). A novel naturally occurring hybrid gene encoded by a plant RNA virus facilitates long distance virus movement. *The EMBO Journal, 14*, 5762–5772.

Di Terlizzi, B., Skrzeczkowski, L. J., Mink, G. I., Scott, S. W., & Zimmerman, M. T. (2001). The RNA 5 of *Prunus necrotic ringspot virus* is a biologically inactive copy of the 3-UTR of the genomic RNA 3. *Archives of Virology, 146*, 825–833.

Door Peeters, A. (2009). *Characterisation of a potential suppressor of RNA silencing of Parietaria mottle virus (PMoV)*. Master thesis. Universidad Politecnica de Valencia.

Edwardson, J. R., & Purciful, D. E. (1974). Relationship of Datura quercina and tobacco streak viruses. *Phytopathology, 64*, 1322–1324.

Fajardo, T. V., Peiro, A., Pallas, V., & Sanchez-Navarro, J. (2013). Systemic transport of Alfalfa mosaic virus can be mediated by the movement proteins of several viruses assigned to five genera of the 30K family. *The Journal of General Virology, 94*, 677–681.

Fenner, F. (1976). Classification and nomenclature of viruses: Second report of the international committee on taxonomy of viruses. *Intervirology, 7*, 1.

Fernandez-Calviño, L., Faulkner, C., & Maule, A. (2011). Plasmodesmata as active conduits for virus cell-to-cell movement. In C. Caranta, M. A. Aranda, M. Tepfer, & J. J. Lopez-Moya (Eds.), *Advances in plant virology* (p. 470). Norwich, UK: Caister Academic Press.

Fiore, N., Fajardo, T. V., Prodan, S., Herranz, M. C., Aparicio, F., Montealegre, J., et al. (2008). Genetic diversity of the movement and coat protein genes of South American isolates of Prunus necrotic ringspot virus. *Archives of Virology, 153*, 909–919.

Fonseca, F., Neto, J. D., Martins, V., & Nolasco, G. (2005). Genomic variability of Prune dwarf virus as affected by agricultural practice. *Archives of Virology*, *150*, 1607–1619.

Francki, R. I. B., Milne, R. G., & Hatta, T. (1985). Ilarvirus group. *Atlas of plant viruses*, Vol. 2, (pp. 81–91). Boca Raton, FL: CRC Press Inc.

Fujita, M., Mise, K., Kajiura, Y., Dohi, K., & Furusawa, I. (1998). Nucleic acid-binding properties and subcellular localization of the 3a protein of brome mosaic bromovirus. *The Journal of General Virology*, *79*, 1273–1280.

Fulton, R. W. (1968). Relationships among the ringspot viruses of *Prunus*. *Tagungsberichte Nr. 97 Deutsche Demokratische Republik Deutsche Akademie Der Landwirtschaftswissenschaten Zu Berlin*, *97*, 123–128.

Galipienso, L., Herranz, M. C., Lopez, C., Pallas, V., & Aramburu, J. (2008). Sequence analysis within the RNA 3 of seven Spanish tomato isolates of Parietaria mottle virus (PMoV-T) reveals important structural differences with the Parietaria isolates (PMoV). *European Journal of Plant Pathology*, *120*, 125–135.

Ge, X., Scott, S. W., & Zimmerman, M. T. (1997). The complete sequence of the genomic RNAs of spinach latent virus. *Archives of Virology*, *142*, 1213–1226.

Gerola, F. M., Bassi, M., & Belli, G. (1969). An electron microscope study of different plants infected with grapevine fanleaf virus. *Giornale Botanico Italiano*, *103*, 271–290.

Girgis, S. M., Bem, F. P., Dovas, C. I., Sclavounos, A., Avgelis, A. D., Tsagris, M., et al. (2009). Characterisation of a novel ilarvirus causing grapevine angular mosaic disease. *European Journal of Plant Pathology*, *125*, 203–221.

Glasa, M., Betinová, E., Kúdela, O., & Subr, Z. (2002). Biological and molecular characterization of Prunus necrotic ringspot virus isolates and possible approaches to their phylogenetic typing. *The Annals of Applied Biology*, *140*, 279–283.

Goldbach, R., Le Gall, O., & Wellinck, J. (1991). Alpha-like viruses in plants. *Seminars in Virology*, *2*, 19–25.

Gomez, G., & Pallas, V. (2001). Identification of an in vitro ribonucleoprotein complex between a viroid RNA and a phloem protein from cucumber plants. *Molecular Plant-Microbe Interactions*, *14*, 910–913.

Gonsalves, D., & Fulton, R. W. (1977). Activation of prunus necrotic ringspot virus and rose mosaic virus by RNA 4 components of some ilarviruses. *Virology*, *81*, 398–407.

Gonsalves, D., & Garnsey, S. M. (1975). Infectivity of heterologous RNA-protein mixtures from alfalfa mosaic, citrus leaf rugose, citrus variegation, and tobacco streak viruses. *Virology*, *67*, 319–326.

Gorbalenya, A. E., & Koonin, E. V. (1993). Helicases- amino-acid-sequence comparisons and structure-function-relationships. *Current Opinion in Structural Biology*, *3*, 419–429.

Gorbalenya, A. E., Pringle, F. M., Zeddam, J. L., Luke, B. T., Cameron, C. E., Kalmakoff, J., et al. (2002). The palm subdomain-based active site is internally permuted in viral RNA-dependent RNA polymerases of an ancient lineage. *Journal of Molecular Biology*, *324*, 47–62.

Gottlieb, A. R., & Berbee, J. (1973). Detection of apple mosaic virus on pistils and pollens of white birch and its seed transmission in cowpea. In: *Proceedings of the second international congress plant pathology* (p. 0929).

Gramstat, A., Courtpozanis, A., & Rohde, W. (1990). The 12 kDa protein of potato virus M displays properties of a nucleic acid-binding regulatory protein. *FEBS Letters*, *276*, 34–38.

Greber, R. S., Klose, M. J., Milne, J. R., & Teakle, D. S. (1991). Transmission of Prunus necrotic ringspot virus using plum pollen and thrips. *The Annals of Applied Biology*, *118*, 589–593.

Greber, R. S., Teakle, D. S., & Mink, G. I. (1992). Thrips-facilitated transmission of Prune dwarf and Prunus necrotic ringspot viruses from cherry pollen to cucumber. *Plant Disease*, *76*, 1039–1041.

Gu, B., Liu, C., Lin-Goerke, J., Maley, D. R., Gutshall, L. L., Feltenberger, C. A., et al. (2000). The RNA helicase and nucleotide triphosphatase activities of the bovine viral diarrhea virus NS3 protein are essential for viral replication. *Journal of Virology, 74*, 1794–1800.

Guo, H.-S., & Ding, S. W. (2002). A viral protein inhibits the long-range signaling of activity of the gene silencing signal. *The EMBO Journal, 21*, 398–407.

Guogas, L. M., Filman, D. J., Hogle, J. M., & Gehrke, L. (2004). Cofolding organizes *alfalfa mosaic virus* RNA and coat protein for replication. *Science, 306*, 2108–2111.

Halk, E. L., & Fulton, R. W. (1978). Stabilization and particle morphology of prune dwarf virus. *Virology, 91*, 434–443.

Halk, E. L., Hsu, H. T., Aebig, J., & Franke, J. (1984). Production of monoclonal antibodies against three ilarviruses and alfalfa mosaic virus and their use in serotyping. *Phytopathology, 74*, 367–372.

Hammond, R. W. (2003). Phylogeny of isolates of Prunus necrotic ringspot virus from the Ilarvirus ring test and identification of group-specific features. *Archives of Virology, 148*, 1195–1210.

Herranz, M. C., Al, Rwahnih M., Sánchez-Navarro, J. A., Elena, S. F., Choueiri, E., Myrta, A., et al. (2008). Low genetic variability in the coat and movement proteins of American plum line pattern virus isolates from different geographic origins. *Archives of Virology, 153*, 367–373.

Herranz, M.C., Niehl, A. Rosales, M. Fiore, N., Zamorano, A. Granell and Pallas, V. (2013). A remarkable synergistic effect at the transcriptomic level in peach fruits doubly infected by Prunus necrotic ringspot virus and Peach latent mosaic viroid. *Virology Journal, 10*, 164.

Herranz, M. C., & Pallas, V. (2004). RNA-binding properties and mapping of the RNA-binding domain from the movement protein of Prunus necrotic ringspot virus. *The Journal of General Virology, 85*, 761–768.

Herranz, M. C., Pallas, V., & Aparicio, F. (2012). Multifunctional roles for the N-terminal basic motif of *Alfalfa mosaic virus* coat protein: Nucleolar/cytoplasmic shuttling, modulation of RNA-binding activity and virion formation. *Molecular Plant-Microbe Interactions, 25*, 1093–1103.

Herranz, M. C., Sanchez-Navarro, J. A., Sauri, A., Mingarro, I., & Pallas, V. (2005). Mutational analysis of the RNA-binding domain of the Prunus necrotic ringspot virus (PNRSV) movement protein reveals its requirement for cell-to-cell movement. *Virology, 339*, 31–41.

Houser-Scott, F., Baer, M. L., Liem, K. F., Cai, J. M., & Gehrke, L. (1994). Nucleotide sequence and structural determinants of specific binding of coat protein or coat protein peptides to the 3' untranslated region of alfalfa mosaic virus RNA 4. *Journal of Virology, 68*, 2194–2205.

Houwing, C. J., & Jaspars, E. M. J. (1980). Preferential binding of 3'-terminal fragments of alfalfa mosaic virus RNA 4 to virions. *Biochemistry, 19*, 5261–5264.

Hull, R. (2002). *Matthews' plant virology*, (Vol. 4). New York: Academic Press.

Ibrahim, A., Hutchens, H. M., Berg, R. H., & Loesch-Fries, L. S. (2012). Alfalfa mosaic virus replicase proteins, P1 and P2, localize to the tonoplast in the presence of virus RNA. *Virology, 433*, 449–461.

Imed, A., Boscia, D., Boari, A., Saldarelli, P., Digiaro, M., & Savino, V. (1997). Comparison of apple mosaic virus isolates from prunus trees and production of specific monoclonal antibodies. *EPPO Bulletin, 27*, 563–564.

James, D., Varga, A., Leippi, L., Godkin, S., & Masters, C. (2010). Sequence analysis of RNA 2 and RNA 3 of lilac leaf chlorosis virus: A putative new member of the genus *Ilarvirus*. *Archives of Virology, 155*, 993–998.

Jaspars, E. M. J. (1988). A core promoter hairpin is essential for subgenomic RNA synthesis in alfalfa mosaic alfamovirus and is conserved on other *Bromoviridae*. *Virus Genes, 17*, 233–242.

Johansen, E., Edwards, M. C., & Hampton, R. O. (1994). Seed transmission of viruses—Current perspectives. *Annual Review of Phytopathology, 32*, 363–386.

Johnson, J. (1936). Tobacco streak, a virus disease. *Phytopathology, 26*, 285–292.

Jones, A. T. (1985). Serological relationship between hydrangea mosaic (HyMV) and Elm mottle (EmotV) viruses. In: *Annual report of Scottish Horticultural Crops Research Institute for 1984* (p. 190).

Jones, A. T., & Mayo, M. A. (1973). Purification and properties of Elm mottle virus. *The Annals of Applied Biology, 75*, 347–357.

Jovel, J., Walker, M., & Sanfaçon, H. (2007). Recovery of *Nicotiana benthamiana* plants from a necrotic response induced by a nepovirus is associated with RN silencing but not with reduced virus titer. *Journal of Virology, 81*, 12285–12297.

Kim, S. H., Kalinina, N. O., Andreev, I., Ryabov, E. V., Fitzgerald, A. G., Taliansky, M. E., et al. (2004). The C-terminal 33 amino acids of the cucumber mosaic virus 3a protein affect virus movement, RNA binding and inhibition of infection and translation. *The Journal of General Virology, 85*, 221–230.

King, A. M. Q., Adams, M. J., Carstens, E. B., & Lefkowitz, E. J. (2012). In *Ninth report of the international committee on taxonomy of viruses*, London: Elsevier.

Klose, M. J., Sdoodee, R., Teakle, D. S., Milne, J. R., Greber, R. S., & Walter, G. H. (1996). Transmission of three strains of tobacco streak ilarvirus by different thrips species using virus-infected pollen. *Journal of Phytopathology, 144*, 281–284.

Koonin, E. V. (1991). The phylogeny of RNA-dependent RNA polymerases of positive-strand RNA viruses. *The Journal of General Virology, 72*(Pt. 9), 2197–2206.

Lakshmi, V., Hallan, V., Ram, R., Ahmed, N., Zaidi, A., & Varma, A. (2011). Diversity of Apple mosaic virus isolates in India based on coat protein and movement protein genes. *Indian Journal of Virology, 22*, 44–49.

Lehoczky, J., Boscia, D., Martelli, G. P., Burgyán, J., Castellano, M. A., Beczner, L., et al. (1987). Occurrence of the line pattern hitherto unknown virus disease of grapevine in Hungary (in Hungarian). *Horticulture, 19*, 61–79.

Li, Q. B., & Palukaitis, P. (1996). Comparison of the nucleic acid- and NTP-binding properties of the movement protein of cucumber mosaic cucumovirus and tobacco mosaic tobamovirus. *Virology, 216*, 71–79.

Li, Y. I., Shih, T. W., Hsu, Y. H., Han, Y. T., Huang, Y. L., & Meng, M. (2001). The helicase-like domain of plant potexvirus replicase participates in formation of RNA 5' cap structure by exhibiting RNA 5'-triphosphatase activity. *Journal of Virology, 75*, 12114–12120.

Lucas, W. J. (2006). Plant viral movement proteins: Agents for cell-to-cell trafficking of viral genomes. *Virology, 344*, 169–184.

Lucy, A. P., Guo, H. S., Li, W. X., & Ding, S. W. (2000). Suppression of post-transcriptional gene silencing by a plant viral protein localized in the nucleus. *The EMBO Journal, 19*, 1672–1680.

MacFarlane, S. A., & McDavin, W. J. (2009). Genome activation by raspberry bushy dwarf virus coat protein. *The Journal of General Virology, 90*, 747–753.

Maliogka, V. I., Dovas, C. I., & Katis, N. I. (2007). Demarcation of ilarviruses based on the phylogeny of RNA2-encoded RdRp and generic ramped annealing RT-PCR. *Archives of Virology, 152*, 1687–1698.

Marcos, J. F., Vilar, M., Perez-Paya, E., & Pallas, V. (1999). In vivo detection, RNA-binding properties and characterization of the RNA-binding domain of the p7 putative movement protein from carnation mottle carmovirus (CarMV). *Virology, 255*, 354–365.

Maroon-Lango, C. J., Aebig, J., & Hammond, J. (2006). Molecular and biological characterization of a novel ilarvirus in bacopa. *Phytopathology, 96*, S73.

Martinez-Gil, L., Sanchez-Navarro, J. A., Cruz, A., Pallas, V., Perez-Gil, J., & Mingarro, I. (2009). Plant virus cell-to-cell movement is not dependent on the transmembrane disposition of its movement protein. *Journal of Virology, 83*, 5535–5543.

Matthews, R. E. F. (1991). *Plant virology*. New York: Academic Press.
Maule, A. J., & Wang, D. W. (1996). Seed transmission of plant viruses: A lesson in biological complexity. *Trends in Microbiology*, *4*, 153–158.
Melcher, U. (2000). The '30K' superfamily of viral movement proteins. *The Journal of General Virology*, *81*, 257–266.
Menzel, W., Hamacher, J., Weissbrodt, S., & Winter, S. (2012). Complete nucleotide sequence of the RNA 3 of Bacopa chlorosis virus and production of polyclonal antibodies to a recombinant coat protein. *Journal of Phytopathology*, *160*, 163–165.
Mink, G. I. (1993). Pollen-transmitted and seed-transmitted viruses and viroids. *Annual Review of Phytopathology*, *31*, 375–402.
Mink, G. I., & Aichele, M. D. (1984). Detection of prunus necrotic ringspot and prune dwarf viruses in prunus seed and seedlings by enzyme-linked immunosorbent-assay. *Plant Disease*, *68*, 378–381.
Murphy, F. A., Fauquet, C. M., Bishop, D. H. L., Ghabrial, S. A., Jarvis, A. W., Martelli, G. P., et al. (1995). Virus taxonomy: Classification and nomenclature of viruses. In *Sixth report of the international committee on taxonomy of viruses* (pp. 450–457), Wein, NY: Springer, Arch. Virol. [Suppl. 10].
Nagano, H., Okuno, T., Mise, K., & Furusawa, I. (1997). Deletion of the C-terminal 33 amino acids of cucumber mosaic virus movement protein enables a chimeric brome mosaic virus to move from cell to cell. *Journal of Virology*, *71*, 2270–2276.
Nagy, P. D. (2008). Yeast as a model host to explore plant virus-host interactions. *Annual Review of Phytopathology*, *46*, 217–242.
Nemeth, M. (1986). *Virus, mycoplasma and Rickettsia diseases of fruit trees*. Dordrecht: Martinus NijHoff, 841 pp.
Oliver, J. E., Freer, J., Andersen, R. L., Cox, K. D., Robinson, T. L., & Fuchs, M. (2009). Genetic diversity of Prunus necrotic ringspot virus isolates within a cherry orchard in New York. *Plant Disease*, *93*, 599–606.
Olsthoorn, R. C. L., Mertens, S., Brederode, T. B., & Bol, J. F. (1999). A conformational switch at the 3'end of a plant virus RNA regulates viral replication. *The EMBO Journal*, *18*, 4856–4864.
Pallas, V., Aparicio, F., Herranz, M. C., Amari, K., Sanchez-Pina, M. A., Myrta, A., et al. (2012). Ilarviruses of *Prunus* spp.: A continued concern for fruit trees. *Phytopathology*, *102*, 1108–1120.
Pallas, V., & Garcia, J. A. (2011). How do plant viruses induce disease? Interactions and interference with host components. *The Journal of General Virology*, *92*, 2691–2705.
Pallas, V., Genoves, A., Sánchez-Pina, M. A., & Navarro, J. A. (2011). Systemic movement of viruses via the plant phloem. In C. Caranta, M. G. Aranda, M. Tepfer, & J. J. López-Moya (Eds.), *Recent advances in plant virology* (pp. 75–101). Norfolk, UK: Caister Academic Press.
Pallas, V., Sanchez-Navarro, J. A., & Diez, J. (1999). In vitro evidence for RNA binding properties of the coat protein of prunus necrotic ringspot ilarvirus and their comparison to related and unrelated viruses. *Archives of Virology*, *144*, 797–803.
Petrillo, J. E., Rocheleau, G., Kelley-Clarke, B., & Gehrke, L. (2005). Evaluation of the conformational switch model for Alfalfa mosaic virus RNA replication. *Journal of Virology*, *79*, 5743–5751.
Poch, O., Sauvaget, I., Delarue, M., & Tordo, N. (1989). Identification of four conserved motifs among the RNA-dependent polymerase encoding elements. *The EMBO Journal*, *8*, 3867–3874.
Postnikova, O. A., & Nemchinov, L. G. (2012). Comparative analysis of microarray data in Arabidopsis transcriptome during compatible interactions with plant viruses. *Virology Journal*, *9*, 101.
Reusken, C. B. E. M., & Bol, J. F. (1996). Structural elements of the 3'-terminal coat protein binding site in alfalfa mosaic virus RNAs. *Nucleic Acids Research*, *24*, 2660–2665.

Reusken, C. B. E. M., Neeleman, L., & Bol, J. F. (1994). The 3'-untranslated region of alfalfa mosaic virus RNA 3 contains at least two independent binding sites for viral coat protein. *Nucleic Acids Research*, *22*, 1346–1353.

Ritzenthaler, C., & Hofmann, C. (2007). Tubule-guide movement of plant viruses. In E. Waigmann & M. Heinlein (Eds.), *Viral transport in plants* (pp. 63–83). Berlin: Springer.

Robert-Seilaniantz, A., Grant, M., & Jones, J. D. G. (2011). Hormone crosstalk in plant disease and defense: More than just jasmonate-salycylate antagonism. *Annual Review of Phytopathology*, *49*, 317–343.

Rozanov, M. N., Koonin, E. V., & Gorbalenya, A. E. (1992). Conservation of the putative methyltransferase domain: A hallmark of the 'Sindbis-like' supergroup of positive-strand RNA viruses. *The Journal of General Virology*, *73*, 2129–2134.

Sánchez-Navarro, J. A., & Bol, J. F. (2001). Role of the Alfalfa mosaic virus movement protein and coat protein in virus transport. *Molecular Plant-Microbe Interactions*, *14*, 1051–1062.

Sánchez-Navarro, J. A., Fajardo, T., Zicca, S., Pallas, V., & Stavolone, L. (2010). Caulimoviridae tubule-guided transport is dictated by movement protein properties. *Journal of Virology*, *84*, 4109–4112.

Sánchez-Navarro, J. A., Herranz, M. C., & Pallas, V. (2006). Cell-to-cell movement of Alfalfa mosaic virus can be mediated by the movement proteins of Ilar-, bromo-, cucumo-, tobamo- and comoviruses and does not require virion formation. *Virology*, *346*, 66–73.

Sánchez-Navarro, J., Miglino, R., Ragozzino, A., & Bol, J. F. (2001). Engineering of alfalfa mosaic virus RNA 3 into an expression vector. *Archives of Virology*, *146*, 923–939.

Sánchez-Navarro, J. A., & Pallás, V. (1994). Nucleotide sequence of apple mosaic ilarvirus RNA 4. *The Journal of General Virology*, *75*, 1441–1445.

Sánchez-Navarro, J. A., & Pallás, V. (1997). Phylogenetical relationships in the ilarviruses: Nucleotide sequence of prunus necrotic ringspot RNA 3. *Archives of Virology*, *142*, 749–763.

Sánchez-Navarro, J. A., Reusken, C. B. E. M., Bol, J. F., & Pallas, V. (1997). Replication of alfalfa mosaic virus RNA 3 with movement and coat protein genes replaced by corresponding genes of Prunus necrotic ringspot ilarvirus. *The Journal of General Virology*, *78*, 3171–3176.

Schmitt, R. A., Williams, H., & Nyland, G. (1977). Virus diseases can decrease peach yields. *Cling Peach Quarterly*, *13*, 17–19.

Scott, S. W., & Zimmerman, M. T. (2001). *American plum line pattern virus* is a distinct ilarvirus. *Acta Horticulturae*, *550*, 221–227.

Scott, S. W., & Zimmerman, M. T. (2006). The complete nucleotide sequence of the genome of *Humulus japonicus latent virus*. *Archives of Virology*, *151*, 1683–1687.

Scott, S. W., & Zimmerman, M. T. (2008). Partial nucleotide sequences of the RNA 1 and RNA 2 of lilac ring mottle confirm that this virus should be considered a member of subgroup 2 of the genus *Ilarvirus*. *Archives of Virology*, *153*, 2169–2172.

Scott, S. W., Zimmerman, M. T., & Ge, X. (2003). Viruses in subgroup 2 of the genus Ilarvirus share both serological relationships and characteristics at the molecular level. *Archives of Virology*, *148*, 2063–2075.

Scott, S. W., Zimmerman, M. T., Yilmaz, S., Bachman, E. J., & Zehr, E. I. (2001). The interaction between Prunus necrotic ringspot virus and Prune dwarf virus in peach stunt disease. *Acta Horticulturae*, *550*, 229–236.

Shepherd, R. J., Francki, R. I. B., Hirth, L., Hollings, M., Inouye, T., MacLeod, R., et al. (1976). New groups of plant viruses approved by the International Committee on Taxonomy of Viruses. *Intervirology*, *6*, 181–184.

Shiel, P. J., Alrefai, R. H., Domier, L. L., Korban, S. S., & Berger, P. H. (1995). The complete nucleotide sequence of *Apple mosaic virus* RNA-3. *Archives of Virology*, *140*, 1247–1256.

Shimura, H., Masuta, C., Yoshida, N., Sueda, K, & Suzuki, M. (2013). The 2b protein of *Asparagus virus* 2 functions as an RNA silencing suppressor against systemic silencing to prove functional synteny with related cucumoviruses. *Virology*, In press.

Siegel, R. W., Adkins, S., & Kao, C. C. (1997). Sequence-specific recognition of a subgenomic RNA promoter by a viral RNA polymerase. *Proceedings of the National Academy of Science United States of America*, *94*, 11238–11243.

Silva, C., Tereso, S., Nolasco, G., & Oliveira, M. M. (2003). Cellular location of Prune dwarf virus in almond sections by in situ reverse transcription-polymerase chain reaction. *Phytopathology*, *93*, 278–285.

Silvestre, R., Untiveros, M., & Cuellar, W. J. (2011). First report of *Potato yellowing virus* (Genus *Ilarvirus*) in *Solanum phureja* from Ecuador. *Plant Disease*, *95*, 355.

Smith, C. H., Roosien, J., van Vloten-Doting, L., & Jaspars, E. M. J. (1981). Evidence that alfalfa mosaic virus infection starts with three RNA-protein complexes. *Virology*, *112*, 169–173.

Spiegel, S., Martin, R. R., Leggett, F., Terborg, M., & Postman, J. (1993). Characterization and geographical distribution of a new ilarvirus from *Fragaria chiloensis*. *Phytopathology*, *83*, 991–995.

Stefanac, Z., & Wrischer, M. (1983). Spinach latent virus: Some properties and comparison of two isolates. *Acta Botanica Croatica*, *42*, 1–9.

Stewart, F. C. (1910). Notes on New York plant diseases. *New York Agricultural Experiment Station Geneva Technical Bulletin*, *328*, 305–404.

Swanson, M. M., Ansel-McKinney, P., Houser-Scott, F., Yusibov, V., Loesch-Fries, L. S., & Gehrke, L. (1998). Viral coat protein peptides with limited sequence homology bind similar domains of alfalfa mosaic virus and tobacco streak virus RNAs. *Journal of Virology*, *72*, 3227–3234.

Tamura, K., Peterson, D., Peterson, N., Stecher, G., Nei, M., & Kumar, S. (2011). MEGA5: Molecular evolutionary genetics analysis using maximum likelihood, evolutionary distance, and maximum parsimony methods. *Molecular Biology and Evolution*, *28*, 2731–2739.

Tan, R., & Frankel, A. D. (1995). Structural variety of arginine-rich RNA-binding peptides. *Proceedings of the National Academy of Science United States of America*, *92*, 5282–5286.

Taschner, P. E., Van der Kuyl, A. C., Neeleman, L., & Bol, J. F. (1991). Replication of an incomplete alfalfa mosaic virus genome in plants transformed with viral replicase genes. *Virology*, *181*, 445–450.

Thomas, B. J., Barton, R. J., & Tuszynski, A. (1983). Hydrangea mosaic-virus, a new ilarvirus from hydrangea-macrophylla (Saxifragaceae). *The Annals of Applied Biology*, *103*, 261–270.

Tzanetakis, I. E., & Martin, R. R. (2005). New features in the genus *Ilarvirus* revealed by the nucleotide sequence of *Fragaria chiloensis latent virus*. *Virus Research*, *112*, 32–37.

Tzanetakis, I. E., Martin, R. R., & Scott, S. W. (2010). Genomic sequences of blackberry chlorotic ringspot virus and strawberry necrotic shock virus and the phylogeny of viruses in subgroup 1 of the genus *Ilarvirus*. *Archives of Virology*, *155*, 557–561.

Ueki, S., & Citovsky, V. (2007). Spread throughout the plant: Systemic transport of viruses. In E. Waigmann & M. Heinlein (Eds.), *Viral transport in plants* (pp. 85–118). Berlin: Springer.

Uyeda, I., & Mink, G. I. (1981). Properties of asparagus virus-Ii, a new member of the ilarvirus group. *Phytopathology*, *71*, 1264–1269.

Uyeda, I., & Mink, G. I. (1983). Relationships among some ilarviruses: Proposed revision of subgroup A. *Phytopathology*, *73*, 47–50.

Uyemoto, G., & Scott, S. W. (1992). Important diseases of *Prunus* caused by viruses and other graft-transmissible pathogens in California and South Carolina. *Plant Disease*, *76*, 5–11.

Van der Heijden, M. W., Carette, J. E., Reinhoud, P. J., Haegi, A., & Bol, J. F. (2001). Alfalfa mosaic virus replicase proteins P1 and P2 interact and colocalize at the vacuolar membrane. *Journal of Virology*, *75*, 1879–1887.

Van der Meer, F. A., Huttinga, H., & Maat, D. Z. (1976). Lilac ring mottle virus: Isolation from lilac, some properties, and relation to lilac ringspot disease. *The Netherlands Journal of Plant Pathology*, *82*, 67–80.

Van der Vossen, E. A., Neeleman, L., & Bol, J. F. (1993). Role of the 5′ leader sequence of alfalfa mosaic virus RNA 3 in replication and translation of the viral RNA. *Nucleic Acids Research*, *21*, 1361–1367.

Van der Vossen, E. A., Neeleman, L., & Bol, J. F. (1994). Early and late functions of alfalfa mosaic virus coat protein can be mutated separately. *Virology*, *202*, 891–903.

Van Regenmortel, M. V. H., Ackermann, H.-W., Calisher, C. H., Dietzgen, R. G., Horzinek, M. C., Keil, G. M., et al. (2013). Virus species polemics: 14 senior virologists oppose a proposed change to the ICTV definition of virus species. *Archives of Virology*, *158*, 1115–1119.

Van Rossum, C. M. A., Garcia, M. L., & Bol, J. F. (1996). Accumulation of alfalfa mosaic virus RNAs 1 and 2 requires the encoded proteins in cis. *Journal of Virology*, *70*, 5100–5105.

Van Vloten-Doting, L. (1975). Coat protein is required for infectivity of tobacco streak virus: Biological equivalence of the coat proteins of tobacco streak and alfalfa mosaic viruses. *Virology*, *65*, 215–225.

Van Vloten Doting, L., Francki, R. I. B., Fulton, R. W., Kaper, J. M., & Lane, L. C. (1981). *Tricornoviridae*—A proposed family of viruses with tripartite, single-stranded RNA genomes. *Intervirology*, *15*, 198.

Vance, V. B., Berger, P. H., Carrington, J. C., Hunt, A. G., & Shi, X. M. (1995). 5' Proximal potyviral sequences mediate potato virus X/potyviral synergistic disease in transgenic tobacco. *Virology*, *206*, 583–590.

Vaskova, D., Petrzik, K., & Karesova, R. (2000). Variability and molecular typing of the woody-tree infecting Prunus necrotic ringspot ilarvirus. *Archives of Virology*, *145*, 699–709.

Vaskova, D., Petrzik, K., & Spak, J. (2000). Molecular variability of the capsid protein of the Prune dwarf virus. *European Journal of Plant Pathology*, *106*, 573–580.

Velankar, S. S., Soultanas, P., Dillingham, M. S., Subramanya, H. S., & Wigley, D. B. (1999). Crystal structures of complexes of PcrA DNA helicase with a DNA substrate indicate an inchworm mechanism. *Cell*, *97*, 75–84.

Vilar, M., Esteve, V., Pallas, V., Marcos, J. F., & Perez-Paya, E. (2001). Structural properties of carnation mottle virus p7 movement protein and its RNA-binding domain. *The Journal of Biological Chemistry*, *276*, 18122–18129.

Vlot, A. C., Laros, S. M., & Bol, J. F. (2003). Coordinate replication of alfalfa mosaic virus RNAs 1 and 2 involves cis- and trans-acting functions of the encoded helicase-like and polymerase-like domains. *Journal of Virology*, *77*, 10790–10798.

Vlot, A. C., Menard, A., & Bol, J. F. (2002). Role of the alfalfa mosaic virus methyltransferase-like domain in negative-strand RNA synthesis. *Journal of Virology*, *76*, 11321–11328.

Waigmann, E., Ueki, S., Trutnyeva, K., & Citovsky, V. (2004). The ins and outs of nondestructive cell-to-cell and systemic movement of plant viruses. *Critical Reviews in Plant Sciences*, *23*, 195–250.

Walter, M. H., Kaiser, W. J., Klein, R. E., & Wyatt, S. D. (1992). Association between tobacco streak ilarvirus seed transmission and anther tissue infection in bean. *Phytopathology*, *82*, 412–415.

Walter, M. H., Wyau, S. D., & Kaiser, W. J. (1995). Comparison of the RNAs and some physicochemical properties of the seed-transmitted tobacco streak virus isolate

Mel 40 and the infrequently seed-transmitted isolate Mel F. *Phytopathology, 85*, 1394–1399.

Wang, X. F., Lee, W. M., Watanabe, T., Schwartz, M., Janda, M., & Ahlquist, P. (2005). Brome mosaic virus 1a nucleoside triphosphatase/helicase domain plays crucial roles in recruiting RNA replication templates. *Journal of Virology, 79*, 13747–13758.

Wellink, J., Van Lent, J. W. M., Verver, J., Sijen, T., Goldbach, R. W., & Van Kammen, A. B. (1993). The cowpea mosaic virus M RNA-encoded 48-kilodalton protein is responsible for induction of tubular structures in protoplasts. *Journal of Virology, 67*, 3660–3664.

Whitham, S. A., Quan, S., Chang, H. S., Cooper, B., Estes, B., Zhu, T., et al. (2003). Diverse RNA viruses elicit the expression of common sets of genes in susceptible Arabidopsis thaliana plants. *The Plant Journal, 33*, 271–283.

Xin, H. W., & Ding, S. W. (2003). Identification and molecular characterization of a naturally occurring RNA virus mutant defective in the initiation of host recovery. *Virology, 317*, 253–262.

Xin, H. W., Ji, L. H., Scott, S. W., Symons, R. H., & Ding, S. W. (1998). Ilarviruses encode a cucumovirus-like 2b gene that is absent in other genera within the *Bromoviridae*. *Journal of Virology, 72*, 6956–6959.

Yusibov, V., & Loesch-Fries, L. S. (1998). Functional significance of three basic N-terminal amino acids of alfalfa mosaic virus coat protein. *Virology, 242*, 1–5.

Zheng, H. Q., Wang, G. L., & Zhang, L. (1997). Alfalfa mosaic virus movement protein induces tubules in plant protoplasts. *Molecular Plant-Microbe Interactions, 10*, 1010–1014.

Zuidema, D., Bierhuizen, M. F. A., & Jaspars, E. M. J. (1983). Removal of the N-terminal part of alfalfa mosaic virus coat protein interferes with the specific binding to RNA 1 and genome activation. *Virology, 129*, 255–260.

CHAPTER SIX

Genetic Variation and HIV-Associated Neurologic Disease

Satinder Dahiya[1], Bryan P. Irish[1], Michael R. Nonnemacher, Brian Wigdahl[2]

Department of Microbiology and Immunology, Center for Molecular Virology and Translational Neuroscience, Institute for Molecular Medicine and Infectious Disease, Drexel University College of Medicine, Philadelphia, Pennsylvania, USA

[1]These authors contributed equally to this work

[2]Corresponding author: e-mail address: brian.wigdahl@drexelmed.edu

Contents

Abstract

HIV-associated neurologic disease continues to be a significant complication in the era of highly active antiretroviral therapy. A substantial subset of the HIV-infected population shows impaired neuropsychological performance as a result of HIV-mediated neuroinflammation and eventual central nervous system (CNS) injury. CNS compartmentalization of HIV, coupled with the evolution of genetically isolated populations in the CNS, is responsible for poor prognosis in patients with AIDS, warranting further investigation and possible additions to the current therapeutic strategy. This chapter

Advances in Virus Research, Volume 87
ISSN 0065-3527
http://dx.doi.org/10.1016/B978-0-12-407698-3.00006-5

reviews key advances in the field of neuropathogenesis and studies that have highlighted how molecular diversity within the HIV genome may impact HIV-associated neurologic disease. We also discuss the possible functional implications of genetic variation within the viral promoter and possibly other regions of the viral genome, especially in the cells of monocyte–macrophage lineage, which are arguably key cellular players in HIV-associated CNS disease.

1. INTRODUCTION

Human immunodeficiency virus type 1 (HIV-1) infects the central nervous system (CNS) initiating a cascade of neuroinflammation and eventually CNS injury. Despite the success of highly active antiretroviral therapy (HAART), neurocognitive impairment (NCI) continues to affect a significant proportion of infected patients. Although the incidence of HIV-1-associated dementia (HAD) has decreased, the overall prevalence of HIV-1-associated neurological disorders (HAND) has increased in the HAART era, primarily because the incidence of subtle forms of HIV-1-associated cognitive impairment has increased. In resource-limited settings, especially in the developing world, poor access to antiretroviral medication results in a much more severe prognosis for HIV-related CNS complications in late-stage HIV infection. HIV enters the nervous system within the first few weeks after initial systemic infection (Pilcher et al., 2001; Schacker, Collier, Hughes, Shea, & Corey, 1996), initiating a cascade of neuroinflammation and eventual CNS invasion and subsequent injury. CNS compartmentalization, including the cerebrospinal fluid (CSF), of HIV species may begin within the first year of infection. Thus, the CNS may be a potential independent site of HIV replication. Genetic variation within the HIV genome and associated selective pressures may lead to an increase in the prevalence of specialized variants that find a niche and begin evolving in the early stages of the disease. This review discusses the key features of HAND, the implications of the molecular and genetic diversity of the HIV-1 genome for HIV disease, and the importance of cells of the monocyte–macrophage lineage in the overall neuropathogenesis of HIV-1.

2. OVERVIEW OF HIV-1 CNS PATHOGENESIS

Entry of HIV-1 into the brain results in a chain of events leading to CNS disease and neurologic impairment. The virus must first circumvent the blood–brain barrier (BBB), a selectively permeable barrier separating the CNS from the peripheral circulation (Fig. 6.1). One route of entry into the CNS involves transit of HIV-1 across the BBB by means of infected cells trafficking from the periphery into the brain. This "Trojan horse" method of entry likely involves infected circulating monocytes carrying HIV-1 into the brain in the form of integrated provirus or infectious viral particles (Haase, 1986). Alternatively, HIV may also traffic into the CNS by lymphocytes that harbor viruses that replicate in macrophages, or as a cell-free virus entering through the endothelial cells or across cells of the choroid plexus (Collman et al., 1992; Spudich & Gonzalez-Scarano,

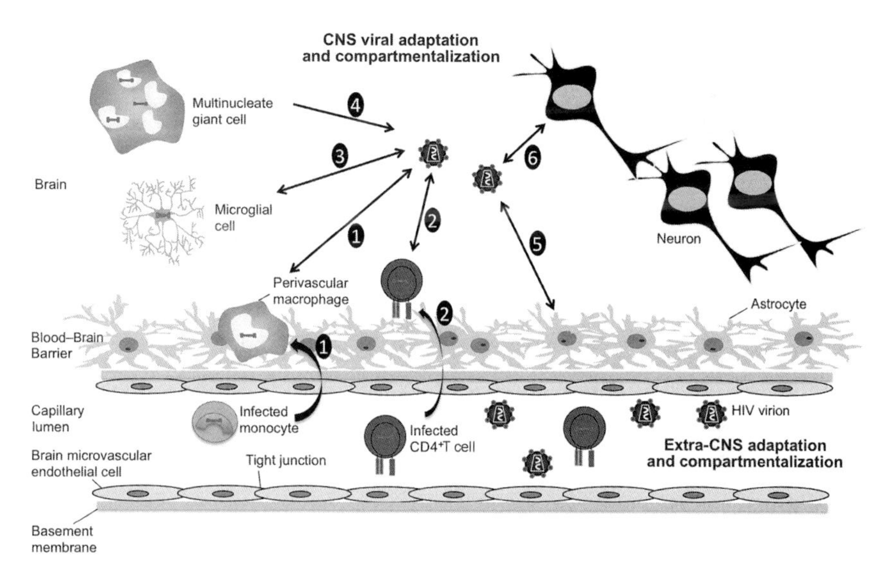

Figure 6.1 Model of HIV trafficking across the BBB and its impact on the CNS. (1) Infected, activated monocytes transport HIV across the BBB through a "Trojan Horse" model, where they differentiate into perivascular macrophages. Infected perivascular macrophages then result in production of HIV within the CNS. (2) To a lesser extent, infected $CD4^{+}T$ cells also serve to carry HIV across the BBB and may also contribute to HIV pools within the CNS. (3) HIV produced in the CNS can result in infection of resident microglial cells. (4) The presence of multinucleate giant cells is an important feature of HIV-related brain pathology; these are produced as a result of cell-to-cell fusion. (5) Astrocytes are known to harbor HIV, but it is well established that they do not result in a productive infection. (6) The viral evolution and adaptation within the CNS adversely affect the physiology of neurons via a variety of mediators including ROS, nitrous oxide (NO), MMPs, and viral proteins that exhibit neurotoxic properties resulting in compromised neurologic functions. (For color version of this figure, the reader is referred to the online version of this chapter.)

2012). Broad systemic infection and immune system activation may exacerbate this process, when infected (Hickey, 1999) and possibly uninfected cells within the CNS release chemotactic mediators into circulation, thereby drawing more activated cells harboring HIV-1 into the brain. This process may establish a positive feedback mechanism of viral entry and subsequent neuroinflammation (Fontaine, Poudrier, & Roger, 2011; Liu, Tang, McArthur, Scott, & Gartner, 2000; Yadav & Collman, 2009). HIV-1 infection of cells of the monocyte–macrophage lineage also induces increased expression of adhesion molecules on vascular endothelial cells, facilitating HIV-1 transit across the BBB (Blodget et al., 2012; Nottet et al., 1996; Rappaport et al., 1999). Infected macrophages induce greater expression of the adhesion molecules E-selectin and vascular cell adhesion molecule-1 (VCAM-1) on the surface of brain endothelial cells than do uninfected macrophages, suggesting that immune cell activation of monocytic cells following HIV-1 infection of the CNS likely plays a key role in facilitating trans-endothelial migration across the BBB (Miller et al., 2012; Nottet et al., 1996; Persidsky et al., 1997; Rappaport et al., 1999). Cells of the monocyte–macrophage lineage are the only cells in the CNS that routinely are shown to express HIV RNA or protein, although other cell types, such as astrocytes, have been shown to harbor HIV sequences but do not show a robust expression of HIV RNA or proteins (Spudich & Gonzalez-Scarano, 2012; Wiley, Schrier, Nelson, Lampert, & Oldstone, 1986). Among the different macrophage subtypes, perivascular macrophages are highly infected in the brains of HIV-1-infected individuals (Kim et al., 2006). Initially, it was thought that perivascular macrophages could not contribute to the long-term presence of HIV-1 in the brain owing to their rapid turnover rate, but reports have suggested that this cell population can harbor virus for long periods and can, therefore, serve as a reservoir for HIV-1, as previously reviewed (Spudich & Gonzalez-Scarano, 2012).

Neurologic disease rarely manifests prior to the onset of immune system dysfunction (McArthur et al., 1997), and patients who do not show early signs of neurologic impairment typically progress through the asymptomatic stage of infection without experiencing a decline in neurologic status (Gannon, Khan, & Kolson, 2011; Selnes et al., 1990). The disconnect between initial infection of the CNS and the presentation of associated neurologic impairment may be explained by an initial immune system clearance of the virus followed by a reseeding of the CNS by HIV-1 at later time points in disease after the immune system has been functionally degraded. This is supported by evidence of increased levels of HIV-1-specific immunoglobulins within the CSF, intrathecally produced anti-HIV-1 antibodies, and increased numbers of HIV-1-specific $CD8^+$ cytotoxic T lymphocytes (CTLs) (Krebs, Ross, McAllister, & Wigdahl, 2000). Recently, using the simian immunodeficiency virus model, it was shown that the intrathecal immune responses correlate inversely with the macrophage-tropic strains in the CNS (Selnes et al., 1990). An alternative explanation for the late onset of neurologic disease could be the selective infection of the CNS with a less neurovirulent strain of HIV-1, which results in a less cytotoxic but more chronic dissemination of the virus within the CNS (Krebs et al., 2000). Further explanation for the late-stage onset of neurological impairment comes from phylogenetic studies of HIV-1 gp160 sequences isolated from multiple tissue compartments as well as multiple compartments within the brain of HIV-1-infected patients (Liu et al., 2000). Recently, it was shown that HIV-1 R5 envelope (Env) sequences evolve with an increased positive charge and that this R5 subset evolves independently from highly macrophage-tropic variants with low-charge gp120s (Gonzalez-Perez et al., 2012). This study and others (Duncan & Sattentau, 2011; Peters, Duenas-Decamp, Sullivan, & Clapham, 2007) highlight the observations that HIV-1 R5 Envs evolve with very distinct properties at different sites in the body and are driven by powerful tissue-specific evolutionary pressures. Additional variables such as the effects of aging on brain, long-term CNS toxicity of HAART, and the impact of drugs of abuse need to be evaluated in detail to provide us with a conclusive model of these correlates of HIV disease.

Others have suggested that infected lymphocytes may be involved in the "Trojan horse" entry of HIV-1 into the CNS (Sloand et al., 1992; Weidenheim, Epshteyn, & Lyman, 1993). Similar to the case with cells of the monocyte–macrophage lineage, HIV-1 infection of $CD4^+$ T cells results in upregulation of cellular adhesion molecules, including leukocyte function antigen-1 and very late antigen-4, which interact with vascular endothelial cell ligands intercellular adhesion molecule-1 (ICAM-1) and VCAM-1, thereby facilitating binding of lymphocytes to the endothelial cell surface (Sloand et al., 1992; Weidenheim et al., 1993). Lymphocytes may also secrete specific enzymes that degrade the basement membrane of endothelial cells, allowing the migration of HIV-1-infected T cells across the BBB (Sloand et al., 1992). However, owing to the relatively low frequency of lymphocytes infiltrating into the CNS, their contribution to HIV-1 trafficking to the brain and subsequent establishment and maintenance of infection remain a topic of debate.

3. CLINICAL DIAGNOSES OF HIV-1 CNS INFECTION

HIV-1 infection of the CNS can result in numerous motor and cognitive deficiencies (Cosenza, Zhao, Si, & Lee, 2002; Robertson, Liner, & Heaton, 2009; Wiley et al., 1986; Williams et al., 2001). Two distinctly recognizable conditions resulting from HIV-1 infection of the CNS are HAD and the less severe, subsyndromic condition called minor cognitive motor disorder (MCMD) (Cherner et al., 2002; Gartner, 2000; McArthur et al., 2003; Minagar et al., 2008; Williams & Hickey, 2002). Patients with either of these conditions are classified collectively as having NCI. Prior to the introduction of HAART in the industrialized world, ~20–30% of HIV-1-infected individuals developed HAD (Childs et al., 1999). The onset of HAD corresponds with high plasma viral loads, and although a reduced incidence of HAD has been observed with patients on HAART, the longer life expectancy of HIV patients has increased the prevalence of the disease (Childs et al., 1999). With the widespread use of HAART, MCMD has become more common (Cherner et al., 2002; McArthur et al., 2003). In the HAART era, it is estimated that ~10% of HIV-infected adults develop HAD; however, MCMD may be several times more common, involving as many as 30% of the HIV-infected population (Cherner et al., 2002; Sacktor et al., 2002). Furthermore, the clinical presentation of MCMD has been associated with neuropathological changes characteristic of HIV encephalitis, and MCMD is associated with a worse overall prognostic outlook (Cherner et al., 2002; McArthur et al., 2003; Mothobi & Brew, 2012; Sacktor et al., 2002). One means of explaining the development of MCMD is that the low-level viral replication associated with successful HAART regimens may lead to slowly progressing neurodegeneration. This is consistent with the longer life spans of patients receiving HAART, and possibly with the inability of certain antiretroviral drugs to effectively penetrate into the brain (Letendre et al., 2004). A recent comparative study (Heaton et al., 2011) of HIV-associated NCI before and after the introduction of HAART concludes that although HAD (the most severe form of HAND) appears to be less common in the HAART era, the long-term benefits of therapy with respect to milder forms of HAND remain in question because their prevalence appears to be increasing (Heaton et al., 2011; McArthur & Brew, 2010). To achieve an optimal comparison of HAND in a longitudinal framework would require development of more consistent clinical definitions of neurologic disease, clinical predictors, and a better-defined characterization of comorbid conditions. Study and compilation of all these variables would be greatly facilitated by developing larger and more representative HIV/AIDS cohorts. Nevertheless, we can safely conclude that the beneficial effects of HAART on neurological complications associated with HIV have been less than complete and that continuous efforts are needed to improve prognosis. Additional variables that need more precise definition in relation to neurologic complications of HIV include drugs of abuse, age, viral strains, genetic variation within the HIV genome during the course of disease, and timing of HAART initiation.

HAD cannot be defined in terms of a single disease entity but must rather be characterized in terms of a broad collection of symptoms encompassing cognitive, motor, and behavioral deficiencies corresponding to the presence of actively replicating HIV-1 within the CNS. HAD is a subcortical dementia, which differentiates it from dementia induced by Alzheimer's disease, that presents clinically as a progressive decline in neurocognitive function (Kolson, Lavi, & Gonzalez-Scarano, 1998; Wendelken & Valcour, 2012). It may also include loss of memory, diminished ability to concentrate, psychomotor retardation, and frequent headache. Early in the course, HAD patients typically experience complications involving mental slowing, impaired motor control and lack of coordination, and behavioral alterations such as apathy, social withdrawal, and personality changes (Price, 1994). In late-stage disease, HAD patients exhibit severe, clinically recognizable cognitive, motor, and behavior deficits. Severe cases of HAD, which have become increasingly rare with the advent of effective combination antiretroviral therapeutics, may manifest with almost absolute mutism, incontinence, and severe, debilitating dementia (del Palacio, Alvarez, & Munoz-Fernandez, 2012; Price, 1994).

Clinical assessment of HIV-1-associated CNS disease requires surrogate biomarkers because brain and spinal cord are relatively inaccessible. HIV-1 RNA measurement in the CSF is one of the most practical means of examining CNS viral load (Marra, Maxwell, Collier, Robertson, & Imrie, 2007; Spudich et al., 2005). CSF markers of immune activation and inflammation have also been used as indicators of disease activity, including CCL2/monocyte chemotactic protein-1 (MCP1) and CXCL10/IP10 (chemokines that facilitate ingress of macrophages and lymphocytes across the BBB) (Chang, Ernst, St Hillaire, & Conant, 2004; Conant et al., 1998), β_2-microglobulin and neopterin (Brew et al., 1992; Brew, Dunbar, Pemberton, & Kaldor, 1996; Enting et al., 2000), quinolic acid (Heyes et al., 2001), arachidonic acid metabolites (Genis et al., 1992), and oxidative stress markers

(Schifitto et al., 2009) (Fig. 6.2). Reduced levels of *N*-acetylaspartate, which indicate decreased neuronal function, and elevated levels of choline, which indicate inflammation and membrane turnover, can also be utilized for overt HAD assessment using magnetic resonance spectroscopy (Chang, 1995; Meyerhoff et al., 1994).

Several pathologies are associated with HAD (Rosenblum, 1990), including white matter pallor, multinucleated-cell encephalitis (associated with the multinucleated giant cells, or MNGCs, or syncytia), and vacuolar myelopathy (Price, 1994). MNGCs are observed in only about half of all HAD patients on postmortem examination (Kato, Hirano, Llena, & Dembitzer, 1987; Wiley & Achim, 1994), and they are comprised of resident CNS mononuclear phagocytes often concentrated around blood vessels (Price, 1994). The presence of MNGCs in infected patients indicates active HIV-1 replication within the CNS (Budka, 1991; Budka et al., 1987). MNGC encephalitis occurs in the subcortical regions of the brain and is often associated with the presence of gliosis and white matter pallor

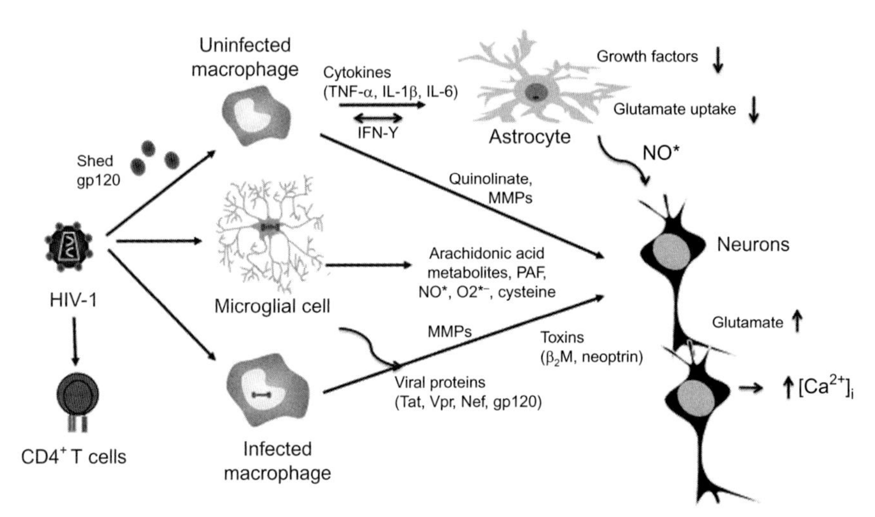

Figure 6.2 Potential pathways and mediators of CNS damage. Subsequent to neuroinvasion, HIV-1-infected perivascular macrophages and brain microglial cells are likely to be the major producers of infectious virus and neurotoxic cellular and viral proteins such as gp120, Tat, and Nef. The extent of CNS dysfunction observed during HIV-1 infection is likely due to both host and viral factors. CNS damage may occur through increased viral replication within the CNS, production of viral neurotoxic proteins, and release of toxins including NO, TNF-α, and quinolinic acid, all of which target neurons, astrocytes, endothelial cells, and oligodendrocytes (not shown). (For color version of this figure, the reader is referred to the online version of this chapter.)

(Chrysikopoulos, Press, Grafe, Hesselink, & Wiley, 1990; Epstein et al., 1984). In addition to these diagnostic approaches, functional magnetic resonance imaging and diffusion tensor imaging have been utilized to assess the changes in brain hemodynamics including alterations in cerebral blood flow, blood oxygen level dependence, and white matter morphometric changes (Ances et al., 2010; Spudich & Gonzalez-Scarano, 2012).

The onset and progression of HIV-1-associated neurodegeneration and subsequent decline in cognitive ability are likely dependent on multiple host and viral factors, some yet to be characterized (Fig. 6.3). This chapter

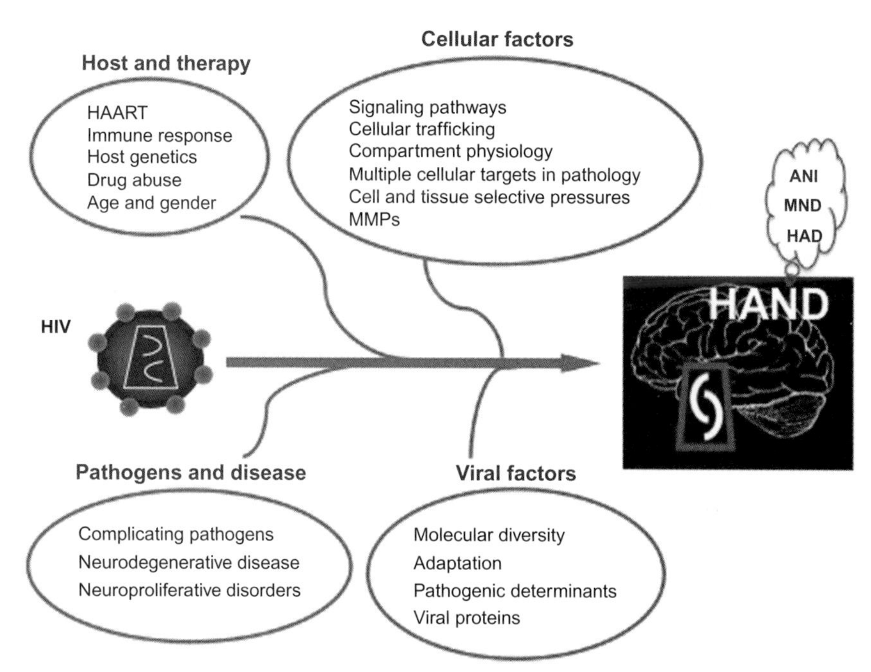

Figure 6.3 The dynamic interplay of multiple viral and host factors contributes to both the onset and severity of HIV-1-associated neurocognitive impairment. An individual's HAART (highly active antiretroviral drug) and/or drugs of abuse status, as well as the presence of comorbidities resulting from opportunistic pathogens, greatly influence overall disease course and the establishment of central nervous system (CNS) pathology. One major complicating factor with respect to HIV-1-induced neurodegeneration is the ability of HIV-1 to adapt and evolve specific genetic variants in response to host immune pressures and subsequently to increase in its capacity to replicate within specific tissue compartments such as the CNS. This ability of HIV to evolve within the CNS eventually manifests as a variety of neurologic symptoms including asymptomatic neurocognitive impairment (ANI), mild neurocognitive disorder (MND), and HIV-associated dementia. (For color version of this figure, the reader is referred to the online version of this chapter.)

reviews the relationship between genetic variation within the HIV-1 genome and host and the onset and progression of HAND.

4. THE EVOLUTION AND ADAPTATION OF HIV-1 AND THE ESTABLISHMENT OF MOLECULAR DIVERSITY

HIV-1 molecular (and consequent phenotypic) diversity is present at the population level as well as among infected individuals. The rapidly evolving nature of HIV-1 results from several factors, including the error-prone nature of reverse transcriptase, selective pressures from the host and from antiretroviral therapy, replication dynamics, and genomic recombination. HIV-1 molecular heterogeneity is most often characterized within the context of specific point mutations, which present as synonymous (non-amino acid altering) and nonsynonymous (amino acid altering) variations (Rambaut, Posada, Crandall, & Holmes, 2004). However, this genetic diversity may also manifest as polymorphic nucleotides throughout the viral genome resulting from insertion, deletion, and recombinatorial events occurring within a viral population (Rambaut et al., 2004). The end consequence is altered viral protein structure and function, as well as changes within noncoding nucleic acid sequences, such as the viral promoter, the long terminal repeat (LTR), which are critical viral components that may dramatically alter the course of viral gene expression.

The clinical trajectory of HIV disease is generally well characterized. Substantial variability occurs with respect to the rate of disease progression among infected individuals; however, the role host factors play in HIV disease progression cannot be discounted (Lackner, Lederman, & Rodriguez, 2012; Lemey et al., 2007). The asymptomatic phase of infection is highly variable and can range from several months to more than 20 years (Lemey et al., 2007). This differential rate of disease progression likely results from the dynamic relationship between virus and host and from specific selective pressures placed upon the virus within a given host (Lackner et al., 2012; Lemey et al., 2007). External factors such as other exogenously acquired infectious diseases may also affect disease progression. Following HIV-1 infection, both a humoral and a cell-mediated immune response are mounted by the host to combat the virus, but without avail, both are eventually defeated as HIV evolves and adapts, enabling it to efficiently replicate (Frost et al., 2005; Richman, Wrin, Little, & Petropoulos, 2003).

This process ultimately leads to the destruction of the host's immune system and the establishment of multiple opportunistic diseases, which define the clinical progression to AIDS (Frost et al., 2005; Richman et al., 2003). The humoral immune response involves the production of neutralizing antibodies, which exert strong selective pressure on the HIV envelope gene *(env)* but do not effectively control viral replication (Cecilia, Kleeberger, Munoz, Giorgi, & Zolla-Pazner, 1999). Recent structural studies of the HIV-1 Env proteins have fueled interest in rational antibody design using candidate-induced antibodies to Env, as previously reviewed (Bonsignori et al., 2012; Walker & Burton, 2010). The CD8$^+$ T-cell response likely serves in a protective capacity during HIV-1 infection, and evidence suggests that at least partial control of virus replication *in vivo* can be associated with the appearance of CTLs (Koup et al., 1994) and that the rate of disease progression is critically dependent on HLA class I alleles (Carrington, Dean, Martin, & O'Brien, 1999; Trachtenberg et al., 2003). It has been suggested that although the majority of the killing of HIV-1-infected cells results from the CTL response, small differences in CTL killing efficiency may be clinically relevant and may correspond with altered disease course (Asquith, Edwards, Lipsitch, & McLean, 2006).

HIV-1 possesses an enormous potential for evolutionary change, a consequence which is loss of the host's immune response to effectively control viral replication (Lemey et al., 2007). A genetically diverse viral population exhibiting a high rate of mutation and recombination, in concert with rapid replication dynamics, facilitates the propagation of infection to a large population of HIV-1-susceptible cells and enables the virus to readily adapt to constantly changing physiological conditions within each individual host (Lemey et al., 2007). Multiple amino acid alterations have been shown to occur within the hypervariable region of HIV-1 Env, changes that allow the virus to evade the host's humoral immune response but that do not negatively impact viral entry into target cells (Frost et al., 2005). HIV-1-specific CD8$^+$ T cells more efficiently target other regions of the virus such as *gag* and *nef* (Addo et al., 2003; Cao, McNevin, Malhotra, & McElrath, 2003; Lichterfeld et al., 2004), especially early in infection. Although viral evolution facilitates the establishment of escape variants capable of evading the CTL response, studies have suggested that evolution of these escape (or partial escape) mutants during chronic HIV-1 infection may occur with, however, an associated loss in replicative fitness (Ganusov et al., 2011; Leslie et al., 2004; Lewis, Dagarag, Khan, Ali, & Yang, 2012; Martinez-Picado et al., 2006).

The HIV-1-specific immune response is becoming increasingly understood. The effect of viral evolution and establishment of genetic diversity on the course and outcome of HIV-1-associated immune and nervous system disease, however, remain unresolved. One issue that is generally accepted is the critical nature of viral evolution with respect to our understanding of the dynamic relationship between virus and host during chronic or persistent infection. Consistent patterns of HIV-1 evolution have been observed throughout the course of infection (Shankarappa et al., 1999); however, investigations of viral diversity and mean divergence from the founder strain of HIV-1 within patients with differing rates of disease progression often conflict with one another (Ganeshan, Dickover, Korber, Bryson, & Wolinsky, 1997). Studies aimed at distinguishing between adaptive and selective neutral mutations have led some to believe that delayed disease progression may be associated with increased positive selection of sites within and accelerated adaptation rates of HIV-1 Env (Ross & Rodrigo, 2002; Williamson, 2003). It remains a topic of debate, however, whether viral adaptation results from, or is the consequence of, differential rates of HIV-1-associated disease progression (Lemey et al., 2007). Moreover, the dynamics of antiretroviral drugs also determine the path of HIV evolution and play a role in therapy outcome. Mathematical simulation studies have established that simulation of clinical trials with new and untested HIV treatment protocols could be used as a potent tool in selecting novel antiretroviral combinations (Hill, Rosenbloom, & Nowak, 2012; Rosenbloom, Hill, Rabi, Siliciano, & Nowak, 2012). Although this simulation approach may have biases toward success and may be limited with respect to the incorporation of all possible sequence variants that would render the compound combination inferior in clinical trials, the simulation can still be used as a preliminary step to strengthen a proposed antiretroviral combination prior to testing in preliminary efficacy analyses.

Studies analyzing the ratio of synonymous (nonamino acid changing) to nonsynonymous (amino acid changing) mutations within HIV-1 have provided insight into the mechanisms governing viral adaptation and the establishment of genetic diversity (Seo, Thorne, Hasegawa, & Kishino, 2002). Fluctuations in the rate of synonymous substitutions have been suggested to reflect changes in mutation rates, whereas nonsynonymous substitution rates may also be affected by changes in selective pressure and effective population size (Lemey et al., 2007). Studies of HIV-1 evolution have been based on the assumption that the rate of synonymous changes is relatively constant among HIV-1-infected patients, an assumption that has been questioned (Ganeshan et al., 1997; Lemey et al., 2007). The rate of synonymous change depends on multiple factors including viral generation time, which may vary considerably among individuals (Lemey et al., 2007). In addition, it does not take into account the potential impact of synonymous changes on *cis*-acting effects between interactions of the genome with virion components that may alter the overall virion maturation and infectivity. Viral replication rates may depend on the activation state of the host's immune system (Silvestri & Feinberg, 2003), the physiological environment (Martinez-Picado et al., 2006), and the environmental conditions existing within latently infected cell populations (Kelly, 1996; Kelly & Morrow, 2003). An investigation utilizing a new computational technique to estimate absolute rates of synonymous and nonsynonymous mutations and characterize how these rates change over time has suggested that the trajectory of HIV-1-associated disease progression among infected individuals may be predicted by the rate of synonymous mutations and that nonsynonymous mutations evolve as a consequence of differential antibody selective pressure. This approach builds on previous relaxed-clock methodology (Drummond et al., 2006), and by comparing evolutionary rates for specific branches within HIV-1 phylogenies, potentially biasing effects of deleterious polymorphisms are corrected for (Lemey et al., 2007). Using this method, a previously unidentified association between the rate of silent HIV-1 evolution and the rate of disease progression has been discovered, demonstrating that host immune mechanisms associated with HIV-1 pathogenesis may also play a role in modulating viral replication and ultimately place restrictions on HIV-1 evolution (Lemey et al., 2007). Most investigations that have compared the rate of nonsynonymous to synonymous changes have concluded that evolution of *pol*, *env*, or *nef* in brain isolates is adaptive in nature (Gray et al., 2011; Huang, Alter, & Wooley, 2002; Spudich & Gonzalez-Scarano, 2012). The challenge now is to understand the selective pressures driving these adaptive changes. To this end, understanding the specific features of the immune response within the CNS during this adaptation period will be important.

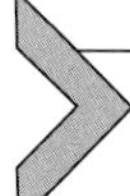

5. MOLECULAR DIVERSITY OF HIV-1 ENV AND NEUROLOGIC DISEASE

HIV-1 Env exists as a trimer in the virion and includes the surface glycoprotein transmembrane subunits gp120 and gp41. Initial viral attachment and subsequent entry into host cells are catalyzed by a high-affinity interaction between gp120 and the host cellular surface antigen CD4. This interaction results in gp120 undergoing a conformational change, exposing the coreceptor-binding site (Doms, 2000). The dynamic interaction between CD4-bound gp120 and the coreceptor then results in additional conformational changes culminating in the structural rearrangement of gp41 and facilitation of virus fusion and entry (Doms, 2000). The chemokine receptor CCR5 is the primary HIV-1 coreceptor utilized by the virus for infection of monocytic phagocytes, and viruses containing CCR5-utilizing envelopes (R5) are the predominant HIV-1 variants isolated from infected brain (Albright et al., 1999; Gorry et al., 2002; Smit et al., 2001). Brain microglia and tissue macrophages are known to express lower levels of CD4 and CCR5 on their surfaces than do peripheral blood $CD4^{+}$ T cells (Lee, Sharron, Montaner, Weissman, & Doms, 1999; Lewin et al., 1996; Wang et al., 2002). CD4 and CCR5 by HIV-1 are interdependent on one another, with one becoming more critical as the other becomes limiting (Doms & Moore, 2000). HIV Env proteins exhibiting increased tropism for cells of the monocyte–macrophage lineage also possess the ability to utilize low levels of both CD4 and CCR5 for fusion and entry, suggesting that decreased dependence on these surface molecules may represent an adaptation preferentially favoring viral replication within the CNS (Gorry et al., 2002; Gray et al., 2005; Martín, LaBranche, & González-Scarano, 2001; Peters et al., 2004; Thomas et al., 2007). However, the mechanistic details concerning how HIV-1 acquires this capacity to enter cells with limiting receptor and coreceptor levels are unclear. However, many of these studies focused on studying isolates derived from postmortem tissues and, therefore, would reflect an end-stage disease phenotype. Moreover, the CNS environment is that of an "immunological privileged" site with a low penetration of antibodies, thus allowing propagation of sequence configurations that would promote neutralization. In fact, isolates from brain are often sensitive to neutralizing antisera (Dunfee et al., 2007; Martin-Garcia, Cocklin, Chaiken, & Gonzalez-Scarano, 2005; Spudich & Gonzalez-Scarano, 2012).

The amino acid sequence of HIV Env determines its affinity for CD4 binding, and sequence alterations within the envelope protein may affect cellular tropism of HIV-1 by changing coreceptor utilization (Hoffman, Stephens, Narayan, & Doms, 1998; Wang et al., 1998). Efficient infection of the brain by HIV-1 is predicated on macrophage tropism, with the majority of viruses isolated from the brain of HIV-1-infected patients representing viruses that preferentially utilize CCR5 for attachment and entry into host cells (Albright et al., 1999; Gorry et al., 2001; Reddy et al., 1996). Studies have suggested that viruses isolated from the brain of HIV-1-infected patients may have higher affinity for CCR5 than do CCR5-utilizing viruses isolated from other tissue compartments, indicating a decreased dependence on CD4 expression by target cells (Gorry et al., 2002; Martín et al., 2001; Shieh, Martin, Baltuch, Malim, & Gonzalez-Scarano, 2000). The chemokine receptor CCR3 has also been implicated in the infection of microglia, and it has been shown that targeting either CCR5 or CCR3 protected both microglia and monocyte-derived macrophages (MDMs) from many strains of HIV-1 (Agrawal et al., 2009; He et al., 1997; Martin-Garcia, Cao, Varela-Rohena, Plassmeyer, & Gonzalez-Scarano, 2006). Additionally, CXCR4 (X4) utilizing viruses and dual-tropic (X4R5) viruses are rarely found in the brain, despite reports suggesting their ability to induce neuronal damage (Chan et al., 1999; Gorry et al., 2001; Reddy et al., 1996).

Studies have demonstrated that brain-derived HIV-1 sequences from HAD patients differ from those isolated from non-HAD patients, suggesting that specific HIV envelope sequences may be associated with clinical onset of dementia (Power et al., 1994; van Marle & Power, 2005). Furthermore, comparative molecular and biological analyses of HIV-1 isolates derived from both HAD and non-HAD patients have suggested that differences in viral tropism can discriminate between patients with and without dementia, indicating that neurotropic HIV-1 variants evolve independently within the brain and contribute to neuropathogenesis (Smit et al., 2001). Several studies have demonstrated that HIV-1 strains derived from patients with HAD differ from those derived from patients without dementia, primarily within the V1 and V3 region of gp120. Variation within the V1 region, in addition to the V2 region, has been associated with altered replication efficiency in macrophages (Toohey, Wehrly, Nishio, Perryman, & Chesebro, 1995), and the V3 region of Env has been implicated as a primary determinant of macrophage tropism and subsequent cytopathicity (Chesebro, Wehrly, Nishio, & Perryman, 1992; Hwang, Boyle, Lyerly, & Cullen, 1991; Korber, MacInnes, Smith, & Myers, 1994; Rossi et al., 2008). Single amino acid alterations within this region have been shown to change viral tropism and entry (McKnight et al., 1995). In addition, association has been made between specific V3 sequences and HAD (Chang et al., 1998; Korber et al., 1994; Power et al., 1994). HAD patients exhibit impaired serological responses against CCR5-dependent HIV-1 strains and increased molecular diversity within the V3 region of Env, suggesting the emergence of viral mutants that may preferentially infect the brain and mediate neurodegeneration (Van Marle et al., 2002). One specific Env variant present in the CD4-binding site of gp120, N283, has been identified at high frequency in brain-derived viruses of HAD patients (Dunfee et al., 2006). N283 has been shown to increase the affinity of gp120 for CD4 by decreasing the dissociation rate between these two molecules, thus enabling HIV-1 to utilize lower levels of CD4 for binding and entry and subsequently enhancing viral replication within macrophages and microglia (Dunfee et al., 2006). The N283 HIV-1 Env variant is found significantly more often in brain-derived Envs from HAD patients (41%) compared with non-HAD patients (8%), suggesting that this macrophage-tropic HIV-1 variant may be specifically associated with neurologic disease (Dunfee et al., 2006). Studies have shown that brain-derived HIV-1 isolates differ from those typically found in systemic circulation (Epstein et al., 1991; Korber, Kunstman, et al., 1994; Wong et al., 1997), and comparisons between peripheral blood-derived viruses and those derived from multiple brain compartments in the same patient have suggested segregated evolution of viral strains present within differing brain regions (Chang et al., 1998). Several studies have shown that HIV-1 proteins may be directly neurotoxic (Dahiya, Nonnemacher, & Wigdahl, 2012; Li et al., 2012; Liu et al., 2000; Meucci et al., 1998; Muller, Schroder, Ushijima, Dapper, & Bormann, 1992) (also see sections below). Individual HIV-1 isolates have been shown to exhibit differential induction of neuronal apoptosis, and this induction is independent of viral replication capacity (Ohagen et al., 1999). Recently, HIV-1 infection was shown to upregulate cathepsin B in macrophages and to reduce cystatin–cathepsin interactions, eventually leading to neuronal apoptosis (Rodriguez-Franco et al., 2012). Also, it was shown that gp120 induced caspase-3-dependent neuronal apoptosis by enhancing A-type transient outward K^+ currents via CXCR4-protein kinase C signaling (Xu et al., 2011). The V3 region of the HIV-1 Env, in addition to conferring increased tropism and subsequent infection of macrophages and microglial cells (Strizki et al., 1996), has been shown to impact the release of neurotoxic molecules following infection of macrophages (Cunningham et al., 1997; Kaul, Garden, & Lipton, 2001; Khanna et al., 2000; Power et al., 1998) and may, itself, be neurotoxic (Pattarini, Pittaluga, & Raiteri, 1998).

6. GENETIC DIVERSITY WITHIN HIV-1 Tat, Vpr, AND THE LTR, AND ITS CONTRIBUTION TO THE ONSET AND SEVERITY OF HIV-1-ASSOCIATED NEUROLOGICAL SYSTEM DISEASE

6.1. HIV-1 transactivator protein Tat

HIV-1 transcription involves an early, Tat-independent and a late, Tat-dependent phase, and transactivation of the viral genome is a critical step in the viral replication cycle, as previously reviewed (Dahiya et al., 2012; Li et al., 2012). The presence of Tat has been shown to increase LTR-mediated transcriptional activity by several hundredfold, and in the absence of Tat, viral replication falls to nearly undetectable levels (Doppler, Schalasta, Amtmann, & Sauer, 1992; Green, Ishino, & Loewenstein, 1989; Rice & Mathews, 1988). Tat is a unique transcription factor in that it binds to the "UCU" bulge of the transactivation response element (TAR), a *cis*-acting RNA enhancer element contained within the 5′ end of all viral transcripts (Brady & Kashanchi, 2005; Rappaport et al., 1999). The interaction of HIV-1 Tat with TAR RNA increases viral transcription and elongation (Raha, Cheng, & Green, 2005; Selby, Bain, Luciw, & Peterlin, 1989). Specifically, HIV-1 Tat is known to promote the binding of P-TEFb (cyclin T1 and cdk9) to the TAR region located within the viral promoter, immediately downstream of the transcriptional initiation site, and the interaction of Tat with P-TEFb and the TAR element results in hyperphosphorylation of the C-terminal domain and subsequent increased processivity of RNA polymerase II (pol II) (Raha et al., 2005; Zhou et al., 2000). The Tat-P-TEFb crystal structure has shown that Tat forms extensive contacts with both the CycT1 and Cdk9 subunits in P-TEFb, resulting in a conformational change and constitutive activation of the enzyme (Tahirov et al., 2010). HIV-1 Tat may also be involved with the formation of the transcriptional preinitiation complex (Dahiya et al., 2012; Raha et al., 2005).

In addition to the HIV-1 LTR, Tat is known to upregulate several other viral as well as cellular genes. Within the CNS, Tat has been shown to stimulate HIV-1 LTR-mediated viral gene expression in the absence of TAR (Taylor & Khalili, 1994), an activity that may result from its ability to enhance the activity of cytokines such as tumor necrosis factor-α (TNF-α) (Sawaya et al., 1998). TNF-α also has the ability to activate the HIV-1 LTR via activation of cytoplasmic nuclear factor kappa B (NF-κB) (Nabel, Rice, Knipe, & Baltimore, 1988; Sawaya et al., 1998), and this positive feedback mechanism may lead to constitutive TNF-α and HIV-1 Tat synthesis by infected glial and microglial cell populations within the brain, ultimately resulting in paracrine dysregulation and damage to neighboring neurons and astrocytes. Similarly, evidence generated from stable expression studies has indicated that HIV-1 Tat may inhibit TNF-α-induced repression of TNF receptor p55, thereby resulting in the amplification of TNF-α activity (Chiao et al., 2001).

HIV-1 Tat can also be secreted from infected cells, including infected macrophages, microglia, and astrocytes and may consequently be taken up by neighboring, uninfected cells (Ensoli et al., 1993; Verhoef, Klein, & Berkhout, 1996). Tat protein has been detected within the brain of infected individuals, and the uptake of Tat by CNS cells has toxic consequences, resulting in large part from neuronal apoptosis (Hudson et al., 2000; Nath et al., 1996). Extracellular Tat can enter neurons via endocytosis through interaction with the low-density lipoprotein receptor-related protein present on the neuronal surface (Vendeville et al., 2004). Recently, it was shown that Tat could bind to the promoters of the phosphatase and the tensin homologue and protein phosphatase 2A (PP2A), eventually resulting in apoptosis of HIV-1-infected $CD4^+$ T cells (Kim, Kukkonen, Gupta, & Aldovini, 2010). Furthermore, Tat has been shown to be transported along anatomical pathways within the brain, indicating that the neurotoxic effects of HIV-1 Tat may occur in regions far removed from the site of active infection (Bruce-Keller et al., 2003). Interestingly, secreted or extracellular Tat has been shown to function as a specific CXCR4 antagonist, selecting against X4-utilizing viruses, and thereby greatly influencing the development and progression of HIV-1 disease (Xiao et al., 2000), specifically within the CNS where R5 viruses are thought to play the predominant role in pathogenesis.

The neuropathologic properties associated with HIV-1 Tat stem from its ability to either directly or indirectly induce apoptosis, upregulate cytokines and chemokines, and interact with matrix metalloproteinases (MMPs). The ability of HIV-1 Tat to upregulate TNF-α and interleukin-1β (IL-1β) has been associated with increased expression of cell adhesion molecules on endothelial cells. Likewise, Tat-induced upregulation of MCP-1 has been shown to exacerbate neuroinvasion, facilitating the loss of BBB integrity, a pathological hallmark of late-stage HAND (Avison et al., 2004; Mayne et al., 1998; Nath, Conant, Chen, Scott, & Major, 1999). Although HIV-1 Tat and various MMPs, including MMP-1, -2, and -9, are known to be independently cytotoxic to cells within the CNS, studies have suggested that the dynamic interaction between Tat and MMPs may be neuroprotective (Johnston et al., 2001; Zhang et al., 2003). Specifically, MMP-1 has been shown to selectively cleave HIV-1 Tat and thereby neutralize its neurotoxic potential (Rumbaugh et al., 2006).

The molecular diversity of HIV-1 Tat protein isolated from brains of patients infected with different HIV-1 clades has been examined, as previously reviewed (Li et al., 2012). Studies examining Tat proteins representative of HIV-1 subtypes B, C, and BF recombinants have demonstrated important structural and functional differences (Siddappa et al., 2006; Turk et al., 2006). BF recombinant HIV-1 isolates, from Argentina, appear to have a replicative advantage over subtype B isolates, possibly owing to the differential ability of Tat to interact with the LTR, and subtype C Tat protein has been shown to be more highly ordered than subtype B Tat. In addition, subtype C Tat protein has been demonstrated to be consistently inferior to subtype B Tat in biological assays with respect to its ability to promote viral proliferation, induce TNF-α and IL-6 expression, and upregulate chemokine coreceptor expression (Siddappa et al., 2006). However, studies have also shown that HIV-1 subtype C Tat exhibits greater transcriptional activity in the Jurkat $CD4^+$ T-cell line when compared with subtypes B and E and that this higher level of transactivation is not LTR sequence dependent but results from variations in the C Tat sequence at amino acid residues 57 (Arg in B and E and Ser in C) and 63 (Glu in B, E, and C), which are within and close to the basic domain, respectively (Kurosu et al., 2002). Phylogenetic analyses of Tat sequences from patients with and without HAD have shown clustering of sequences with respect to clinical diagnosis of neurological impairment as well as tissue of origin (Bratanich et al., 1998; Mayne et al., 1998). Nonsynonymous versus synonymous mutation rates among brain-derived Tat sequences isolated from patients with NCI were shown to be significantly greater than those isolated from patients without clinical evidence of neurologic disease (Bratanich et al., 1998). Collectively, these studies suggest that differing selective pressures act on individual HIV-1 genes within the CNS and that these differing pressures may influence both the development and subsequent severity of NCI. Comparisons of matched brain- and spleen-derived Tat sequences have suggested that greater sequence homology exists among brain-derived Tat clones than what is observed between brain- and spleen-derived clones (Mayne et al., 1998). Additionally, significant sequence heterogeneity exists within brain-derived Tat in domains associated with viral replication and intracellular transport (Mayne et al., 1998). Importantly, HIV-1 Tat derived from HAD patients has been associated with greater neuronal death both *in vitro* and *in vivo* compared with Tat from non-HAD patients, and this characteristic has been attributed, in part, to enhanced MMP-2 expression induced by brain-derived HIV-1 Tat (Johnston et al., 2001). Interestingly, however, these same brain-derived Tat isolates also appear to be limited in their ability to enhance viral gene expression despite their increased activation of host transcriptional machinery (Silva et al., 2003). However, one must remember that these viral gene activation studies were performed with a viral regulatory region that was derived from a non-CNS tissue source and may, therefore, not be naturally compatible with respect to optimal LTR activation by a Tat protein selected for CNS replication. This is particularly important because previous studies (Burdo, Gartner, Mauger, & Wigdahl, 2004; Hogan, Nonnemacher, Krebs, Henderson, & Wigdahl, 2003; Hogan, Stauff, et al., 2003) demonstrated that LTRs derived from the CNS are likely to be structurally and functionally different from LTRs derived from other tissue sources and that colinear Tat and LTR combinations may result in more efficient LTR activation (Li et al., 2011). Nonetheless, taken together, these reports suggest that genetic diversity of HIV-1 Tat very likely contributes to the establishment and severity of HAND.

6.2. HIV-1 Vpr

Viral protein r (Vpr) is a 96-amino acid accessory protein that is packaged into the HIV-1 virion via its association with the p6 domain of HIV-1 Gag (Emerman & Malim, 1998). Vpr is a multifunctional protein affecting both early and late stages of the HIV-1 viral life cycle and is associated with the nuclear localization and import of the HIV-1 preintegration complex (Lu, Spearman, & Ratner, 1993; Mahalingam, Collman, Patel, Monken, & Srinivasan, 1995; Mahalingam, Khan, Jabbar, et al., 1995; Mahalingam, Khan, Murali, et al., 1995). Lacking a true nuclear localization sequence, Vpr is known to localize to the nucleus when expressed *in vitro* (Lu et al., 1993; Mahalingam, Khan, Jabbar, et al., 1995; Mahalingam, Khan, Murali, et al., 1995). Therefore, Vpr likely facilitates nuclear localization via its interaction with cellular proteins involved in nuclear import, possibly karyopherin α and β (Lu et al., 1993; Mahalingam, Collman, et al., 1995; Mahalingam, Khan, Jabbar, et al., 1995; Mahalingam, Khan, Murali, et al., 1995). Vpr has been suggested by several studies to induce cell cycle arrest in HIV-1-infected cells, with Vpr-expressing cells accumulating in the G_2 phase (Ayyavoo et al., 1997; Jowett et al., 1995). The efficacy of

Vpr-induced transactivation of the LTR has been shown to correlate with the induction of G_2 arrest in host cells (Ardon et al., 2006; DeHart et al., 2007; Zimmerman et al., 2004). Interestingly, cell cycle arrest of infected cells has been shown to increase HIV-1 LTR transcriptional activity, independent of Vpr (Cohen, Dehni, Sodroski, & Haseltine, 1990). Vpr has also been shown by several studies to transactivate the HIV-1 LTR by a variety of mechanisms (Cohen et al., 1990; Goh et al., 1998; McAllister et al., 2000; Sawaya et al., 1999). Vpr has been shown to activate the LTR via interaction with HIV-1 Tat and to indirectly increase LTR activity through its interaction with the transcription factor p300 (Felzien et al., 1998; Sawaya et al., 1999). Studies have suggested that Vpr may be involved in ternary complexes with Sp and the LTR, and investigations have indicated that Vpr is able to directly interact with the HIV-1 LTR via its binding to C/EBP-binding sites I and II (Burdo, Gartner, et al., 2004; Hogan, Nonnemacher, et al., 2003).

Studies of Vpr binding to HIV-1 LTR sequences encompassing the ATF/CREB, C/EBP site I, and the promoter-distal NF-κB site have suggested that the Vpr preferentially interacts with sequences spanning C/EBP site I and the adjacent NF-κB-binding site (Burdo, Nonnemacher, et al., 2004). This result in addition to the established proximity of the these two binding sites indicates that Vpr and NF-κB binding may be mutually exclusive; however, the downstream NF-κB element would still be available for binding independent of Vpr interactions upstream (Burdo, Gartner, et al., 2004). Importantly, it has been demonstrated that HIV-1 Vpr induces IL-8 production in monocytes through activation of both NF-κB and NF-IL-6 (C/EBP), and elevated levels of IL-8 are thought to be responsible for certain clinical manifestations observed among AIDS patients throughout the course of disease (Roux, Alfieri, Hrimech, Cohen, & Tanner, 2000). This finding suggests that HIV-1 LTR activity is likely influenced by a complex and dynamic balance between Vpr and members of the C/EBP transcription factor family, as well as NF-κB isoforms (Burdo, Nonnemacher, et al., 2004). Additionally, both Vpr and C/EBP factors are known to be required for efficient HIV-1 replication within cells of myeloid lineage. Studies have suggested that Vpr-regulated promoter activation may be enhanced as a result of increased binding of NF-κB and C/EBP factors to their respective binding sites (Roux et al., 2000), and other studies have suggested that Vpr may also mediate promoter activity via direct binding to C/EBP sites and other adjacent binding sequences (Burdo, Nonnemacher, et al., 2004; Hogan, Nonnemacher, et al., 2003). Evidence supports the concept that sequence-dependent interactions between Vpr and C/EBP site I may occur in the context of neurologic disease. Electrophoretic mobility shift (EMS) analyses have revealed a direct association between Vpr and HIV-1 LTR sequences, which include C/EBP site I, the promoter-distal NF-κB site, and the upstream ATF–CREB-binding site (Burdo, Nonnemacher, et al., 2004; Hogan, Nonnemacher, et al., 2003). This relationship was shown to be sequence-specific with respect to C/EBP site I (Burdo, Nonnemacher, et al., 2004). The 3T C/EBP-binding site variant, described earlier, which binds C/EBP factors with low relative affinity, has also been shown to be the C/EBP site I variant that binds Vpr with the highest relative affinity (Hogan, Nonnemacher, et al., 2003). Importantly, the affinity of C/EBP-binding sites for Vpr is associated with HAD, with high-affinity sites being more prevalent in HAD patients (Burdo, Nonnemacher, et al., 2004).

Cell types within the CNS that are capable of supporting productive HIV-1 infection are limited to macrophages and microglia; however, neuropathological abnormalities associated with cognitive impairment (MCMD and HAD) are thought to result, in large part, from neuronal dropout and apoptosis of neurons (Gelbard et al., 1995; Ohagen et al., 1999; Petito, 1995). Astrocytes do not support a high-level productive HIV-1 replication, potentially involving a defect in Rev function (Gorry et al., 1999; Messam & Major, 2000; Neumann et al., 1995; Tornatore, Chandra, Berger, & Major, 1994). Therefore, an indirect mechanism leading to apoptosis of neurons may exist, and HIV-1 Vpr is one of the viral gene products implicated in this process. Recently, it was shown that the effect of Vpr on neuronal death is in part via released proinflammatory factors. In this study, supernatants from Vpr-deleted HIV-1 mutant-infected MDMs contained lower concentrations of IL-1β, IL-8, and TNF-α and showed reduced neurotoxicity compared with wild-type HIV-1-infected MDM supernatants (Guha et al., 2012).

Importantly, free Vpr has been identified in the serum of HIV-1-infected patients and in the CSF of HIV/AIDS patients with neurological disease (Levy, Refaeli, & Weiner, 1995). Studies have linked Vpr to the induction of apoptosis of T cells, and although the HIV-1 envelope glycoprotein gp120 and Tat regulatory protein have been most commonly associated with cellular death during HIV-1 infection, virion-encapsulated Vpr may also be involved in CNS cell death *in vivo*. Experiments using extracellular Vpr have demonstrated that Vpr is able to bind promonocytic and lymphoid cells and increase permissiveness to HIV-1 replication (Levy et al., 1995).

Extracellular Vpr has been shown to associate directly with the plasmalemma of cultured rat hippocampal neurons and causes a large inward sodium current, depolarization, and cell death (Levy et al., 1995; Piller, Jans, Gage, & Jans, 1998). In addition, HIV-1 Vpr has been shown to potently induce apoptosis both in the undifferentiated neuronal precursor cell line NT-2 and in mature human neurons (Patel, Mukhtar, & Pomerantz, 2000). Thus, based on the cytotoxic and neurotoxic effects of extracellular Vpr, one may postulate that cell-free Vpr likely contributes to cellular depletion within lymphoid, peripheral blood, and CNS tissue compartments. Furthermore, the fact that extracellular Vpr is present in the serum and CSF of AIDS patients with neurologic disease suggests that extracellular Vpr may play a significant role in AIDS pathology and HIV-1-associated neurologic complications (Levy et al., 1995; Piller et al., 1998). Moreover, Vpr has been implicated in modulating the host glucocorticoid receptor to affect transcription from the LTR as well as other host genes (Refaeli, Levy, & Weiner, 1995). Vpr has also been shown to transactivate promoters containing glucocorticoid-responsive elements. This modulation is most likely via direct interaction with the glucocorticoid receptor, with Vpr acting as a coactivator of glucocorticoid receptor (Kino et al., 1999). Vpr has also been shown to induce oxidative stress in microglial cells via the hypoxia-inducible factor pathway (Deshmane et al., 2009) and has been shown to interact with ANT, PP2A, and HAX-1, which have been shown to play important roles in pathways that culminate in neuronal degeneration (Na et al., 2011; Zhao, Li, & Bukrinsky, 2011). Nonetheless, extensive studies related to the role Vpr plays in HIV pathogenesis have established it as a crucial accessory protein with a multitude of functions spread across different stages of the viral life cycle.

6.3. HIV-1 LTR activity within cells of the monocyte–macrophage lineage

The HIV-1 LTR is approximately 640 bp long and consists of the U3, R, and U5 segmented regions. The U3 region is further divided into the modulatory, enhancer, and core regions, which facilitate the interaction of both viral and cellular proteins involved with regulating viral gene expression (Cullen, 1991; Pereira, Bentley, Peeters, Churchill, & Deacon, 2000). With HIV-1 subtype B, the core region contains the TATAA box and a GC-rich sequence, which facilitates binding of members of the Sp family of transcription factors. The TATAA box binds TBP (TATAA-binding protein), in addition to other cellular proteins involved with the pol II transcriptional complex (Jones & Peterlin, 1994). The enhancer element is located immediately upstream of the core region and is associated primarily with the presence of two 10-bp NF-κB-binding sites (Nabel & Baltimore, 1987). The modulatory region, which consists of sequences located upstream of the NF-κB-binding sites, contains numerous transcription factor-binding sites specific for factors including C/EBP, ATF/CREB, LEF-1, NF-AT, and many others (Krebs et al., 2000). Studies of the modulatory region have revealed that this region is rich in *cis*-acting-binding elements, which serve to both repress and activate the HIV-1 LTR (Pereira et al., 2000). Furthermore, the interaction of viral proteins, specifically Vpr and Tat, with the LTR provides an additional element of complexity to the regulation of viral gene expression.

One of the primary regulators of HIV-1 LTR activity in all susceptible host cell populations, including cells of the monocyte–macrophage lineage, is NF-κB (Asin, Bren, Carmona, Solan, & Paya, 2001). Several studies have been aimed at determining the dependence of NF-κB family members with respect to transcriptional activation of the HIV-1 LTR in T cells and its subsequent effect on reactivation of HIV-1 from latency (Chen, Feinberg, & Baltimore, 1997; Folks et al., 1986; Ross, Buckler-White, Rabson, Englund, & Martin, 1991). Depending on the type of T cell examined, and differences in the experimental approaches employed, results from these studies have been conflicting. Generally, HIV-1 LTR NF-κB-binding sites are indispensable with respect to viral replication in $CD4^+$ T-cell lines (Alcami et al., 1995; Chen et al., 1997). Studies involving human monocytic cells and transformed human monocyte and macrophage cell lines have centered, in large part, on determining how monocytic differentiation affects HIV-1 expression and how HIV-1 infection results in NF-κB activation (Griffin, Leung, Folks, Kunkel, & Nabel, 1989; Raziuddin et al., 1991; Schuitemaker et al., 1992). Interestingly, differentiated macrophages already contain a constitutive nuclear pool of NF-κB, and what role this preexisting pool of NF-κB plays with respect to modulating HIV-1 gene expression remains unclear (Asin et al., 2001). However, studies have suggested that preexisting NF-κB heterodimers within the nuclei of these cells play a role in transcriptional initiation following infection by HIV-1 (Asin et al., 2001). Overall, studies have concluded that NF-κB *cis*-acting elements within the HIV-1 LTR are critical for efficient LTR activity and subsequent viral gene expression both within $CD4^+$ T cells and in cells of monocyte–macrophage lineage.

The importance of NF-κB can also be observed in the difference in LTR activity of different HIV-1 subtypes. HIV-1 subtype B is the predominant subtype in North America and in Europe, whereas subtypes C and E are most prevalent in other parts of the world, including east Asia, Thailand, India, and southern Africa (Janssens, Buve, & Nkengasong, 1997; Novitsky et al., 1999; Ping et al., 1999). One of the most striking differences between HIV-1 subtype C and other HIV-1 subtypes, such as B and E, resides in the LTR. Subtype C viruses have been shown to contain three NF-κB-binding sites within the enhancer element of the LTR, whereas subtypes B and E have only two and one, respectively (Gao et al., 1996; Kurosu et al., 2001; Montano et al., 2000). The functional consequence of this difference was revealed by transient transfection assay in HeLa cells, which showed that subtype C LTRs have higher promoter/enhancer activity compared with subtypes B and E. Subtype C does appear to be transmitted more efficiently than other subtypes (Essex, 1999), and specific genetic biological differences such as increased NF-κB-binding sites may play a role in this increased transmission efficiency.

Investigations of the transcriptional regulation of the HIV-1 LTR in cells of the monocyte–macrophage lineage have also focused on the role of C/EBP and Sp transcription factors with respect to regulation of viral gene expression within the CNS. C/EBP factors have been shown to be critically involved in the regulation of monocyte-specific gene expression (Matsusaka et al., 1993; Pope et al., 1994; Tanaka et al., 1995). C/EBPβ has been shown to bind at least two sites within the HIV-1 LTR and has been demonstrated to activate viral transcription in transient expression analyses (Ross et al., 2001; Tesmer, Rajadhyaksha, Babin, & Bina, 1993). Studies have revealed that at least one intact C/EBP site is required for HIV-1 replication in monocytic cells; however, this is not required for replication in T cells (Henderson & Calame, 1997; Henderson, Connor, & Calame, 1996; Henderson, Zou, & Calame, 1995).

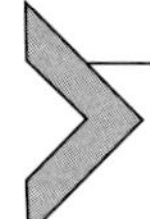

7. SEQUENCE VARIATION OF SPECIFIC TRANSCRIPTION FACTOR-BINDING SITES WITHIN THE HIV-1 LTR AND ITS CORRELATION WITH NERVOUS SYSTEM DISEASE

The relationship between LTR genetic diversity and HIV-1 disease is complex. Several reports have suggested that LTR sequence variation may alter promoter activity in varying cell types (Henderson & Calame, 1997; Henderson et al., 1996, 1995; Krebs, Mehrens, Pomeroy, Goodenow, & Wigdahl, 1998;

McAllister et al., 2000). Numerous investigations have reported sequence variation within LTRs isolated from infected patients (Hogan, Nonnemacher, et al., 2003; Hogan, Stauff, et al., 2003; Michael, D'Arcy, Ehrenberg, & Redfield, 1994; Nonnemacher, Irish, Liu, Mauger, & Wigdahl, 2004; Ross et al., 2001). Comparative analysis of LTRs isolated from both peripheral blood mononuclear cells and brain across a population of individuals has also revealed compartmentalization of the specific LTR variants and showed that LTRs isolated from the CNS are more closely related to previously characterized brain-derived LTRs than to LTRs isolated from other physiological compartments (Hogan, Nonnemacher, et al., 2003; Hogan, Stauff, et al., 2003; Michael et al., 1994; Nonnemacher et al., 2004; Ross et al., 2001). These naturally occurring sequence alterations within the LTR, which appear to arise as a result of tissue-specific selective pressures, may have a profound impact on the ability of the LTR to support HIV-1 infection by differentially modulating the ability of critical transcription factors to bind the LTR (Hogan, Stauff, et al., 2003). Although several studies have suggested that there is no correlation between LTR sequence variation and altered viral tropism and replication (Pomerantz, Feinberg, Andino, & Baltimore, 1991; Schuitemaker et al., 1993; Velpandi, Nagashunmugam, Otsuka, Cartas, & Srinivasan, 1992), studies involving the analysis of two different LTR variants have demonstrated that increased LTR activity based on transient expression studies corresponds to increased viral replication (Golub, Li, & Volsky, 1991; McAllister et al., 2000). Studies utilizing an HIV-1 LAI infectious molecular clone revealed that when the native high-affinity NF-κB-proximal Sp-binding site III was replaced with a low-affinity site, replication within Jurkat $CD4^+$ T cells was markedly decreased, whereas little effect on replication was observed within U-937 monocytic cells (McAllister et al., 2000). These results were consistent with previously published results (Zeichner, Hirka, Andrews, & Alwine, 1992; Zeichner, Kim, & Alwine, 1991a, 1991b), demonstrating that mutations introduced into the HIV-1 LTR that resulted in altered transient expression activity also resulted in similar alteration in viral replication potential when the same mutations were placed into the context of a replication-competent virus. Transient transfection analyses of HIV-1 LAI LTR-luciferase constructs in the Jurkat T-cell line have suggested that the large reduction in viral replication within these cells, caused by low-affinity Sp-binding site III variants, may be the result of reduced basal, Vpr, and Tat-mediated LTR activity (McAllister et al., 2000). When examined within the context of an HIV-1 YU-2 LTR-luciferase construct, the naturally low-affinity Sp-binding site III was replaced with a high-affinity site, which resulted in increased basal YU-2 LTR activity in

Jurkat T cells and reduced LTR activity in U-937 monocytic cells (McAllister et al., 2000). In addition, LTRs derived from HIV-1-infected patients have been shown to differentially regulate transient expression in a cell type-specific manner, a finding that is reinforced by studies involving cell type-specific reporter gene expression directed by a brain-derived LTR in transgenic mouse CNS tissue (Corboy, Buzy, Zink, & Clements, 1992; McAllister et al., 2000; Michael et al., 1994).

Several investigations have compared LTR sequences from HIV-1-infected long-term nonprogressors (LTNPs) and rapid progressors, and in each of these studies, no direct relationship could be established between LTR sequence variation and disease progression (Zhang et al., 1997). Furthermore, transient expression analyses performed in both cell lines and primary monocytes demonstrated no simple correlation between promoter length and rapidity of disease course (Zhang et al., 1997). However, two LTNPs were shown to harbor virus that exhibited what could be defined as a defective LTR. G to A hypermutations were observed throughout the promoter region of LTRs derived from one LTNP (Zhang et al., 1997), and another LTNP was shown to harbor virus with multiple insertions and deletions across the LTR (Rousseau, Abrams, Lee, Urbano, & King, 1997), both indicating defects within the 5′ LTR structure and suggesting that impaired functionality of the HIV-1 LTR may correspond to long-term nonprogression in a subset of HIV-1-infected patients. Other reports have postulated that LTRs with increased activity may correlate with greater viral infectivity and propagation throughout high-risk populations. HIV-1 subtype C LTRs, which have been show to contain three NF-κB-binding sites, exhibit enhanced LTR activation when compared with LTRs containing only one or two of these sites. Additionally, studies have shown that subtype C viruses also produce increased levels of p24, thus indicating greater replication rates than viruses representative of other HIV-1 subtypes (Naghavi, Schwartz, Sonnerborg, & Vahlne, 1999).

Studies have demonstrated that specific HIV-1 LTR C/EBP-binding site sequence configurations may be preferentially compartmentalized in the brain of infected patients and that these LTR variants exhibit enhanced LTR-mediated transcriptional activity (Hogan, Stauff, et al., 2003; Ross et al., 2001). Transient expression studies have suggested that an NF-κB-proximal C/EBP site I that binds C/EBP factors with high relative affinity results in increased basal as well as IL-6-induced LTR activity (Ross et al., 2001). Investigations have revealed that specific HIV-1 LTR C/EBP configurations preferentially encountered in the brain exhibit enhanced

LTR-specific activity (Ross et al., 2001). A high relative affinity 6G C/EBP site I (T to G change at nucleotide position 6) was commonly found in brain-derived LTRs but was infrequently encountered in peripheral blood-derived LTRs, as demonstrated by analyses of variations at each nucleotide position within C/EBP site I (Hogan, Stauff, et al., 2003). A differential level of conservation was also observed at C/EBP site II. Analyses of overall conservation of each site demonstrated that C/EBP site II was highly conserved in LTRs derived from brain and less conserved among those derived from the peripheral blood compartment. Overall, these studies demonstrated that brain-derived LTRs contain two high relative affinity C/EBP-binding sites and suggest that these sites may play a particular role in LTR-directed transcription with respect to CNS disease.

Studies have also suggested a direct correlation between specific C/EBP sequence variants and HAD. Sequence analysis of C/EBP-binding sites I and II using peripheral blood-derived HIV-1 LTR sequences was reported in three studies (Estable et al., 1996; Kirchhoff, Greenough, Hamacher, Sullivan, & Desrosiers, 1997; Michael et al., 1994). Because these published reports used different classification systems to describe disease severity, LTRs were designated as belonging to one of three groups prior to analysis. HIV-1 LTR sequences from asymptomatic patients with nonprogressing or stage I disease were assigned to disease severity group 1 (DSG 1) (Hogan, Stauff, et al., 2003). DSG 2 was comprised of patients characterized as having slow-progressing, stage II or III disease (Hogan, Stauff, et al., 2003). DSG 3 consisted of patients who were originally classified as having progressing, stage IV HIV-1 disease (Hogan, Stauff, et al., 2003). At C/EBP-binding site I, a 3T (C to T change at nucleotide position 3) configuration was observed at low prevalence within LTRs isolated from the peripheral blood of HIV-1-infected patients early in disease, and at relatively high prevalence from patients with late-stage disease (Hogan, Stauff, et al., 2003). The prevalence of the 3T C/EBP site I variant was not identified among LTRs from DSG1 patients and increased from approximately 8% of all DSG 2 LTRs to nearly 50% of all DSG 3 LTRs (Hogan, Stauff, et al., 2003). Within C/EBP site II, the consensus B (conB) configuration increased significantly throughout disease progression, whereas the prevalence of the 6G and 4C (T to C change at nucleotide position 4) variants decreased (Hogan, Stauff, et al., 2003). The conB configuration at C/EBP site II increased in prevalence from approximately 24% of all DSG 1 LTRs to approximately 93% of all DSG 3 LTRs. Conversely, the 4C C/EBP site II sequence variant decreased from approximately 28% in DSG 1 to approximately 8% in DSG 2 and was completely absent among DSG 3 LTRs (Hogan, Stauff, et al., 2003). Likewise, the 6G C/EBP site sequence variant decreased in prevalence from nearly 35% in DSG 1 to approximately 1% in DSG 3 LTRs. Interestingly, in this as well as similar studies, described in more detail below, involving analysis of Sp transcription factor-binding sites I, II, and III, both NF-κB-binding sites I and II were shown to be highly conserved in the conB configuration throughout disease progression (Hogan, Stauff, et al., 2003; Nonnemacher et al., 2004).

With respect to the impact of the genetic variants in C/EBP sites I and II and CNS disease, the 3T C/EBP site I variant was also observed in 25% of all brain-derived LTRs from patients diagnosed with dementia but was absent in brain-derived LTRs from patients without dementia (Hogan, Stauff, et al., 2003). The 3T C/EBP site I sequence configuration has been shown to have low relative affinity for C/EBP factors (Burdo, Gartner, et al., 2004; Hogan, Stauff, et al., 2003). Taken together, these results suggest that the 3T C/EBP site I configuration may provide a valuable tool in evaluating the likelihood of HIV-1-infected patients developing HAD. Similar to the observations made concerning the 3T C/EBP site I variant, the 6G and 4C C/EBP site II variants were observed in approximately 10% and 7% of HAD patients, respectively, and neither the 6G nor the 4C variant was found in LTRs derived from patients without dementia (Hogan, Stauff, et al., 2003). One study examining the regional distribution of HIV-1 LTRs containing the 6G and 4C C/EBP site II sequence variants within the brains of patients with HAD revealed statistically significant differences, with the high-affinity 6G C/EBP site II accumulating in the midfrontal gyrus and the low-affinity 4C C/EBP site II accumulating in the cerebellum (Burdo, Gartner, et al., 2004). These observations are consistent with reports of viral replication rates within these neuroanatomical regions, with the midfrontal gyrus representing a neuroanatomical region known to exhibit high-level HIV-1 replication (Glass, Fedor, Wesselingh, & McArthur, 1995), whereas viral gene expression within the cerebellum has been shown to occur at very low levels in patients with HAD (Burdo, Gartner, et al., 2004). This suggests that the presence of the 4C C/EBP site II variant within the cerebellum may represent a means by which HIV-1 maintains a silent genome and establishes a latent viral reservoir in the brain.

The three Sp-binding sites that comprise the remaining sequences within the core region of the HIV-1 LTR are also very important to HIV-1 basal and Tat-mediated transactivation (McAllister et al., 2000). Mutation of these binding sites diminishes both basal promoter activity and viral replication (although not in all cell types) (J.J. McAllister and B. Wigdahl, unpublished observation). Specific sequence variations in the NF-κB-proximal Sp site III present within the brain-derived HIV-1 variant, YU-2, result in a failure to interact efficiently with members of the Sp transcription factor family (McAllister et al., 2000) (unpublished observation). This result, combined with the observation that the ratio of Sp1:Sp3 factor binding to Sp site III is increased during monocytic differentiation, suggests that HIV-1 replication within cells of monocyte–macrophage lineage in the brain may be impacted by changes in Sp factor expression that accompany monocytic differentiation as well as alterations of the functional interactions between Sp factors and the NF-κB proximal, G/C-rich Sp-binding site (McAllister et al., 2000) (unpublished observation). Sequence variation within the Sp-binding sites and altered Sp factor recruitment may also impact the ability of HIV-1 Vpr to interact with and subsequently upregulate LTR activity (McAllister et al., 2000 and unpublished observations). Studies aimed at characterizing specific Sp-binding site sequence variants within 348 peripheral blood-derived HIV-1 LTRs isolated from patients with disease ranging in severity from DSG 1 to DSG 3 have demonstrated the presence of a low-affinity 5T (C to T change at nucleotide position 5) variant in Sp site III (Nonnemacher et al., 2004). The 5T Sp site III was shown to increase in prevalence throughout disease progression, with approximately 60% of all DSG 3 LTRs (Nonnemacher et al., 2004). Similar to the 3T C/EBP site I variant, the 5T sequence configuration results in substantially decreased binding affinity for Sp transcription factors, as demonstrated by both EMS (Nonnemacher et al., 2004) and surface plasmon resonance analyses (Nonnemacher and Wigdahl, unpublished observations). Interestingly, when the 3T C/EBP site I containing LTRs were examined for the presence of the 5T Sp site III sequence variant, an absolute correlation was observed with respect to DSG3. Of nine DSG2 LTRs that contained the 3T C/EBP site I sequence variant, six also contained the 5T Sp site III variant; and of 44 DSG3 LTRs that contained the 3T C/EBP site I sequence variant, all 44 also contained the 5T Sp site III variant (unpublished data). Importantly, the 5T Sp site III variant was also observed in 16% of all brain-derived LTRs from patients diagnosed with dementia but was absent in brain-derived LTRs from patients without dementia (unpublished observations). Taking into consideration that decreased binding of C/EBP and Sp factors corresponds with impaired HIV-1 LTR activity, the sequence variation observed within this region is almost certainly relevant to viral pathogenesis and may also influence both the course and severity of immunologic as well neurologic HIV-1 disease progression.

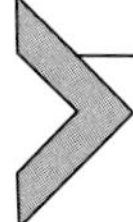

8. HOST GENETIC DETERMINANTS OF HIV-1 INFECTION AND CNS DISEASE

The course of HIV-1 infection is determined by a complex and dynamic interplay between viral and host factors. The very existence of LTNPs, in addition to individuals who have been exposed to HIV but remain uninfected, strongly suggests the existence of predisposing factors that may represent major determinants of clinical disease outcome. Post-seroconversion, progression to AIDS may take as little as 2 years in some individuals, while others may remain symptom-free for more than a decade. To this end, several studies have determined that the marked heterogeneity among infected individuals is governed, at least in part, by host genetic variants that serve to modulate virus replication and antiviral immunity (Carrington & O'Brien, 2003; Fauci, 2003; O'Brien & Nelson, 2004). The most widely recognized host genetic variant to be associated with HIV-1 disease progression is the 32-bp deletion in the coding region of the CCR5 gene (CCR5-Δ32). CCR5-Δ32 is one of the most significant host polymorphism with respect to HIV infection, and it has been shown to effectively block HIV-1 infection in homozygous individuals and to significantly retard disease progression and development of AIDS in heterozygotes (Carrington et al., 1997; Dean et al., 1996; Liu et al., 1996; O'Brien & Nelson, 2004; Samson et al., 1996). Subsequent to the discovery of CCR5-Δ 32, 13 additional host polymorphisms have been identified, and these genetic variants differ widely with respect to their influence on HIV-1-associated disease progression and development of AIDS-defining illnesses (O'Brien & Nelson, 2004). Here, we restrict our discussion to CCR5-Δ32 and CCR2-V64I, which have been linked to HIV-1 neuropathogenesis.

Chemokines and chemokine receptors play a critical role in the pathogenesis and transmission of HIV-1 (Michael, 1999; Paxton & Kang, 1998). The chemokine receptor CCR5 is utilized by R5 strains of HIV-1 to gain entry into host cells, particularly those of the monocyte–macrophage lineage, cells that have been shown to be intricately involved in HAND (Alkhatib et al., 1996; Berger et al., 1998; Choe et al., 1996; Roos et al., 1992). Several single-nucleotide polymorphisms identified within the regulatory region of CCR5 have been implicated in modulating the rate of HIV-1 disease progression (Kostrikis et al., 1999; Martin et al., 1998).

The CCR5-Δ32 gene variant has been shown to result in the production of aberrant CCR5 protein and is known to provide considerable protection against HIV-1 infection in individuals homozygous for the mutation and to result in slowed disease progression in heterozygous patients infected with HIV-1 (Barroga et al., 2000; Liu et al., 1996; Singh et al., 2003). Chemokines and their cognate receptors are also expressed within the brain (Glabinski et al., 1995), and CCR5 has been shown to be the primary coreceptor utilized by brain-derived HIV-1 strains isolated from HAD patients (Albright et al., 1999; Boven, van der Bruggen, van Asbeck, Marx, & Nottet, 1999). This, in addition to resident brain microglia and macrophages being the primary cells supporting replication of R5 HIV-1 strains within the CNS, leads to the conclusion that any host genetic variant that may inhibit or at least decrease the ability of HIV-1 to infect these cell types could have a profound effect on the neuropathogenesis of HIV-1 infection and subsequent neurocognitive abnormalities.

The chemokine receptor CCR2 has been shown to function as a minor coreceptor for HIV-1, and its natural ligand is MCP-1, a β-chemokine, which has been linked to neuropsychological impairment (Conant et al., 1998; Weiss, Cuff, & Berman, 1999). A genetic variant within the promoter region of MCP-1 (−2518-G/A polymorphism) has been associated with increased HIV-1 disease progression and development of HAD (Singh et al., 2004), and homozygosity for an MCP-1 2578-G allele has been associated with a 50% decrease in the risk of becoming infected with HIV-1 (Gonzalez et al., 2002). A specific polymorphism (G to A change at nucleotide position 190) in the coding region of the CCR2 gene results in the expression of isoleucine in place of valine at amino acid position 64 (V64I), and this nonsynonymous mutation has been linked with impaired progression of HIV-1 disease in adults (Kostrikis et al., 1998; van Rij et al., 1998); however, this association has been contested by other studies (Michael et al., 1997; Mulherin et al., 2003). In a study investigating development of neurologic complications in 121 HIV-1 patients, the CCR2-V64I polymorphism correlated with more rapid progression to neuropsychological impairment (Singh et al., 2004). An observation that was further confirmed among individuals possessing the CCR5-wt/wt genotype suggested that heterozygosity for the CCR5-Δ32 genotype could not be used to explain the differential rate of progression to neurologic disease in this patient cohort (Singh et al., 2004). Interestingly, although the CCR2-V64I allele was associated with more rapid onset of neuropsychological impairment, it was shown not to have an impact on the overall rate of

HIV-1-associated disease progression (Singh et al., 2004). Additionally, the CCR2-64 genotype did not correspond to differential HIV-1 plasma viral load or to $CD4^+$ T-cell count, suggesting that this genotype is not directly involved with viral entry or replication but with the host inflammatory response mediated by the binding of MCP-1 to the CCR2 receptor within the CNS (Singh et al., 2004). Moreover, no association was observed between the genotype of CCR2 and HIV disease progression or therapeutic response (Philpott et al., 2004), indicating that HAART therapy benefits may overshadow the advantages imparted by the V64I polymorphism.

More recently, to investigate the virus–host interactions and their role in HIV-1 pathogenesis, genome-wide association studies (GWAS) have been conducted (Table 6.1) that utilized unbiased searches at the genome-wide level to identify yet unidentified genetic factors and cellular pathways associated with HIV-1 infection and these studies have been previously reviewed (van Manen, van 't Wout, & Schuitemaker, 2012). These association studies were greatly facilitated by a number of clinical HIV-1 disease cohorts that have been established over the years. An alternative approach of sequencing the whole exome that selectively sequences the coding regions can prove to be more cost-effective, but the caveat is that many trait-related single-nucleotide polymorphisms fall in intergenic and noncoding regions (Hindorff et al., 2009; Manolio, 2010). Moreover, the gaps in our knowledge of the viral promoter, the LTR, can also be filled by application of comparative genomics and deep sequencing techniques to find key signatures within the LTR that associate with changes in viral pathogenesis and/or alterations in clinical measurements of disease severity such as viral load, $CD4^+$ T-cell count, HAND, and/or other pathologies associated with HIV disease.

Nevertheless, the efforts in these large-scale approaches (Fellay, Shianna, Telenti, & Goldstein, 2010) have enabled studies to examine HIV pathogenesis at a more global level, but the inter-individual variability that plays an important role in disease progression remains to be explored. The shift to whole genome sequencing will further help in accelerating the effort to find signatures that can be potentially used for therapeutic intervention regardless of whether they are of cellular or viral in origin.

9. CONCLUSION

The onset of HIV-1 CNS disease is largely dependent on the trafficking of infected monocytic cells from the periphery across the BBB, where virus is subsequently disseminated to susceptible cell populations within

Table 6.1 Genetic factors and cellular pathways associated with HIV-1 infection identified through genome-wide association studies (GWAS) on HIV pathogenesis

Phenotype	Association	Validation technique	Reference(s)
		Significant and confirmed	
RNA VL set point	HCP5 (rs2395029)	Genotyping, Illumina's HumanHap550 BeadChip	Catano et al. (2008), Fellay et al. (2007), Stern et al. (2012), Telenti and Johnson (2012)
VL controllers	−35 HLAC (rs9264942)	Genotyping, genotype–phenotype associations	Catano et al. (2008), Fellay et al. (2007)
	HLA-B*5703	IFN-γ ELISPOT, tetramer staining, Illumina HumanHap BeadChip	Pereyra et al. (2010)
Long-term nonprogression	HCP5 (rs2395029)	Meta-analysis using Infinium II HumanHap300 BeadChips	Limou et al. (2009)
	CXCR6 (rs2234358)	Illumina Infinium II HumanHap300 BeadChips	Limou et al. (2010)
Progression to AIDS	PARD3B (rs11884476)	Regression analysis	Troyer et al. (2011)
		Significant	
VL controllers	MICA (rs4418214) >300 SNPs in MHC	Genotype association, regression modeling	Pereyra et al. (2010)
		Confirmed	
CD4 T-cell decline	ZNRD1 (rs9261174)	Genotyping, shRNA gene knockdown	Ballana et al. (2010), Fellay et al. (2007)
Progression to AIDS	PROX1 (rs17762192)	Meta-analysis; immunoblotting	Herbeck et al. (2010)
Nevirapine tolerance	CCHCR1 (rs1265112)	Illumina HumanHap550v3 Genotyping BeadChip	Chantarangsu et al. (2011), Lingappa et al. (2011)

Table 6.1 Genetic factors and cellular pathways associated with HIV-1 infection identified through genome-wide association studies (GWAS) on HIV pathogenesis—cont'd

Phenotype	Association	Validation technique	Reference(s)
		Putative	
Long-term nonprogression	C6orf48 (rs9368699)	Meta-analysis using Infinium II HumanHap300	Limou et al. (2009)
Progression to AIDS	PRMT6 (rs4118325)	Illumina HumanHap300 BeadChips	Le Clerc et al. (2009)
	SOX5 (rs1522232)	Illumina HumanHap300 BeadChips	
	AGR3 (rs152363)	Illumina's Infinium HumanHap300 BeadChip	van Manen et al. (2011)
In vitro replication in macrophages	DYRK1A (rs12483205)	Illumina 610 Quad BeadChip, *in vitro* HIV-1 replication assays	Bol et al. (2011)
Mother-to-child transmission	HS3ST3A1 (rs8069770)	Illumina's HumanHap650Y Genotyping BeadChip	Joubert et al. (2010)

Significant and confirmed: p value for association less than 5×10^{-8} and repeated independently to confirm the association with significance. Significant: meets current standard for genome-wide significance in GWAS (has a p value below 5×10^{-8}); needs to be replicated independently. Confirmed: SNP tested to have association with a phenotype but failed to achieve a significant p value to have genome-wide significance (i.e., p value $>5 \times 10^{-8}$). Putative: p value $>5 \times 10^{-8}$ and yet to be confirmed by independent analysis.
Adapted from van Manen et al. (2012).

the brain including microglia and perivascular macrophages. Phylogenetic studies have suggested that cells of the monocyte–macrophage lineage may become infected within the bone marrow, and molecular genetic analyses of HIV-1 LTR, Tat, and Env sequences have revealed that evolutionary events occur both within the periphery and in the CNS (Liu et al., 2000). Current evidence indicates that HIV-1 within the brain may adapt specifically to selective pressures present within that physiological compartment, conferring upon the virus the capacity to utilize lower levels of CD4 and CCR5 for binding and entry (Doms, 2000; Doms & Moore, 2000). This observation, combined with results suggesting the presence of greater molecular diversity among HIV-1 quasispecies isolated from HAD versus non-HAD patients (Power et al., 1994), supports the idea that HIV-1 becomes specifically conditioned for efficient replication within the CNS. To date, no specific brain-adapted, neurovirulent virus has been identified, but lack of these observations may be due to limited experimental capabilities and incomplete patient information. If a brain-specific virus exists, one might postulate that it would result in greater neurologic disease and more severe NCI of its host. However, this also has not been observed, perhaps owing in part to the suppressive effects of combination antiretroviral therapy on reseeding of the virus in the brain owing to effective control of the viral replication in the periphery or more effective penetration of antiretroviral drug combinations into the CNS resulting in improved intra-CNS control of viral gene expression and replication.

The past two decades have witnessed an extraordinary experimental effort centered on elucidating both the causes and effects of HIV-1 infection of the CNS. Through these efforts, a number of viral adaptations and evolutionary events have been observed. Genetic mutations present throughout the HIV-1 genome have been identified and correlated both with peripheral immune and/or CNS disease progression and severity (Burdo, Gartner, et al., 2004; Dunfee et al., 2006; Hogan, Stauff, et al., 2003; Johnston et al., 2001). Multiple amino acid variants have been identified within the hypervariable regions of the HIV-1 Env that facilitate viral evasion of the host's adaptive immune response (Dunfee et al., 2006). Brain-derived HIV-1 Env sequences from HAD patients have been shown to differ from those of non-HAD patients, suggesting that specific HIV-1 Env sequences may correlate with neurotropism, neuroinvasion, CNS infection, and the subsequent onset of NCI (Van Marle et al., 2002). Investigations of the HIV-1 LTR within cells of the monocyte–macrophage lineage have focused heavily on the role of C/EBP and Sp transcription factor families. Specific C/EBP-binding site sequence variants contained within the LTR that are associated with increased LTR-mediated transcriptional activity have been identified within the brains of HAD patients (Burdo, Gartner, et al., 2004; Hogan, Stauff, et al., 2003; Ross et al., 2001). Likewise, sequence variation within Sp transcription factor-binding sites, which results in altered Sp factor recruitment and thus diminished viral replication, may also impact the capacity of HIV-1 Vpr to modulate LTR activity (McAllister et al., 2000; Nonnemacher et al., 2004). Analyses of HIV-1 Tat derived from patients with and without clinical indications of neurological impairment have revealed evidence of phylogenetic clustering with respect to neuropathological status and tissue of origin (Bratanich et al., 1998; Mayne et al., 1998). HIV-1 Tat isolated from the brain of HAD patients has been linked with increased neuronal death (Johnston et al., 2001), and evidence suggests that brain-derived HIV-1 Tat may be limited in its ability to upregulate viral gene expression (Silva et al., 2003).

The knowledge obtained by studying these individual viral evolutionary events has greatly enhanced our understanding of how HIV-1 establishes infection within the CNS and ultimately causes disease; however, the overall value of this information is limited in that the analysis of these individual mutations alone does not allow for a complete appreciation of how the virus is adapting to global environmental changes within a given host or host population. In order to gain an accurate understanding of why HIV-1 selects for specific genetic variants, one must take a step back and determine whether multiple molecular alterations may be coselected for within the same viral genome. For example, the 3T C/EBP site I-binding site sequence variant would, at first glance, seem deleterious to the virus in that it prevents binding of a transcription factor that has been shown to be critical for viral replication with cells of the monocyte lineage. The same can be said for Tat variants isolated from the brains of infected patients, which appear to be more toxic to neuronal cells and less capable of transactivating the HIV-1 LTR. However, on further examination, these and other seemingly nonadvantageous viral structural and functional changes begin to make sense within the bigger picture of HIV-1 adaptation to changing host conditions. Although the 3T C/EBP site I variant binds C/EBP factors with low relative affinity, it has been shown to bind Vpr with high relative affinity, and studies have shown that only one functional C/EBP site is required for viral replication within cells of the monocyte–macrophage lineage (Burdo, Nonnemacher, et al., 2004; Hogan, Nonnemacher, et al., 2003). Interestingly, in all HIV-1 LTRs identified as containing the 3T C/EBP site I configuration, C/EBP-binding site II has been identified as containing a conB sequence configuration, a high-affinity C/EBP-binding site. Why, then, would the virus choose to sacrifice C/EBP binding in exchange for Vpr binding to the LTR? The answer to that question is not completely known, but one explanation may be found in the fact that Vpr binding to the HIV-1 LTR has been correlated with HAD (Burdo, Gartner, et al., 2004; Hogan, Stauff, et al., 2003). Therefore, within the brain, the virus contains a Tat protein that is inefficient with respect to driving viral gene expression, a deficiency that may be compensated for by Vpr binding to and subsequently transactivating the HIV-1 LTR, a process that can occur only after the virus has mutated in such a way as to ameliorate binding of C/EBP factors to C/EBP site I, while maintaining binding integrity at C/EBP site II. A similar story can be told regarding the 5T Sp site III sequence variant. Although this theory has not yet been demonstrated experimentally, perhaps by preventing binding of Sp factors to Sp-binding site III the virus may facilitate greater binding of NF-κB, a transcription factor that has been shown to be absolutely required for viral replication within all HIV-1-susceptible host cell populations (McAllister et al., 2000; Nonnemacher et al., 2004).

Regional distribution analysis of HIV-1 LTR sequence variants has revealed a relationship between the accumulation patterns of specific C/EBP site II variants' viral replication rates (Burdo, Gartner, et al., 2004). The 6G C/EBP site II sequence configuration was shown to localize in the midfrontal gyrus, an area known to support relatively high levels of viral gene expression, whereas the 4C C/EBP site II variant was shown to accumulate in the cerebellum, a region exhibiting low-level viral replication (Burdo, Gartner, et al., 2004). These results, which provide evidence for the regional distribution of specific HIV-1 LTR genotypes within the brains of HAD and non-HAD patients, are consistent with investigations suggesting the neuroanatomical regionalization of HIV-1 Env sequences. It would be of great value to determine whether multiple HIV-1 gene products such as Env, LTR, and Tat evolve independently of one another, thus making the regional distribution of specific sequence variants merely coincidental, or whether HIV-1 evolves in a coordinated manner. Coordinated selection of multiple genetic variants within differing genomic regions could explain, at least in part, the occurrence of particular sequence alterations that have been identified throughout the HIV-1 genome and that correspond with neurologic disease. The idea that multiple HIV-1 genes may adapt and evolve together in response to changing physiological conditions within specific tissue compartments, as well as microenvironments within those compartments, raises the possibility that specific brain-adapted HIV-1 envelopes correspond to specific brain-adapted Tat proteins, which in turn correspond to specific brain-adapted LTR sequences. Further investigation is required to substantiate this concept.

It may be useful to consider both the timing of HIV-1 entry into the CNS and the mechanisms by which it may distribute and accumulate within specific brain regions. As discussed previously, regionalization could very well be the result of brain-specific viral evolution whereby HIV-1 adapts to the particular physiological conditions present within varying CNS microenvironments. Another possibility is that, through the continual trafficking of HIV-1 to the CNS during the course of infection, only those viral species already genetically adapted for survival within particular brain regions remain viable, while all others die out. Another possible explanation of the region-specific distribution of HIV-1 within the brain is that trafficking of HIV-1 occurs primarily during late-stage disease and that viral evolutionary events occur prior to neuroinvasion within the peripheral blood and possibly the bone marrow. This latter possibility is, however, contradictory to evidence suggesting that the 6G and 4C C/EBP site II sequence configurations are found only in the peripheral blood in patients with early-stage disease but are present in the brain in HAD patients with late-stage disease. This finding would indicate that HIV-1 containing these LTR genotypes must invade the CNS during the early stages of disease.

One of the most profound advancements that might be realized by identifying and characterizing HIV-1 genetic mutations is their potential use as diagnostic markers of disease progression and as targets for preventive and therapeutic strategies. Several genetic mutations have been associated with varying degrees of disease severity, including the 3T C/EBP site I and 5T Sp site III variants. These markers of disease progression are currently being evaluated for their potential prognostic value (Burdo, Gartner, et al., 2004; Burdo, Nonnemacher, et al., 2004; Hogan, Nonnemacher, et al., 2003; Hogan, Stauff, et al., 2003; Nonnemacher et al., 2004). Studies are ongoing to develop an HIV-1 sequence database that will allow for the identification of additional sequence variants that may also serve as molecular markers of HIV-1 immune and nervous system disease. Developing an understanding of viral as well as host genetic heterogeneity may also prove useful in the prevention and treatment of HIV-1 infection and disease. This factor is being utilized in microbicide development, where the knowledge of how HIV-1 selectively adapts to more readily and efficiently infect and replicate within specific cellular compartments may generate effective pharmacologic barriers to viral entry and infection. Of course, enhancing our appreciation of how HIV-1 adapts for replication within the CNS will greatly facilitate the development of the next generation of antiretroviral therapeutics that can more effectively target compartmentalized virus within the brain and other tissue reservoirs.

It is clear that individual genetic mutations cannot be fully understood in isolation and that specific genomic mutations may very well be coselected for within specific viruses and viral quasispecies in response to local and global physiological changes within the host's peripheral circulation as well as specific tissue compartments such as the brain. Based on the available data, there is little debate as to whether genetic diversity of HIV-1 contributes to the onset and severity of HIV-1-associated neurologic disease; however, further investigation is required to definitively determine the impact that the genomic heterogeneity of HIV-1, and that of the host, has on the development of neuropathology and the onset of NCI.

ACKNOWLEDGMENTS

This work was supported in part by funds from the Public Health Service, National Institutes of Health, through grants from the National Institute of Neurological Disorders and Stroke (NS32092 to B. W.) and the National Institute of Drug Abuse (DA19807 to B. W.). M. R. N. was also supported by faculty development funds provided by the Department of Microbiology and Immunology and the Institute for Molecular Medicine and Infectious Disease.

REFERENCES

Addo, M. M., Yu, X. G., Rathod, A., Cohen, D., Eldridge, R. L., Strick, D., et al. (2003). Comprehensive epitope analysis of human immunodeficiency virus type 1 (HIV-1)-specific T-cell responses directed against the entire expressed HIV-1 genome demonstrate broadly directed responses, but no correlation to viral load. *Journal of Virology*, *77*, 2081.

Agrawal, L., Maxwell, C. R., Peters, P. J., Clapham, P. R., Liu, S. M., Mackay, C. R., et al. (2009). Complexity in human immunodeficiency virus type 1 (HIV-1) co-receptor usage: Roles of CCR3 and CCR5 in HIV-1 infection of monocyte-derived macrophages and brain microglia. *The Journal of General Virology*, *90*, 710.

Albright, A. V., Shieh, J. T., Itoh, T., Lee, B., Pleasure, D., O'Connor, M. J., et al. (1999). Microglia express CCR5, CXCR4, and CCR3, but of these, CCR5 is the principal coreceptor for human immunodeficiency virus type 1 dementia isolates. *Journal of Virology*, *73*, 205.

Alcami, J., Lain de Lera, T., Folgueira, L., Pedraza, M. A., Jacque, J. M., Bachelerie, F., et al. (1995). Absolute dependence on kappa B responsive elements for initiation and Tat-mediated amplification of HIV transcription in blood CD4 T lymphocytes. *The EMBO Journal*, *14*, 1552.

Alkhatib, G., Combadiere, C., Broder, C. C., Feng, Y., Kennedy, P. E., Murphy, P. M., et al. (1996). CC CKR5: A RANTES, MIP-1alpha, MIP-1beta receptor as a fusion cofactor for macrophage-tropic HIV-1. *Science*, *272*, 1955.

Ances, B. M., Vaida, F., Yeh, M. J., Liang, C. L., Buxton, R. B., Letendre, S., et al. (2010). HIV infection and aging independently affect brain function as measured by functional magnetic resonance imaging. *The Journal of Infectious Diseases*, *201*, 336.

Ardon, O., Zimmerman, E. S., Andersen, J. L., DeHart, J. L., Blackett, J., & Planelles, V. (2006). Induction of G2 arrest and binding to cyclophilin A are independent phenotypes of human immunodeficiency virus type 1 Vpr. *Journal of Virology*, *80*, 3694.

Asin, S., Bren, G. D., Carmona, E. M., Solan, N. J., & Paya, C. V. (2001). NF-kappaB cis-acting motifs of the human immunodeficiency virus (HIV) long terminal repeat regulate HIV transcription in human macrophages. *Journal of Virology*, *75*, 11408.

Asquith, B., Edwards, C. T., Lipsitch, M., & McLean, A. R. (2006). Inefficient cytotoxic T lymphocyte-mediated killing of HIV-1-infected cells in vivo. *PLoS Biology*, *4*, e90.

Avison, M. J., Nath, A., Greene-Avison, R., Schmitt, F. A., Bales, R. A., Ethisham, A., et al. (2004). Inflammatory changes and breakdown of microvascular integrity in early human immunodeficiency virus dementia. *Journal of Neurovirology, 10*, 223.

Ayyavoo, V., Mahalingam, S., Rafaeli, Y., Kudchodkar, S., Chang, D., Nagashunmugam, T., et al. (1997). HIV-1 viral protein R (Vpr) regulates viral replication and cellular proliferation in T cells and monocytoid cells in vitro. *Journal of Leukocyte Biology, 62*, 93.

Ballana, E., Senserrich, J., Pauls, E., Faner, R., Mercader, J. M., Uyttebroeck, F., et al. (2010). ZNRD1 (zinc ribbon domain-containing 1) is a host cellular factor that influences HIV-1 replication and disease progression. *Clinical Infectious Diseases, 50*, 1022.

Barroga, C. F., Raskino, C., Fangon, M. C., Palumbo, P. E., Baker, C. J., Englund, J. A., et al. (2000). The CCR5Delta32 allele slows disease progression of human immunodeficiency virus-1-infected children receiving antiretroviral treatment. *The Journal of Infectious Diseases, 182*, 413.

Berger, E. A., Doms, R. W., Fenyo, E. M., Korber, B. T., Littman, D. R., Moore, J. P., et al. (1998). A new classification for HIV-1. *Nature, 391*, 240.

Blodget, E., Shen, C., Aldrovandi, G., Rollie, A., Gupta, S. K., Stein, J. H., et al. (2012). Relationship between microbial translocation and endothelial function in HIV infected patients. *PLoS One, 7*, e42624.

Bol, S. M., Moerland, P. D., Limou, S., van Remmerden, Y., Coulonges, C., van Manen, D., et al. (2011). Genome-wide association study identifies single nucleotide polymorphism in DYRK1A associated with replication of HIV-1 in monocyte-derived macrophages. *PLoS One, 6*, e17190.

Bonsignori, M., Alam, S. M., Liao, H. X., Verkoczy, L., Tomaras, G. D., Haynes, B. F., et al. (2012). HIV-1 antibodies from infection and vaccination: Insights for guiding vaccine design. *Trends in Microbiology, 20*, 532–539.

Boven, L. A., van der Bruggen, T., van Asbeck, B. S., Marx, J. J., & Nottet, H. S. (1999). Potential role of CCR5 polymorphism in the development of AIDS dementia complex. *FEMS Immunology and Medical Microbiology, 26*, 243.

Brady, J., & Kashanchi, F. (2005). Tat gets the "green" light on transcription initiation. *Retrovirology, 2*, 69.

Bratanich, A. C., Liu, C., McArthur, J. C., Fudyk, T., Glass, J. D., Mittoo, S., et al. (1998). Brain-derived HIV-1 tat sequences from AIDS patients with dementia show increased molecular heterogeneity. *Journal of Neurovirology, 4*, 387.

Brew, B. J., Bhalla, R. B., Paul, M., Sidtis, J. J., Keilp, J. J., Sadler, A. E., et al. (1992). Cerebrospinal fluid beta 2-microglobulin in patients with AIDS dementia complex: An expanded series including response to zidovudine treatment. *AIDS, 6*, 461.

Brew, B. J., Dunbar, N., Pemberton, L., & Kaldor, J. (1996). Predictive markers of AIDS dementia complex: CD4 cell count and cerebrospinal fluid concentrations of beta 2-microglobulin and neopterin. *The Journal of Infectious Diseases, 174*, 294.

Bruce-Keller, A. J., Chauhan, A., Dimayuga, F. O., Gee, J., Keller, J. N., & Nath, A. (2003). Synaptic transport of human immunodeficiency virus-Tat protein causes neurotoxicity and gliosis in rat brain. *The Journal of Neuroscience, 23*, 8417.

Budka, H. (1991). Neuropathology of human immunodeficiency virus infection. *Brain Pathology, 1*, 163.

Budka, H., Costanzi, G., Cristina, S., Lechi, A., Parravicini, C., Trabattoni, R., et al. (1987). Brain pathology induced by infection with the human immunodeficiency virus (HIV). A histological, immunocytochemical, and electron microscopical study of 100 autopsy cases. *Acta Neuropathologica (Berlin), 75*, 185.

Burdo, T. H., Gartner, S., Mauger, D., & Wigdahl, B. (2004). Region-specific distribution of human immunodeficiency virus type 1 long terminal repeats containing specific configurations of CCAAT/enhancer-binding protein site II in brains derived from demented and nondemented patients. *Journal of Neurovirology, 10*(Suppl. 1), 7–14.

Burdo, T. H., Nonnemacher, M., Irish, B. P., Choi, C. H., Krebs, F. C., Gartner, S., et al. (2004). High-affinity interaction between HIV-1 Vpr and specific sequences that span the C/EBP and adjacent NF-kappaB sites within the HIV-1 LTR correlate with HIV-1-associated dementia. *DNA and Cell Biology*, *23*, 261.

Cao, J., McNevin, J., Malhotra, U., & McElrath, M. J. (2003). Evolution of CD8+ T cell immunity and viral escape following acute HIV-1 infection. *Journal of Immunology*, *171*, 3837.

Carrington, M., Dean, M., Martin, M. P., & O'Brien, S. J. (1999). Genetics of HIV-1 infection: Chemokine receptor CCR5 polymorphism and its consequences. *Human Molecular Genetics*, *8*, 1939.

Carrington, M., Kissner, T., Gerrard, B., Ivanov, S., O'Brien, S. J., & Dean, M. (1997). Novel alleles of the chemokine-receptor gene CCR5. *American Journal of Human Genetics*, *61*, 1261.

Carrington, M., & O'Brien, S. J. (2003). The influence of HLA genotype on AIDS. *Annual Review of Medicine*, *54*, 535.

Catano, G., Kulkarni, H., He, W., Marconi, V. C., Agan, B. K., Landrum, M., et al. (2008). HIV-1 disease-influencing effects associated with ZNRD1, HCP5 and HLA-C alleles are attributable mainly to either HLA-A10 or HLA-B*57 alleles. *PLoS One*, *3*, e3636.

Cecilia, D., Kleeberger, C., Munoz, A., Giorgi, J. V., & Zolla-Pazner, S. (1999). A longitudinal study of neutralizing antibodies and disease progression in HIV-1-infected subjects. *The Journal of Infectious Diseases*, *179*, 1365.

Chan, S. Y., Speck, R. F., Power, C., Gaffen, S. L., Chesebro, B., & Goldsmith, M. A. (1999). V3 recombinants indicate a central role for CCR5 as a coreceptor in tissue infection by human immunodeficiency virus type 1. *Journal of Virology*, *73*, 2350.

Chang, L. (1995). In vivo magnetic resonance spectroscopy in HIV and HIV-related brain diseases. *Reviews in the Neurosciences*, *6*, 365.

Chang, L., Ernst, T., St Hillaire, C., & Conant, K. (2004). Antiretroviral treatment alters relationship between MCP-1 and neurometabolites in HIV patients. *Antiviral Therapy*, *9*, 431.

Chang, J., Jozwiak, R., Wang, B., Ng, T., Ge, Y. C., Bolton, W., et al. (1998). Unique HIV type 1 V3 region sequences derived from six different regions of brain: Region-specific evolution within host-determined quasispecies. *AIDS Research and Human Retroviruses*, *14*, 25.

Chantarangsu, S., Mushiroda, T., Mahasirimongkol, S., Kiertiburanakul, S., Sungkanuparph, S., Manosuthi, W., et al. (2011). Genome-wide association study identifies variations in 6p21.3 associated with nevirapine-induced rash. *Clinical Infectious Diseases*, *53*, 341.

Chen, B. K., Feinberg, M. B., & Baltimore, D. (1997). The kappaB sites in the human immunodeficiency virus type 1 long terminal repeat enhance virus replication yet are not absolutely required for viral growth. *Journal of Virology*, *71*, 5495.

Cherner, M., Masliah, E., Ellis, R. J., Marcotte, T. D., Moore, D. J., Grant, I., et al. (2002). Neurocognitive dysfunction predicts postmortem findings of HIV encephalitis. *Neurology*, *59*, 1563.

Chesebro, B., Wehrly, K., Nishio, J., & Perryman, S. (1992). Macrophage-tropic human immunodeficiency virus isolates from different patients exhibit unusual V3 envelope sequence homogeneity in comparison with T-cell-tropic isolates: Definition of critical amino acids involved in cell tropism. *Journal of Virology*, *66*, 6547.

Chiao, C., Bader, T., Stenger, J. E., Baldwin, W., Brady, J., & Barrett, J. C. (2001). HIV type 1 Tat inhibits tumor necrosis factor alpha-induced repression of tumor necrosis factor receptor p55 and amplifies tumor necrosis factor alpha activity in stably tat-transfected HeLa Cells. *AIDS Research and Human Retroviruses*, *17*, 1125.

Childs, E. A., Lyles, R. H., Selnes, O. A., Chen, B., Miller, E. N., Cohen, B. A., et al. (1999). Plasma viral load and CD4 lymphocytes predict HIV-associated dementia and sensory neuropathy. *Neurology, 52*, 607.

Choe, H., Farzan, M., Sun, Y., Sullivan, N., Rollins, B., Ponath, P. D., et al. (1996). The beta-chemokine receptors CCR3 and CCR5 facilitate infection by primary HIV-1 isolates. *Cell, 85*, 1135.

Chrysikopoulos, H. S., Press, G. A., Grafe, M. R., Hesselink, J. R., & Wiley, C. A. (1990). Encephalitis caused by human immunodeficiency virus: CT and MR imaging manifestations with clinical and pathologic correlation. *Radiology, 175*, 185.

Cohen, E. A., Dehni, G., Sodroski, J. G., & Haseltine, W. A. (1990). Human immunodeficiency virus vpr product is a virion-associated regulatory protein. *Journal of Virology, 64*, 3097.

Collman, R., Balliet, J. W., Gregory, S. A., Friedman, H., Kolson, D. L., Nathanson, N., et al. (1992). An infectious molecular clone of an unusual macrophage-tropic and highly cytopathic strain of human immunodeficiency virus type 1. *Journal of Virology, 66*, 7517.

Conant, K., Garzino-Demo, A., Nath, A., McArthur, J. C., Halliday, W., Power, C., et al. (1998). Induction of monocyte chemoattractant protein-1 in HIV-1 Tat-stimulated astrocytes and elevation in AIDS dementia. *Proceedings of the National Academy of Sciences of the United States of America, 95*, 3117.

Corboy, J. R., Buzy, J. M., Zink, M. C., & Clements, J. E. (1992). Expression directed from HIV long terminal repeats in the central nervous system of transgenic mice. *Science, 258*, 1804.

Cosenza, M. A., Zhao, M. L., Si, Q., & Lee, S. C. (2002). Human brain parenchymal microglia express CD14 and CD45 and are productively infected by HIV-1 in HIV-1 encephalitis. *Brain Pathology, 12*, 442.

Cullen, B. R. (1991). Regulation of gene expression in the human immunodeficiency virus type 1. *Advances in Virus Research, 40*, 1.

Cunningham, A. L., Naif, H., Saksena, N., Lynch, G., Chang, J., Li, S., et al. (1997). HIV infection of macrophages and pathogenesis of AIDS dementia complex: Interaction of the host cell and viral genotype. *Journal of Leukocyte Biology, 62*, 117.

Dahiya, S., Nonnemacher, M. R., & Wigdahl, B. (2012). Deployment of the human immunodeficiency virus type 1 protein arsenal: Combating the host to enhance viral transcription and providing targets for therapeutic development. *The Journal of General Virology, 93*, 1151.

Dean, M., Carrington, M., Winkler, C., Huttley, G. A., Smith, M. W., Allikmets, R., et al. (1996). Genetic restriction of HIV-1 infection and progression to AIDS by a deletion allele of the CKR5 structural gene. Hemophilia growth and development study, multicenter AIDS cohort study, multicenter hemophilia cohort study, San Francisco city cohort, ALIVE study. *Science, 273*, 1856.

DeHart, J. L., Zimmerman, E. S., Ardon, O., Monteiro-Filho, C. M., Arganaraz, E. R., & Planelles, V. (2007). HIV-1 Vpr activates the G2 checkpoint through manipulation of the ubiquitin proteasome system. *Virology Journal, 4*, 57.

del Palacio, M., Alvarez, S., & Munoz-Fernandez, M. A. (2012). HIV-1 infection and neurocognitive impairment in the current era. *Reviews in Medical Virology, 22*, 33.

Deshmane, S. L., Mukerjee, R., Fan, S., Del Valle, L., Michiels, C., Sweet, T., et al. (2009). Activation of the oxidative stress pathway by HIV-1 Vpr leads to induction of hypoxia-inducible factor 1alpha expression. *The Journal of Biological Chemistry, 284*, 11364.

Doms, R. W. (2000). Beyond receptor expression: The influence of receptor conformation, density, and affinity in HIV-1 infection. *Virology, 276*, 229.

Doms, R. W., & Moore, J. P. (2000). HIV-1 membrane fusion: Targets of opportunity. *The Journal of Cell Biology, 151*, F9.

Doppler, C., Schalasta, G., Amtmann, E., & Sauer, G. (1992). Binding of NF-kB to the HIV-1 LTR is not sufficient to induce HIV-1 LTR activity. *AIDS Research and Human Retroviruses*, *8*, 245.

Drummond, N. S., Vilar, F. J., Naisbitt, D. J., Hanson, A., Woods, A., Park, B. K., et al. (2006). Drug-specific T cells in an HIV-positive patient with nevirapine-induced hepatitis. *Antiviral Therapy*, *11*, 393.

Duncan, C. J., & Sattentau, Q. J. (2011). Viral determinants of HIV-1 macrophage tropism. *Viruses*, *3*, 2255.

Dunfee, R. L., Thomas, E. R., Gorry, P. R., Wang, J., Taylor, J., Kunstman, K., et al. (2006). The HIV Env variant N283 enhances macrophage tropism and is associated with brain infection and dementia. *Proceedings of the National Academy of Sciences of the United States of America*, *103*, 15160.

Dunfee, R. L., Thomas, E. R., Wang, J., Kunstman, K., Wolinsky, S. M., & Gabuzda, D. (2007). Loss of the N-linked glycosylation site at position 386 in the HIV envelope V4 region enhances macrophage tropism and is associated with dementia. *Virology*, *367*, 222.

Emerman, M., & Malim, M. H. (1998). HIV-1 regulatory/accessory genes: Keys to unraveling viral and host cell biology. *Science*, *280*, 1880.

Ensoli, B., Buonaguro, L., Barillari, G., Fiorelli, V., Gendelman, R., Morgan, R. A., et al. (1993). Release, uptake, and effects of extracellular human immunodeficiency virus type 1 Tat protein on cell growth and viral transactivation. *Journal of Virology*, *67*, 277.

Enting, R. H., Foudraine, N. A., Lange, J. M., Jurriaans, S., van der Poll, T., Weverling, G. J., et al. (2000). Cerebrospinal fluid beta2-microglobulin, monocyte chemotactic protein-1, and soluble tumour necrosis factor alpha receptors before and after treatment with lamivudine plus zidovudine or stavudine. *Journal of Neuroimmunology*, *102*, 216.

Epstein, L. G., Kuiken, C., Blumberg, B. M., Hartman, S., Sharer, L. R., Clement, M., et al. (1991). HIV-1 V3 domain variation in brain and spleen of children with AIDS: Tissue-specific evolution within host-determined quasispecies. *Virology*, *180*, 583.

Epstein, L. G., Sharer, L. R., Cho, E. S., Myenhofer, M., Navia, B., & Price, R. W. (1984). HTLV-III/LAV-like retrovirus particles in the brains of patients with AIDS encephalopathy. *AIDS Research*, *1*, 447.

Essex, M. (1999). Human immunodeficiency viruses in the developing world. *Advances in Virus Research*, *53*, 71.

Estable, M. C., Bell, B., Merzouki, A., Montaner, J. S., O'Shaughnessy, M. V., & Sadowski, I. J. (1996). Human immunodeficiency virus type 1 long terminal repeat variants from 42 patients representing all stages of infection display a wide range of sequence polymorphism and transcription activity. *Journal of Virology*, *70*, 4053.

Fauci, A. S. (2003). HIV and AIDS: 20 Years of science. *Nature Medicine*, *9*, 839.

Fellay, J., Shianna, K. V., Ge, D., Colombo, S., Ledergerber, B., Weale, M., et al. (2007). A whole-genome association study of major determinants for host control of HIV-1. *Science*, *317*, 944.

Fellay, J., Shianna, K. V., Telenti, A., & Goldstein, D. B. (2010). Host genetics and HIV-1: The final phase? *PLoS Pathogens*, *6*, e1001033.

Felzien, L. K., Woffendin, C., Hottiger, M. O., Subbramanian, R. A., Cohen, E. A., & Nabel, G. J. (1998). HIV transcriptional activation by the accessory protein, VPR, is mediated by the p300 co-activator. *Proceedings of the National Academy of Sciences of the United States of America*, *95*, 5281.

Folks, T., Powell, D. M., Lightfoote, M. M., Benn, S., Martin, M. A., & Fauci, A. S. (1986). Induction of HTLV-III/LAV from a nonvirus-producing T-cell line: Implications for latency. *Science*, *231*, 600.

Fontaine, J., Poudrier, J., & Roger, M. (2011). Short communication: Persistence of high blood levels of the chemokines CCL2, CCL19, and CCL20 during the course of HIV infection. *AIDS Research and Human Retroviruses*, *27*, 655.

Frost, S. D., Wrin, T., Smith, D. M., Kosakovsky Pond, S. L., Liu, Y., Paxinos, E., et al. (2005). Neutralizing antibody responses drive the evolution of human immunodeficiency virus type 1 envelope during recent HIV infection. *Proceedings of the National Academy of Sciences of the United States of America, 102*, 18514.

Ganeshan, S., Dickover, R. E., Korber, B. T., Bryson, Y. J., & Wolinsky, S. M. (1997). Human immunodeficiency virus type 1 genetic evolution in children with different rates of development of disease. *Journal of Virology, 71*, 663.

Gannon, P., Khan, M. Z., & Kolson, D. L. (2011). Current understanding of HIV-associated neurocognitive disorders pathogenesis. *Current Opinion in Neurology, 24*, 275.

Ganusov, V. V., Goonetilleke, N., Liu, M. K., Ferrari, G., Shaw, G. M., McMichael, A. J., et al. (2011). Fitness costs and diversity of the cytotoxic T lymphocyte (CTL) response determine the rate of CTL escape during acute and chronic phases of HIV infection. *Journal of Virology, 85*, 10518.

Gao, F., Robertson, D. L., Morrison, S. G., Hui, H., Craig, S., Decker, J., et al. (1996). The heterosexual human immunodeficiency virus type 1 epidemic in Thailand is caused by an intersubtype (A/E) recombinant of African origin. *Journal of Virology, 70*, 7013.

Gartner, S. (2000). HIV infection and dementia. *Science, 287*, 602.

Gelbard, H. A., James, H. J., Sharer, L. R., Perry, S. W., Saito, Y., Kazee, A. M., et al. (1995). Apoptotic neurons in brains from paediatric patients with HIV-1 encephalitis and progressive encephalopathy. *Neuropathology and Applied Neurobiology, 21*, 208.

Genis, P., Jett, M., Bernton, E. W., Boyle, T., Gelbard, H. A., Dzenko, K., et al. (1992). Cytokines and arachidonic metabolites produced during human immunodeficiency virus (HIV)-infected macrophage-astroglia interactions: Implications for the neuropathogenesis of HIV disease. *The Journal of Experimental Medicine, 176*, 1703.

Glabinski, A. R., Tani, M., Aras, S., Stoler, M. H., Tuohy, V. K., & Ransohoff, R. M. (1995). Regulation and function of central nervous system chemokines. *International Journal of Developmental Neuroscience, 13*, 153.

Glass, J. D., Fedor, H., Wesselingh, S. L., & McArthur, J. C. (1995). Immunocytochemical quantitation of human immunodeficiency virus in the brain: Correlations with dementia. *Annals of Neurology, 38*, 755.

Goh, W. C., Rogel, M. E., Kinsey, C. M., Michael, S. F., Fultz, P. N., Nowak, M. A., et al. (1998). HIV-1 Vpr increases viral expression by manipulation of the cell cycle: A mechanism for selection of Vpr in vivo. *Nature Medicine, 4*, 65.

Golub, E. I., Li, G. R., & Volsky, D. J. (1991). Induction of dormant HIV-1 by sodium butyrate: Involvement of the TATA box in the activation of the HIV-1 promoter. *AIDS, 5*, 663.

Gonzalez, E., Rovin, B. H., Sen, L., Cooke, G., Dhanda, R., Mummidi, S., et al. (2002). HIV-1 infection and AIDS dementia are influenced by a mutant MCP-1 allele linked to increased monocyte infiltration of tissues and MCP-1 levels. *Proceedings of the National Academy of Sciences of the United States of America, 99*, 13795.

Gonzalez-Perez, M. P., O'Connell, O., Lin, R., Sullivan, W. M., Bell, J., Simmonds, P., et al. (2012). Independent evolution of macrophage-tropism and increased charge between HIV-1 R5 envelopes present in brain and immune tissue. *Retrovirology, 9*, 20.

Gorry, P. R., Bristol, G., Zack, J. A., Ritola, K., Swanstrom, R., Birch, C. J., et al. (2001). Macrophage tropism of human immunodeficiency virus type 1 isolates from brain and lymphoid tissues predicts neurotropism independent of coreceptor specificity. *Journal of Virology, 75*, 10073.

Gorry, P. R., Howard, J. L., Churchill, M. J., Anderson, J. L., Cunningham, A., Adrian, D., et al. (1999). Diminished production of human immunodeficiency virus type 1 in astrocytes results from inefficient translation of gag, env, and nef mRNAs despite efficient expression of Tat and Rev. *Journal of Virology, 73*, 352.

Gorry, P. R., Taylor, J., Holm, G. H., Mehle, A., Morgan, T., Cayabyab, M., et al. (2002). Increased CCR5 affinity and reduced CCR5/CD4 dependence of a neurovirulent primary human immunodeficiency virus type 1 isolate. *Journal of Virology*, *76*, 6277.
Gray, L. R., Gabuzda, D., Cowley, D., Ellett, A., Chiavaroli, L., Wesselingh, S. L., et al. (2011). CD4 and MHC class 1 down-modulation activities of nef alleles from brain- and lymphoid tissue-derived primary HIV-1 isolates. *Journal of Neurovirology*, *17*, 82.
Gray, L., Sterjovski, J., Churchill, M., Ellery, P., Nasr, N., Lewin, S. R., et al. (2005). Uncoupling coreceptor usage of human immunodeficiency virus type 1 (HIV-1) from macrophage tropism reveals biological properties of CCR5-restricted HIV-1 isolates from patients with acquired immunodeficiency syndrome. *Virology*, *337*, 384.
Green, M., Ishino, M., & Loewenstein, P. M. (1989). Mutational analysis of HIV-1 Tat minimal domain peptides: Identification of trans-dominant mutants that suppress HIV-LTR-driven gene expression. *Cell*, *58*, 215.
Griffin, G. E., Leung, K., Folks, T. M., Kunkel, S., & Nabel, G. J. (1989). Activation of HIV gene expression during monocyte differentiation by induction of NF-kappa B. *Nature*, *339*, 70.
Guha, D., Nagilla, P., Redinger, C., Srinivasan, A., Schatten, G. P., & Ayyavoo, V. (2012). Neuronal apoptosis by HIV-1 Vpr: Contribution of proinflammatory molecular networks from infected target cells. *Journal of Neuroinflammation*, *9*, 138.
Haase, A. T. (1986). Pathogenesis of lentivirus infections. *Nature*, *322*, 130.
He, J., Chen, Y., Farzan, M., Choe, H., Ohagen, A., Gartner, S., et al. (1997). CCR3 and CCR5 are co-receptors for HIV-1 infection of microglia. *Nature*, *385*, 645.
Heaton, R. K., Franklin, D. R., Ellis, R. J., McCutchan, J. A., Letendre, S. L., Leblanc, S., et al. (2011). HIV-associated neurocognitive disorders before and during the era of combination antiretroviral therapy: Differences in rates, nature, and predictors. *Journal of Neurovirology*, *17*, 3.
Henderson, A. J., & Calame, K. L. (1997). CCAAT/enhancer binding protein (C/EBP) sites are required for HIV-1 replication in primary macrophages but not CD4(+) T cells. *Proceedings of the National Academy of Sciences of the United States of America*, *94*, 8714.
Henderson, A. J., Connor, R. I., & Calame, K. L. (1996). C/EBP activators are required for HIV-1 replication and proviral induction in monocytic cell lines. *Immunity*, *5*, 91.
Henderson, A. J., Zou, X., & Calame, K. L. (1995). C/EBP proteins activate transcription from the human immunodeficiency virus type 1 long terminal repeat in macrophages/monocytes. *Journal of Virology*, *69*, 5337.
Herbeck, J. T., Gottlieb, G. S., Winkler, C. A., Nelson, G. W., An, P., Maust, B. S., et al. (2010). Multistage genomewide association study identifies a locus at 1q41 associated with rate of HIV-1 disease progression to clinical AIDS. *The Journal of Infectious Diseases*, *201*, 618.
Heyes, M. P., Ellis, R. J., Ryan, L., Childers, M. E., Grant, I., Wolfson, T., et al. (2001). Elevated cerebrospinal fluid quinolinic acid levels are associated with region-specific cerebral volume loss in HIV infection. *Brain*, *124*, 1033.
Hickey, W. F. (1999). Leukocyte traffic in the central nervous system: The participants and their roles. *Seminars in Immunology*, *11*, 125.
Hill, A. L., Rosenbloom, D. I., & Nowak, M. A. (2012). Evolutionary dynamics of HIV at multiple spatial and temporal scales. *Journal of Molecular Medicine (Berlin, Germany)*, *90*, 543.
Hindorff, L. A., Sethupathy, P., Junkins, H. A., Ramos, E. M., Mehta, J. P., Collins, F. S., et al. (2009). Potential etiologic and functional implications of genome-wide association loci for human diseases and traits. *Proceedings of the National Academy of Sciences of the United States of America*, *106*, 9362.
Hoffman, T. L., Stephens, E. B., Narayan, O., & Doms, R. W. (1998). HIV type I envelope determinants for use of the CCR2b, CCR3, STRL33, and APJ coreceptors. *Proceedings of the National Academy of Sciences of the United States of America*, *95*, 11360.

Hogan, T. H., Nonnemacher, M. R., Krebs, F. C., Henderson, A., & Wigdahl, B. (2003). HIV-1 Vpr binding to HIV-1 LTR C/EBP cis-acting elements and adjacent regions is sequence-specific. *Biomedicine and Pharmacotherapy*, *57*, 41.

Hogan, T. H., Stauff, D. L., Krebs, F. C., Gartner, S., Quiterio, S. J., & Wigdahl, B. (2003). Structural and functional evolution of human immunodeficiency virus type 1 long terminal repeat CCAAT/enhancer binding protein sites and their use as molecular markers for central nervous system disease progression. *Journal of Neurovirology*, *9*, 55.

Huang, K. J., Alter, G. M., & Wooley, D. P. (2002). The reverse transcriptase sequence of human immunodeficiency virus type 1 is under positive evolutionary selection within the central nervous system. *Journal of Neurovirology*, *8*, 281.

Hudson, L., Liu, J., Nath, A., Jones, M., Raghavan, R., Narayan, O., et al. (2000). Detection of the human immunodeficiency virus regulatory protein tat in CNS tissues. *Journal of Neurovirology*, *6*, 145.

Hwang, S. S., Boyle, T. J., Lyerly, H. K., & Cullen, B. R. (1991). Identification of the envelope V3 loop as the primary determinant of cell tropism in HIV-1. *Science*, *253*, 71.

Janssens, W., Buve, A., & Nkengasong, J. N. (1997). The puzzle of HIV-1 subtypes in Africa. *AIDS*, *11*, 705.

Johnston, J. B., Zhang, K., Silva, C., Shalinsky, D. R., Conant, K., Ni, W., et al. (2001). HIV-1 Tat neurotoxicity is prevented by matrix metalloproteinase inhibitors. *Annals of Neurology*, *49*, 230.

Jones, K. A., & Peterlin, B. M. (1994). Control of RNA initiation and elongation at the HIV-1 promoter. *Annual Review of Biochemistry*, *63*, 717.

Joubert, B. R., Lange, E. M., Franceschini, N., Mwapasa, V., North, K. E., & Meshnick, S. R. (2010). A whole genome association study of mother-to-child transmission of HIV in Malawi. *Genome Medicine*, *2*, 17.

Jowett, J. B., Planelles, V., Poon, B., Shah, N. P., Chen, M. L., & Chen, I. S. (1995). The human immunodeficiency virus type 1 vpr gene arrests infected T cells in the G2+M phase of the cell cycle. *Journal of Virology*, *69*, 6304.

Kato, T., Hirano, A., Llena, J. F., & Dembitzer, H. M. (1987). Neuropathology of acquired immune deficiency syndrome (AIDS) in 53 autopsy cases with particular emphasis on microglial nodules and multinucleated giant cells. *Acta Neuropathology (Berlin, Germany)*, *73*, 287.

Kaul, M., Garden, G. A., & Lipton, S. A. (2001). Pathways to neuronal injury and apoptosis in HIV-associated dementia. *Nature*, *410*, 988.

Kelly, J. K. (1996). Replication rate and evolution in the human immunodeficiency virus. *Journal of Theoretical Biology*, *180*, 359.

Kelly, N. J., & Morrow, C. D. (2003). Yeast tRNA(Phe) expressed in human cells can be selected by HIV-1 for use as a reverse transcription primer. *Virology*, *313*, 354.

Khanna, K. V., Yu, X. F., Ford, D. H., Ratner, L., Hildreth, J. K., & Markham, R. B. (2000). Differences among HIV-1 variants in their ability to elicit secretion of TNF-alpha. *Journal of Immunology*, *164*, 1408.

Kim, W. K., Alvarez, X., Fisher, J., Bronfin, B., Westmoreland, S., McLaurin, J., et al. (2006). CD163 identifies perivascular macrophages in normal and viral encephalitic brains and potential precursors to perivascular macrophages in blood. *The American Journal of Pathology*, *168*, 822.

Kim, N., Kukkonen, S., Gupta, S., & Aldovini, A. (2010). Association of Tat with promoters of PTEN and PP2A subunits is key to transcriptional activation of apoptotic pathways in HIV-infected CD4+ T cells. *PLoS Pathogens*, *6*, e1001103.

Kino, T., Gragerov, A., Kopp, J. B., Stauber, R. H., Pavlakis, G. N., & Chrousos, G. P. (1999). The HIV-1 virion-associated protein vpr is a coactivator of the human glucocorticoid receptor. *The Journal of Experimental Medicine*, *189*, 51.

Kirchhoff, F., Greenough, T. C., Hamacher, M., Sullivan, J. L., & Desrosiers, R. C. (1997). Activity of human immunodeficiency virus type 1 promoter/TAR regions and tat1 genes derived from individuals with different rates of disease progression. *Virology*, *232*, 319.

Kolson, D. L., Lavi, E., & Gonzalez-Scarano, F. (1998). The effects of human immunodeficiency virus in the central nervous system. *Advances in Virus Research*, *50*, 1.

Korber, B. T., Kunstman, K. J., Patterson, B. K., Furtado, M., McEvilly, M. M., Levy, R., et al. (1994). Genetic differences between blood- and brain-derived viral sequences from human immunodeficiency virus type 1-infected patients: Evidence of conserved elements in the V3 region of the envelope protein of brain-derived sequences. *Journal of Virology*, *68*, 7467.

Korber, B. T., MacInnes, K., Smith, R. F., & Myers, G. (1994). Mutational trends in V3 loop protein sequences observed in different genetic lineages of human immunodeficiency virus type 1. *Journal of Virology*, *68*, 6730.

Kostrikis, L. G., Huang, Y., Moore, J. P., Wolinsky, S. M., Zhang, L., Guo, Y., et al. (1998). A chemokine receptor CCR2 allele delays HIV-1 disease progression and is associated with a CCR5 promoter mutation. *Nature Medicine*, *4*, 350.

Kostrikis, L. G., Neumann, A. U., Thomson, B., Korber, B. T., McHardy, P., Karanicolas, R., et al. (1999). A polymorphism in the regulatory region of the CC-chemokine receptor 5 gene influences perinatal transmission of human immunodeficiency virus type 1 to African-American infants. *Journal of Virology*, *73*, 10264.

Koup, R. A., Safrit, J. T., Cao, Y., Andrews, C. A., McLeod, G., Borkowsky, W., et al. (1994). Temporal association of cellular immune responses with the initial control of viremia in primary human immunodeficiency virus type 1 syndrome. *Journal of Virology*, *68*, 4650.

Krebs, F. C., Mehrens, D., Pomeroy, S., Goodenow, M. M., & Wigdahl, B. (1998). Human immunodeficiency virus type 1 long terminal repeat quasispecies differ in basal transcription and nuclear factor recruitment in human glial cells and lymphocytes. *Journal of Biomedical Science*, *5*, 31.

Krebs, F. C., Ross, H., McAllister, J., & Wigdahl, B. (2000). HIV-1-associated central nervous system dysfunction. *Advances in Pharmacology*, *49*, 315.

Kurosu, T., Mukai, T., Auwanit, W., Ayuthaya, P. I., Saeng-Aroon, S., & Ikuta, K. (2001). Variable sequences in the long terminal repeat and Its downstream region of some of HIV Type 1 CRF01_AE recently distributing among Thai carriers. *AIDS Research and Human Retroviruses*, *17*, 863.

Kurosu, T., Mukai, T., Komoto, S., Ibrahim, M. S., Li, Y. G., Kobayashi, T., et al. (2002). Human immunodeficiency virus type 1 subtype C exhibits higher transactivation activity of Tat than subtypes B and E. *Microbiology and Immunology*, *46*, 787.

Lackner, A. A., Lederman, M. M., & Rodriguez, B. (2012). HIV pathogenesis: The host. *Cold Spring Harbor Perspectives in Medicine*, *2*, a007005.

Le Clerc, S., Limou, S., Coulonges, C., Carpentier, W., Dina, C., Taing, L., et al. (2009). Genomewide association study of a rapid progression cohort identifies new susceptibility alleles for AIDS (ANRS Genomewide Association Study 03). *The Journal of Infectious Diseases*, *200*, 1194.

Lee, B., Sharron, M., Montaner, L. J., Weissman, D., & Doms, R. W. (1999). Quantification of CD4, CCR5, and CXCR4 levels on lymphocyte subsets, dendritic cells, and differentially conditioned monocyte-derived macrophages. *Proceedings of the National Academy of Sciences of the United States of America*, *96*, 5215.

Lemey, P., Kosakovsky Pond, S. L., Drummond, A. J., Pybus, O. G., Shapiro, B., Barroso, H., et al. (2007). Synonymous substitution rates predict HIV disease progression as a result of underlying replication dynamics. *PLoS Computational Biology*, *3*, e29.

Leslie, A. J., Pfafferott, K. J., Chetty, P., Draenert, R., Addo, M. M., Feeney, M., et al. (2004). HIV evolution: CTL escape mutation and reversion after transmission. *Nature Medicine*, *10*, 282.

Letendre, S. L., McCutchan, J. A., Childers, M. E., Woods, S. P., Lazzaretto, D., Heaton, R. K., et al. (2004). Enhancing antiretroviral therapy for human immunodeficiency virus cognitive disorders. *Annals of Neurology*, *56*, 416.

Levy, D. N., Refaeli, Y., & Weiner, D. B. (1995). Extracellular Vpr protein increases cellular permissiveness to human immunodeficiency virus replication and reactivates virus from latency. *Journal of Virology*, *69*, 1243.

Lewin, S. R., Sonza, S., Irving, L. B., McDonald, C. F., Mills, J., & Crowe, S. M. (1996). Surface CD4 is critical to in vitro HIV infection of human alveolar macrophages. *AIDS Research and Human Retroviruses*, *12*, 877.

Lewis, M. J., Dagarag, M., Khan, B., Ali, A., & Yang, O. O. (2012). Partial escape of HIV-1 from cytotoxic T lymphocytes during chronic infection. *Journal of Virology*, *86*, 7459.

Li, L., Aiamkitsumrit, B., Pirrone, V., Nonnemacher, M. R., Wojno, A., Passic, S., et al. (2011). Development of co-selected single nucleotide polymorphisms in the viral promoter precedes the onset of human immunodeficiency virus type 1-associated neurocognitive impairment. *Journal of Neurovirology*, *17*, 92.

Li, L., Dahiya, S., Kortagere, S., Aiamkitsumrit, B., Cunningham, D., Pirrone, V., et al. (2012). Impact of Tat genetic variation on HIV-1 disease. *Advances in Virology*, *2012*, 123605.

Lichterfeld, M., Yu, X. G., Cohen, D., Addo, M. M., Malenfant, J., Perkins, B., et al. (2004). HIV-1 Nef is preferentially recognized by CD8 T cells in primary HIV-1 infection despite a relatively high degree of genetic diversity. *AIDS*, *18*, 1383.

Limou, S., Coulonges, C., Herbeck, J. T., van Manen, D., An, P., Le Clerc, S., et al. (2010). Multiple-cohort genetic association study reveals CXCR6 as a new chemokine receptor involved in long-term nonprogression to AIDS. *The Journal of Infectious Diseases*, *202*, 908.

Limou, S., Le Clerc, S., Coulonges, C., Carpentier, W., Dina, C., Delaneau, O., et al. (2009). Genomewide association study of an AIDS-nonprogression cohort emphasizes the role played by HLA genes (ANRS Genomewide Association Study 02). *The Journal of Infectious Diseases*, *199*, 419.

Lingappa, J. R., Petrovski, S., Kahle, E., Fellay, J., Shianna, K., McElrath, M. J., et al. (2011). Genomewide association study for determinants of HIV-1 acquisition and viral set point in HIV-1 serodiscordant couples with quantified virus exposure. *PLoS One*, *6*, e28632.

Liu, R., Paxton, W. A., Choe, S., Ceradini, D., Martin, S. R., Horuk, R., et al. (1996). Homozygous defect in HIV-1 coreceptor accounts for resistance of some multiply-exposed individuals to HIV-1 infection. *Cell*, *86*, 367.

Liu, Y., Tang, X. P., McArthur, J. C., Scott, J., & Gartner, S. (2000). Analysis of human immunodeficiency virus type 1 gp160 sequences from a patient with HIV dementia: Evidence for monocyte trafficking into brain. *Journal of Neurovirology*, *6*(Suppl. 1), S70.

Lu, Y. L., Spearman, P., & Ratner, L. (1993). Human immunodeficiency virus type 1 viral protein R localization in infected cells and virions. *Journal of Virology*, *67*, 6542.

Mahalingam, S., Collman, R. G., Patel, M., Monken, C. E., & Srinivasan, A. (1995). Functional analysis of HIV-1 Vpr: Identification of determinants essential for subcellular localization. *Virology*, *212*, 331.

Mahalingam, S., Khan, S. A., Jabbar, M. A., Monken, C. E., Collman, R. G., & Srinivasan, A. (1995). Identification of residues in the N-terminal acidic domain of HIV-1 Vpr essential for virion incorporation. *Virology*, *207*, 297.

Mahalingam, S., Khan, S. A., Murali, R., Jabbar, M. A., Monken, C. E., Collman, R. G., et al. (1995). Mutagenesis of the putative alpha-helical domain of the Vpr protein of human immunodeficiency virus type 1: Effect on stability and virion incorporation. *Proceedings of the National Academy of Sciences of the United States of America*, *92*, 3794.

Manolio, T. A. (2010). Genomewide association studies and assessment of the risk of disease. *The New England Journal of Medicine, 363*, 166.

Marra, C. M., Maxwell, C. L., Collier, A. C., Robertson, K. R., & Imrie, A. (2007). Interpreting cerebrospinal fluid pleocytosis in HIV in the era of potent antiretroviral therapy. *BMC Infectious Diseases, 7*, 37.

Martin, M. P., Dean, M., Smith, M. W., Winkler, C., Gerrard, B., Michael, N. L., et al. (1998). Genetic acceleration of AIDS progression by a promoter variant of CCR5. *Science, 282*, 1907.

Martín, J., LaBranche, C. C., & González-Scarano, F. (2001). Differential CD4/CCR5 utilization, gp120 conformation, and neutralization sensitivity between envelopes from a microglia-adapted human immunodeficiency virus type 1 and its parental isolate. *Journal of Virology, 75*, 3568.

Martinez-Picado, J., Prado, J. G., Fry, E. E., Pfafferott, K., Leslie, A., Chetty, S., et al. (2006). Fitness cost of escape mutations in p24 Gag in association with control of human immunodeficiency virus type 1. *Journal of Virology, 80*, 3617.

Martin-Garcia, J., Cao, W., Varela-Rohena, A., Plassmeyer, M. L., & Gonzalez-Scarano, F. (2006). HIV-1 tropism for the central nervous system: Brain-derived envelope glycoproteins with lower CD4 dependence and reduced sensitivity to a fusion inhibitor. *Virology, 346*, 169.

Martin-Garcia, J., Cocklin, S., Chaiken, I. M., & Gonzalez-Scarano, F. (2005). Interaction with CD4 and antibodies to CD4-induced epitopes of the envelope gp120 from a microglial cell-adapted human immunodeficiency virus type 1 isolate. *Journal of Virology, 79*, 6703.

Matsusaka, T., Fujikawa, K., Nishio, Y., Mukaida, N., Matsushima, K., Kishimoto, T., et al. (1993). Transcription factors NF-IL6 and NF-kappa B synergistically activate transcription of the inflammatory cytokines, interleukin 6 and interleukin 8. *Proceedings of the National Academy of Sciences of the United States of America, 90*, 10193.

Mayne, M., Bratanich, A. C., Chen, P., Rana, F., Nath, A., & Power, C. (1998). HIV-1 tat molecular diversity and induction of TNF-alpha: Implications for HIV-induced neurological disease. *Neuroimmunomodulation, 5*, 184.

McAllister, J. J., Phillips, D., Millhouse, S., Conner, J., Hogan, T., Ross, H. L., et al. (2000). Analysis of the HIV-1 LTR NF-kappaB-proximal Sp site III: Evidence for cell type-specific gene regulation and viral replication. *Virology, 274*, 262.

McArthur, J. C., & Brew, B. J. (2010). HIV-associated neurocognitive disorders: Is there a hidden epidemic? *AIDS, 24*, 1367.

McArthur, J. C., Haughey, N., Gartner, S., Conant, K., Pardo, C., Nath, A., et al. (2003). Human immunodeficiency virus-associated dementia: An evolving disease. *Journal of Neurovirology, 9*, 205.

McArthur, J. C., McClernon, D. R., Cronin, M. F., Nance-Sproson, T. E., Saah, A. J., St Clair, M., et al. (1997). Relationship between human immunodeficiency virus-associated dementia and viral load in cerebrospinal fluid and brain. *Annals of Neurology, 42*, 689.

McKnight, A., Weiss, R. A., Shotton, C., Takeuchi, Y., Hoshino, H., & Clapham, P. R. (1995). Change in tropism upon immune escape by human immunodeficiency virus. *Journal of Virology, 69*, 3167.

Messam, C. A., & Major, E. O. (2000). Stages of restricted HIV-1 infection in astrocyte cultures derived from human fetal brain tissue. *Journal of Neurovirology, 6*(Suppl. 1), S90.

Meucci, O., Fatatis, A., Simen, A. A., Bushell, T. J., Gray, P. W., & Miller, R. J. (1998). Chemokines regulate hippocampal neuronal signaling and gp120 neurotoxicity. *Proceedings of the National Academy of Sciences of the United States of America, 95*, 14500.

Meyerhoff, D. J., MacKay, S., Poole, N., Dillon, W. P., Weiner, M. W., & Fein, G. (1994). N-acetylaspartate reductions measured by 1H MRSI in cognitively impaired HIV-seropositive individuals. *Magnetic Resonance Imaging*, *12*, 653.

Michael, N. L. (1999). Host genetic influences on HIV-1 pathogenesis. *Current Opinion in Immunology*, *11*, 466.

Michael, N. L., Chang, G., Louie, L. G., Mascola, J. R., Dondero, D., Birx, D. L., et al. (1997). The role of viral phenotype and CCR-5 gene defects in HIV-1 transmission and disease progression. *Nature Medicine*, *3*, 338.

Michael, N. L., D'Arcy, L., Ehrenberg, P. K., & Redfield, R. R. (1994). Naturally occurring genotypes of the human immunodeficiency virus type 1 long terminal repeat display a wide range of basal and Tat-induced transcriptional activities. *Journal of Virology*, *68*, 3163.

Miller, T. I., Borkowsky, W., DiMeglio, L. A., Dooley, L., Geffner, M. E., Hazra, R., et al. (2012). Metabolic abnormalities and viral replication are associated with biomarkers of vascular dysfunction in HIV-infected children. *HIV Medicine*, *13*, 264.

Minagar, A., Commins, D., Alexander, J. S., Hoque, R., Chiappelli, F., Singer, E. J., et al. (2008). NeuroAIDS: Characteristics and diagnosis of the neurological complications of AIDS. *Molecular Diagnosis and Therapy*, *12*, 25.

Montano, M. A., Nixon, C. P., Ndung'u, T., Bussmann, H., Novitsky, V. A., Dickman, D., et al. (2000). Elevated tumor necrosis factor-alpha activation of human immunodeficiency virus type 1 subtype C in Southern Africa is associated with an NF-kappaB enhancer gain-of-function. *The Journal of Infectious Diseases*, *181*, 76.

Mothobi, N. Z., & Brew, B. J. (2012). Neurocognitive dysfunction in the highly active antiretroviral therapy era. *Current Opinion in Infectious Diseases*, *25*, 4.

Mulherin, S. A., O'Brien, T. R., Ioannidis, J. P., Goedert, J. J., Buchbinder, S. P., Coutinho, R. A., et al. (2003). Effects of CCR5-Delta32 and CCR2-64I alleles on HIV-1 disease progression: The protection varies with duration of infection. *AIDS*, *17*, 377.

Muller, W. E., Schroder, H. C., Ushijima, H., Dapper, J., & Bormann, J. (1992). gp120 of HIV-1 induces apoptosis in rat cortical cell cultures: Prevention by memantine. *European Journal of Pharmacology*, *226*, 209.

Na, H., Acharjee, S., Jones, G., Vivithanaporn, P., Noorbakhsh, F., McFarlane, N., et al. (2011). Interactions between human immunodeficiency virus (HIV)-1 Vpr expression and innate immunity influence neurovirulence. *Retrovirology*, *8*, 44.

Nabel, G., & Baltimore, D. (1987). An inducible transcription factor activates expression of human immunodeficiency virus in T cells. *Nature*, *326*, 711.

Nabel, G. J., Rice, S. A., Knipe, D. M., & Baltimore, D. (1988). Alternative mechanisms for activation of human immunodeficiency virus enhancer in T cells. *Science*, *239*, 1299.

Naghavi, M. H., Schwartz, S., Sonnerborg, A., & Vahlne, A. (1999). Long terminal repeat promoter/enhancer activity of different subtypes of HIV type 1. *AIDS Research and Human Retroviruses*, *15*, 1293.

Nath, A., Conant, K., Chen, P., Scott, C., & Major, E. O. (1999). Transient exposure to HIV-1 Tat protein results in cytokine production in macrophages and astrocytes. A hit and run phenomenon. *The Journal of Biological Chemistry*, *274*, 17098.

Nath, A., Psooy, K., Martin, C., Knudsen, B., Magnuson, D. S., Haughey, N., et al. (1996). Identification of a human immunodeficiency virus type 1 Tat epitope that is neuroexcitatory and neurotoxic. *Journal of Virology*, *70*, 1475.

Neumann, M., Felber, B. K., Kleinschmidt, A., Froese, B., Erfle, V., Pavlakis, G. N., et al. (1995). Restriction of human immunodeficiency virus type 1 production in a human astrocytoma cell line is associated with a cellular block in Rev function. *Journal of Virology*, *69*, 2159.

Nonnemacher, M. R., Irish, B. P., Liu, Y., Mauger, D., & Wigdahl, B. (2004). Specific sequence configurations of HIV-1 LTR G/C box array result in altered recruitment of Sp isoforms and correlate with disease progression. *Journal of Neuroimmunology*, *157*, 39.

Nottet, H. S., Persidsky, Y., Sasseville, V. G., Nukuna, A. N., Bock, P., Zhai, Q. H., et al. (1996). Mechanisms for the transendothelial migration of HIV-1-infected monocytes into brain. *Journal of Immunology, 156*, 1284.

Novitsky, V. A., Montano, M. A., McLane, M. F., Renjifo, B., Vannberg, F., Foley, B. T., et al. (1999). Molecular cloning and phylogenetic analysis of human immunodeficiency virus type 1 subtype C: A set of 23 full-length clones from Botswana. *Journal of Virology, 73*, 4427.

O'Brien, S. J., & Nelson, G. W. (2004). Human genes that limit AIDS. *Nature Genetics, 36*, 565.

Ohagen, A., Ghosh, S., He, J., Huang, K., Chen, Y., Yuan, M., et al. (1999). Apoptosis induced by infection of primary brain cultures with diverse human immunodeficiency virus type 1 isolates: Evidence for a role of the envelope. *Journal of Virology, 73*, 897.

Patel, C. A., Mukhtar, M., & Pomerantz, R. J. (2000). Human immunodeficiency virus type 1 Vpr induces apoptosis in human neuronal cells. *Journal of Virology, 74*, 9717.

Pattarini, R., Pittaluga, A., & Raiteri, M. (1998). The human immunodeficiency virus-1 envelope protein gp120 binds through its V3 sequence to the glycine site of N-methyl-D-aspartate receptors mediating noradrenaline release in the hippocampus. *Neuroscience, 87*, 147.

Paxton, W. A., & Kang, S. (1998). Chemokine receptor allelic polymorphisms: Relationships to HIV resistance and disease progression. *Seminars in Immunology, 10*, 187.

Pereira, L. A., Bentley, K., Peeters, A., Churchill, M. J., & Deacon, N. J. (2000). A compilation of cellular transcription factor interactions with the HIV-1 LTR promoter. *Nucleic Acids Research, 28*, 663.

Pereyra, F., Jia, X., McLaren, P. J., Telenti, A., de Bakker, P. I., Walker, B. D., et al. (2010). The major genetic determinants of HIV-1 control affect HLA class I peptide presentation. *Science, 330*, 1551.

Persidsky, Y., Stins, M., Way, D., Witte, M. H., Weinand, M., Kim, K. S., et al. (1997). A model for monocyte migration through the blood–brain barrier during HIV-1 encephalitis. *Journal of Immunology, 158*, 3499.

Peters, P. J., Bhattacharya, J., Hibbitts, S., Dittmar, M. T., Simmons, G., Bell, J., et al. (2004). Biological analysis of human immunodeficiency virus type 1 R5 envelopes amplified from brain and lymph node tissues of AIDS patients with neuropathology reveals two distinct tropism phenotypes and identifies envelopes in the brain that confer an enhanced tropism and fusigenicity for macrophages. *Journal of Virology, 78*, 6915.

Peters, P. J., Duenas-Decamp, M. J., Sullivan, W. M., & Clapham, P. R. (2007). Variation of macrophage tropism among HIV-1 R5 envelopes in brain and other tissues. *Journal of Neuroimmune Pharmacology, 2*, 32.

Petito, C. K. (1995). Mechanisms of cell death in brains of patients with AIDS. *Journal of Neuropathology and Experimental Neurology, 54*, 404.

Philpott, S., Burger, H., Tarwater, P. M., Lu, M., Gange, S. J., Anastos, K., et al. (2004). CCR2 genotype and disease progression in a treated population of HIV type 1-infected women. *Clinical Infectious Diseases, 39*, 861.

Pilcher, C. D., Shugars, D. C., Fiscus, S. A., Miller, W. C., Menezes, P., Giner, J., et al. (2001). HIV in body fluids during primary HIV infection: Implications for pathogenesis, treatment and public health. *AIDS, 15*, 837.

Piller, S. C., Jans, P., Gage, P. W., & Jans, D. A. (1998). Extracellular HIV-1 virus protein R causes a large inward current and cell death in cultured hippocampal neurons: Implications for AIDS pathology. *Proceedings of the National Academy of Sciences of the United States of America, 95*, 4595.

Ping, L. H., Nelson, J. A., Hoffman, I. F., Schock, J., Lamers, S. L., Goodman, M., et al. (1999). Characterization of V3 sequence heterogeneity in subtype C human immunodeficiency virus type 1 isolates from Malawi: Underrepresentation of X4 variants. *Journal of Virology, 73*, 6271.

Pomerantz, R. J., Feinberg, M. B., Andino, R., & Baltimore, D. (1991). The long terminal repeat is not a major determinant of the cellular tropism of human immunodeficiency virus type 1. *Journal of Virology*, *65*, 1041.

Pope, M., Betjes, M. G., Romani, N., Hirmand, H., Cameron, P. U., Hoffman, L., et al. (1994). Conjugates of dendritic cells and memory T lymphocytes from skin facilitate productive infection with HIV-1. *Cell*, *78*, 389.

Power, C., McArthur, J. C., Johnson, R. T., Griffin, D. E., Glass, J. D., Perryman, S., et al. (1994). Demented and nondemented patients with AIDS differ in brain-derived human immunodeficiency virus type 1 envelope sequences. *Journal of Virology*, *68*, 4643.

Power, C., McArthur, J. C., Nath, A., Wehrly, K., Mayne, M., Nishio, J., et al. (1998). Neuronal death induced by brain-derived human immunodeficiency virus type 1 envelope genes differs between demented and nondemented AIDS patients. *Journal of Virology*, *72*, 9045.

Price, R. W. (1994). Understanding the AIDS dementia complex (ADC). The challenge of HIV and its effects on the central nervous system. *Research Publications—Association for Research in Nervous and Mental Disease*, *72*, 1.

Raha, T., Cheng, S. W., & Green, M. R. (2005). HIV-1 Tat stimulates transcription complex assembly through recruitment of TBP in the absence of TAFs. *PLoS Biology*, *3*, e44.

Rambaut, A., Posada, D., Crandall, K. A., & Holmes, E. C. (2004). The causes and consequences of HIV evolution. *Nature Reviews. Genetics*, *5*, 52.

Rappaport, J., Joseph, J., Croul, S., Alexander, G., Del Valle, L., Amini, S., et al. (1999). Molecular pathway involved in HIV-1-induced CNS pathology: Role of viral regulatory protein, Tat. *Journal of Leukocyte Biology*, *65*, 458.

Raziuddin, Mikovits, J. A., Calvert, I., Ghosh, S., Kung, H. F., & Ruscetti, F. W. (1991). Negative regulation of human immunodeficiency virus type 1 expression in monocytes: Role of the 65-kDa plus 50-kDa NF-kappa B dimer. *Proceedings of the National Academy of Sciences of the United States of America*, *88*, 9426.

Reddy, R. T., Achim, C. L., Sirko, D. A., Tehranchi, S., Kraus, F. G., Wong-Staal, F., et al. (1996). Sequence analysis of the V3 loop in brain and spleen of patients with HIV encephalitis. *AIDS Research and Human Retroviruses*, *12*, 477.

Refaeli, Y., Levy, D. N., & Weiner, D. B. (1995). The glucocorticoid receptor type II complex is a target of the HIV-1 vpr gene product. *Proceedings of the National Academy of Sciences of the United States of America*, *92*, 3621.

Rice, A. P., & Mathews, M. B. (1988). Transcriptional but not translational regulation of HIV-1 by the tat gene product. *Nature*, *332*, 551.

Richman, D. D., Wrin, T., Little, S. J., & Petropoulos, C. J. (2003). Rapid evolution of the neutralizing antibody response to HIV type 1 infection. *Proceedings of the National Academy of Sciences of the United States of America*, *100*, 4144.

Robertson, K., Liner, J., & Heaton, R. (2009). Neuropsychological assessment of HIV-infected populations in international settings. *Neuropsychology Review*, *19*, 232.

Rodriguez-Franco, E. J., Cantres-Rosario, Y. M., Plaud-Valentin, M., Romeu, R., Rodriguez, Y., Skolasky, R., et al. (2012). Dysregulation of macrophage-secreted cathepsin B contributes to HIV-1-linked neuronal apoptosis. *PLoS One*, *7*, e36571.

Roos, M. T., Lange, J. M., de Goede, R. E., Coutinho, R. A., Schellekens, P. T., Miedema, F., et al. (1992). Viral phenotype and immune response in primary human immunodeficiency virus type 1 infection. *The Journal of Infectious Diseases*, *165*, 427.

Rosenbloom, D. I., Hill, A. L., Rabi, S. A., Siliciano, R. F., & Nowak, M. A. (2012). Antiretroviral dynamics determines HIV evolution and predicts therapy outcome. *Nat Med*, *18*, 1378–1385.

Rosenblum, M. K. (1990). Infection of the central nervous system by the human immunodeficiency virus type 1. Morphology and relation to syndromes of progressive encephalopathy and myelopathy in patients with AIDS. *Pathology Annual*, *25*(Pt. 1), 117.

Ross, E. K., Buckler-White, A. J., Rabson, A. B., Englund, G., & Martin, M. A. (1991). Contribution of NF-kappa B and Sp1 binding motifs to the replicative capacity of human immunodeficiency virus type 1: Distinct patterns of viral growth are determined by T-cell types. *Journal of Virology*, *65*, 4350.

Ross, H. L., Gartner, S., McArthur, J. C., Corboy, J. R., McAllister, J. J., Millhouse, S., et al. (2001). HIV-1 LTR C/EBP binding site sequence configurations preferentially encountered in brain lead to enhanced C/EBP factor binding and increased LTR-specific activity. *Journal of Neurovirology*, *7*, 235.

Ross, H. A., & Rodrigo, A. G. (2002). Immune-mediated positive selection drives human immunodeficiency virus type 1 molecular variation and predicts disease duration. *Journal of Virology*, *76*, 11715.

Rossi, F., Querido, B., Nimmagadda, M., Cocklin, S., Navas-Martin, S., & Martin-Garcia, J. (2008). The V1-V3 region of a brain-derived HIV-1 envelope glycoprotein determines macrophage tropism, low CD4 dependence, increased fusogenicity and altered sensitivity to entry inhibitors. *Retrovirology*, *5*, 89.

Rousseau, C., Abrams, E., Lee, M., Urbano, R., & King, M. C. (1997). Long terminal repeat and nef gene variants of human immunodeficiency virus type 1 in perinatally infected long-term survivors and rapid progressors. *AIDS Research and Human Retroviruses*, *13*, 1611.

Roux, P., Alfieri, C., Hrimech, M., Cohen, E. A., & Tanner, J. E. (2000). Activation of transcription factors NF-kappaB and NF-IL-6 by human immunodeficiency virus type 1 protein R (Vpr) induces interleukin-8 expression. *Journal of Virology*, *74*, 4658.

Rumbaugh, J., Turchan-Cholewo, J., Galey, D., St Hillaire, C., Anderson, C., Conant, K., et al. (2006). Interaction of HIV Tat and matrix metalloproteinase in HIV neuropathogenesis: A new host defense mechanism. *The FASEB Journal*, *20*, 1736.

Sacktor, N., McDermott, M. P., Marder, K., Schifitto, G., Selnes, O. A., McArthur, J. C., et al. (2002). HIV-associated cognitive impairment before and after the advent of combination therapy. *Journal of Neurovirology*, *8*, 136.

Samson, M., Libert, F., Doranz, B. J., Rucker, J., Liesnard, C., Farber, C. M., et al. (1996). Resistance to HIV-1 infection in caucasian individuals bearing mutant alleles of the CCR-5 chemokine receptor gene. *Nature*, *382*, 722.

Sawaya, B. E., Khalili, K., Rappaport, J., Serio, D., Chen, W., Srinivasan, A., et al. (1999). Suppression of HIV-1 transcription and replication by a Vpr mutant. *Gene Therapy*, *6*, 947.

Sawaya, B. E., Thatikunta, P., Denisova, L., Brady, J., Khalili, K., & Amini, S. (1998). Regulation of TNFalpha and TGFbeta-1 gene transcription by HIV-1 Tat in CNS cells. *Journal of Neuroimmunology*, *87*, 33.

Schacker, T., Collier, A. C., Hughes, J., Shea, T., & Corey, L. (1996). Clinical and epidemiologic features of primary HIV infection. *Annals of Internal Medicine*, *125*, 257.

Schifitto, G., Yiannoutsos, C. T., Ernst, T., Navia, B. A., Nath, A., Sacktor, N., et al. (2009). Selegiline and oxidative stress in HIV-associated cognitive impairment. *Neurology*, *73*, 1975.

Schuitemaker, H., Groenink, M., Meyaard, L., Kootstra, N. A., Fouchier, R. A., Gruters, R. A., et al. (1993). Early replication steps but not cell type-specific signalling of the viral long terminal repeat determine HIV-1 monocytotropism. *AIDS Research and Human Retroviruses*, *9*, 669.

Schuitemaker, H., Kootstra, N. A., Koppelman, M. H., Bruisten, S. M., Huisman, H. G., Tersmette, M., et al. (1992). Proliferation-dependent HIV-1 infection of monocytes occurs during differentiation into macrophages. *The Journal of Clinical Investigation*, *89*, 1154.

Selby, M. J., Bain, E. S., Luciw, P. A., & Peterlin, B. M. (1989). Structure, sequence, and position of the stem-loop in tar determine transcriptional elongation by tat through the HIV-1 long terminal repeat. *Genes and Development*, *3*, 547.

Selnes, O. A., Miller, E., McArthur, J., Gordon, B., Munoz, A., Sheridan, K., et al. (1990). HIV-1 infection: No evidence of cognitive decline during the asymptomatic stages. The multicenter AIDS cohort study. *Neurology*, *40*, 204.

Seo, T. K., Thorne, J. L., Hasegawa, M., & Kishino, H. (2002). A viral sampling design for testing the molecular clock and for estimating evolutionary rates and divergence times. *Bioinformatics*, *18*, 115.

Shankarappa, R., Margolick, J. B., Gange, S. J., Rodrigo, A. G., Upchurch, D., Farzadegan, H., et al. (1999). Consistent viral evolutionary changes associated with the progression of human immunodeficiency virus type 1 infection. *Journal of Virology*, *73*, 10489.

Shieh, J. T., Martin, J., Baltuch, G., Malim, M. H., & Gonzalez-Scarano, F. (2000). Determinants of syncytium formation in microglia by human immunodeficiency virus type 1: Role of the V1/V2 domains. *Journal of Virology*, *74*, 693.

Siddappa, N. B., Venkatramanan, M., Venkatesh, P., Janki, M. V., Jayasuryan, N., Desai, A., et al. (2006). Transactivation and signaling functions of Tat are not correlated: Biological and immunological characterization of HIV-1 subtype-C Tat protein. *Retrovirology*, *3*, 53.

Silva, C., Zhang, K., Tsutsui, S., Holden, J. K., Gill, M. J., & Power, C. (2003). Growth hormone prevents human immunodeficiency virus-induced neuronal p53 expression. *Annals of Neurology*, *54*, 605.

Silvestri, G., & Feinberg, M. B. (2003). Turnover of lymphocytes and conceptual paradigms in HIV infection. *The Journal of Clinical Investigation*, *112*, 821.

Singh, K. K., Barroga, C. F., Hughes, M. D., Chen, J., Raskino, C., McKinney, R. E., et al. (2003). Genetic influence of CCR5, CCR2, and SDF1 variants on human immunodeficiency virus 1 (HIV-1)-related disease progression and neurological impairment, in children with symptomatic HIV-1 infection. *The Journal of Infectious Diseases*, *188*, 1461.

Singh, K. K., Ellis, R. J., Marquie-Beck, J., Letendre, S., Heaton, R. K., Grant, I., et al. (2004). CCR2 polymorphisms affect neuropsychological impairment in HIV-1-infected adults. *Journal of Neuroimmunology*, *157*, 185.

Sloand, E. M., Klein, H. G., Banks, S. M., Vareldzis, B., Merritt, S., & Pierce, P. (1992). Epidemiology of thrombocytopenia in HIV infection. *European Journal of Haematology*, *48*, 168.

Smit, T. K., Wang, B., Ng, T., Osborne, R., Brew, B., & Saksena, N. K. (2001). Varied tropism of HIV-1 isolates derived from different regions of adult brain cortex discriminate between patients with and without AIDS dementia complex (ADC): Evidence for neurotropic HIV variants. *Virology*, *279*, 509.

Spudich, S., & Gonzalez-Scarano, F. (2012). HIV-1-related central nervous system disease: Current issues in pathogenesis, diagnosis, and treatment. *Cold Spring Harbor Perspectives in Medicine*, *2*, a007120.

Spudich, S. S., Nilsson, A. C., Lollo, N. D., Liegler, T. J., Petropoulos, C. J., Deeks, S. G., et al. (2005). Cerebrospinal fluid HIV infection and pleocytosis: Relation to systemic infection and antiretroviral treatment. *BMC Infectious Diseases*, *5*, 98.

Stern, M., Czaja, K., Rauch, A., Rickenbach, M., Gunthard, H. F., Battegay, M., et al. (2012). HLA-Bw4 identifies a population of HIV-infected patients with an increased capacity to control viral replication after structured treatment interruption. *HIV Medicine*, *13*, 589.

Strizki, J. M., Albright, A. V., Sheng, H., O'Connor, M., Perrin, L., & Gonzalez-Scarano, F. (1996). Infection of primary human microglia and monocyte-derived macrophages with human immunodeficiency virus type 1 isolates: Evidence of differential tropism. *Journal of Virology*, *70*, 7654.

Tahirov, T. H., Babayeva, N. D., Varzavand, K., Cooper, J. J., Sedore, S. C., & Price, D. H. (2010). Crystal structure of HIV-1 Tat complexed with human P-TEFb. *Nature*, *465*, 747.

Tanaka, T., Akira, S., Yoshida, K., Umemoto, M., Yoneda, Y., Shirafuji, N., et al. (1995). Targeted disruption of the NF-IL6 gene discloses its essential role in bacteria killing and tumor cytotoxicity by macrophages. *Cell*, *80*, 353.

Taylor, J. P., & Khalili, K. (1994). Activation of HIV-1 transcription by Tat in cells derived from the CNS: Evidence for the participation of NF-kappa B—A review. *Advances in Neuroimmunology*, *4*, 291.

Telenti, A., & Johnson, W. E. (2012). Host genes important to HIV replication and evolution. *Cold Spring Harbor Perspectives in Medicine*, *2*, a007203.

Tesmer, V. M., Rajadhyaksha, A., Babin, J., & Bina, M. (1993). NF-IL6-mediated transcriptional activation of the long terminal repeat of the human immunodeficiency virus type 1. *Proceedings of the National Academy of Sciences of the United States of America*, *90*, 7298.

Thomas, E. R., Dunfee, R. L., Stanton, J., Bogdan, D., Taylor, J., Kunstman, K., et al. (2007). Macrophage entry mediated by HIV Envs from brain and lymphoid tissues is determined by the capacity to use low CD4 levels and overall efficiency of fusion. *Virology*, *360*, 105.

Toohey, K., Wehrly, K., Nishio, J., Perryman, S., & Chesebro, B. (1995). Human immunodeficiency virus envelope V1 and V2 regions influence replication efficiency in macrophages by affecting virus spread. *Virology*, *213*, 70.

Tornatore, C., Chandra, R., Berger, J. R., & Major, E. O. (1994). HIV-1 infection of subcortical astrocytes in the pediatric central nervous system. *Neurology*, *44*, 481.

Trachtenberg, E., Korber, B., Sollars, C., Kepler, T. B., Hraber, P. T., Hayes, E., et al. (2003). Advantage of rare HLA supertype in HIV disease progression. *Nature Medicine*, *9*, 928.

Troyer, J. L., Nelson, G. W., Lautenberger, J. A., Chinn, L., McIntosh, C., Johnson, R. C., et al. (2011). Genome-wide association study implicates PARD3B-based AIDS restriction. *The Journal of Infectious Diseases*, *203*, 1491.

Turk, G., Carobene, M., Monczor, A., Rubio, A. E., Gomez-Carrillo, M., & Salomon, H. (2006). Higher transactivation activity associated with LTR and Tat elements from HIV-1 BF intersubtype recombinant variants. *Retrovirology*, *3*, 14.

van Manen, D., Delaneau, O., Kootstra, N. A., Boeser-Nunnink, B. D., Limou, S., Bol, S. M., et al. (2011). Genome-wide association scan in HIV-1-infected individuals identifying variants influencing disease course. *PLoS One*, *6*, e22208.

van Manen, D., van 't Wout, A. B., & Schuitemaker, H. (2012a). Genome-wide association studies on HIV susceptibility, pathogenesis and pharmacogenomics. *Retrovirology*, *9*, 70.

van Marle, G., & Power, C. (2005). Human immunodeficiency virus type 1 genetic diversity in the nervous system: Evolutionary epiphenomenon or disease determinant? *Journal of Neurovirology*, *11*, 107.

Van Marle, G., Rourke, S. B., Zhang, K., Silva, C., Ethier, J., Gill, M. J., et al. (2002). HIV dementia patients exhibit reduced viral neutralization and increased envelope sequence diversity in blood and brain. *AIDS*, *16*, 1905.

van Rij, R. P., de Roda Husman, A. M., Brouwer, M., Goudsmit, J., Coutinho, R. A., & Schuitemaker, H. (1998). Role of CCR2 genotype in the clinical course of syncytium-inducing (SI) or non-SI human immunodeficiency virus type 1 infection and in the time to conversion to SI virus variants. *The Journal of Infectious Diseases*, *178*, 1806.

Velpandi, A., Nagashunmugam, T., Otsuka, T., Cartas, M., & Srinivasan, A. (1992). Structure-function studies of HIV-1: Influence of long terminal repeat U3 region sequences on virus production. *DNA and Cell Biology*, *11*, 369.

Vendeville, A., Rayne, F., Bonhoure, A., Bettache, N., Montcourrier, P., & Beaumelle, B. (2004). HIV-1 Tat enters T cells using coated pits before translocating from acidified endosomes and eliciting biological responses. *Molecular Biology of the Cell*, *15*, 2347.

Verhoef, K., Klein, A., & Berkhout, B. (1996). Paracrine activation of the HIV-1 LTR promoter by the viral Tat protein is mechanistically similar to trans-activation within a cell. *Virology*, *225*, 316.

Walker, L. M., & Burton, D. R. (2010). Rational antibody-based HIV-1 vaccine design: Current approaches and future directions. *Current Opinion in Immunology*, *22*, 358.

Wang, Z. X., Berson, J. F., Zhang, T. Y., Cen, Y. H., Sun, Y., Sharron, M., et al. (1998). CXCR4 sequences involved in coreceptor determination of human immunodeficiency virus type-1 tropism. Unmasking of activity with M-tropic Env glycoproteins. *The Journal of Biological Chemistry, 273*, 15007–15015.

Wang, J., Crawford, K., Yuan, M., Wang, H., Gorry, P. R., & Gabuzda, D. (2002). Regulation of CC chemokine receptor 5 and CD4 expression and human immunodeficiency virus type 1 replication in human macrophages and microglia by T helper type 2 cytokines. *The Journal of Infectious Diseases, 185*, 885.

Weidenheim, K. M., Epshteyn, I., & Lyman, W. D. (1993). Immunocytochemical identification of T-cells in HIV-1 encephalitis: Implications for pathogenesis of CNS disease. *Modern Pathology, 6*, 167.

Weiss, J. M., Cuff, C. A., & Berman, J. W. (1999). TGF-beta downmodulates cytokine-induced monocyte chemoattractant protein (MCP)-1 expression in human endothelial cells. A putative role for TGF-beta in the modulation of TNF receptor expression. *Endothelium, 6*, 291.

Wendelken, L. A., & Valcour, V. (2012). Impact of HIV and aging on neuropsychological function. *Journal of Neurovirology, 18*, 256.

Wiley, C. A., & Achim, C. (1994). Human immunodeficiency virus encephalitis is the pathological correlate of dementia in acquired immunodeficiency syndrome. *Annals of Neurology, 36*, 673.

Wiley, C. A., Schrier, R. D., Nelson, J. A., Lampert, P. W., & Oldstone, M. B. (1986). Cellular localization of human immunodeficiency virus infection within the brains of acquired immune deficiency syndrome patients. *Proceedings of the National Academy of Sciences of the United States of America, 83*, 7089.

Williams, K. C., Corey, S., Westmoreland, S. V., Pauley, D., Knight, H., deBakker, C., et al. (2001). Perivascular macrophages are the primary cell type productively infected by simian immunodeficiency virus in the brains of macaques: Implications for the neuropathogenesis of AIDS. *The Journal of Experimental Medicine, 193*, 905.

Williams, K. C., & Hickey, W. F. (2002). Central nervous system damage, monocytes and macrophages, and neurological disorders in AIDS. *Annual Review of Neuroscience, 25*, 537.

Williamson, S. (2003). Adaptation in the env gene of HIV-1 and evolutionary theories of disease progression. *Molecular Biology and Evolution, 20*, 1318.

Wong, J. K., Ignacio, C. C., Torriani, F., Havlir, D., Fitch, N. J., & Richman, D. D. (1997). In vivo compartmentalization of human immunodeficiency virus: Evidence from the examination of pol sequences from autopsy tissues. *Journal of Virology, 71*, 2059.

Xiao, H., Neuveut, C., Tiffany, H. L., Benkirane, M., Rich, E. A., Murphy, P. M., et al. (2000). Selective CXCR4 antagonism by Tat: Implications for in vivo expansion of coreceptor use by HIV-1. *Proceedings of the National Academy of Sciences of the United States of America, 97*, 11466.

Xu, C., Liu, J., Chen, L., Liang, S., Fujii, N., Tamamura, H., et al. (2011). HIV-1 gp120 enhances outward potassium current via CXCR4 and cAMP-dependent protein kinase A signaling in cultured rat microglia. *Glia, 59*, 997.

Yadav, A., & Collman, R. G. (2009). CNS inflammation and macrophage/microglial biology associated with HIV-1 infection. *Journal of Neuroimmune Pharmacology, 4*, 430.

Zeichner, S. L., Hirka, G., Andrews, P. W., & Alwine, J. C. (1992). Differentiation-dependent human immunodeficiency virus long terminal repeat regulatory elements active in human teratocarcinoma cells. *Journal of Virology, 66*, 2268.

Zeichner, S. L., Kim, J. Y., & Alwine, J. C. (1991a). Analysis of the human immunodeficiency virus long terminal repeat by in vitro transcription competition and linker scanning mutagenesis. *Gene Expression, 1*, 15.

Zeichner, S. L., Kim, J. Y., & Alwine, J. C. (1991b). Linker-scanning mutational analysis of the transcriptional activity of the human immunodeficiency virus type 1 long terminal repeat. *Journal of Virology*, *65*, 2436.

Zhang, L., Huang, Y., Yuan, H., Chen, B. K., Ip, J., & Ho, D. D. (1997). Genotypic and phenotypic characterization of long terminal repeat sequences from long-term survivors of human immunodeficiency virus type 1 infection. *Journal of Virology*, *71*, 5608.

Zhang, K., McQuibban, G. A., Silva, C., Butler, G. S., Johnston, J. B., Holden, J., et al. (2003). HIV-induced metalloproteinase processing of the chemokine stromal cell derived factor-1 causes neurodegeneration. *Nature Neuroscience*, *6*, 1064.

Zhao, R. Y., Li, G., & Bukrinsky, M. I. (2011). Vpr-host interactions during HIV-1 viral life cycle. *Journal of Neuroimmune Pharmacology*, *6*, 216.

Zhou, M., Halanski, M. A., Radonovich, M. F., Kashanchi, F., Peng, J., Price, D. H., et al. (2000). Tat modifies the activity of CDK9 to phosphorylate serine 5 of the RNA polymerase II carboxyl-terminal domain during human immunodeficiency virus type 1 transcription. *Molecular and Cellular Biology*, *20*, 5077.

Zimmerman, E. S., Chen, J., Andersen, J. L., Ardon, O., Dehart, J. L., Blackett, J., et al. (2004). Human immunodeficiency virus type 1 Vpr-mediated G2 arrest requires Rad17 and Hus1 and induces nuclear BRCA1 and gamma-H2AX focus formation. *Molecular and Cellular Biology*, *24*, 9286.

INDEX

Note: Page numbers followed by "*f*" indicate figures, and "*t*" indicate tables.

I

J

W